Partnerships in Healthcare: Transforming Relational Process

Partnerships in Healthcare: Transforming Relational Process

Edited by:

Anthony L. Suchman
Richard J. Botelho
Patricia Hinton-Walker

UNIVERSITY OF ROCHESTER PRESS

First published 1998

University of Rochester Press
668 Mt. Hope Avenue
Rochester, New York, 14620, USA
and at P.O. Box 9, Woodbridge, Suffolk IP12 3DF, UK

ISBN 1–878822–80–2

Library of Congress Cataloging-in-Publication Data

Partnerships in healthcare : transforming relational process / edited by Anthony L. Suchman, Patricia Hinton Walker, Richard J. Botelho.
p. cm.
Includes index.
ISBN 1–878822–80–2 (alk. paper)
1. Medical cooperation. 2. Medical care. I. Suchman, Anthony L., 1954– . II. Walker, Patricia Hinton. III. Botelho, Richard J., 1953–
[DNLM: 1. Partnership Practice—organization & administration—congresses. 2. Delivery of Health Care, Integrated—organization & administration—congresses. 3. Interprofessional Relations—congresses. 4. Professional-Patient Relations—congresses. 5. Community-Institutional Relations—congresses. W 92 P273 1998]
RA394.P34 1998
362.1′068—dc21
DNLM/DLC
for Library of Congress 97–45065
CIP

British Library Cataloguing-in-Publication Data

A catalogue record for this book is available from the British Library

This publication is printed on acid-free paper
Printed in the United States of America
Typeset by Cornerstone Composition Services

*For Lynne Feldman, Alexandra and Julian Suchman,
Nancy and Analivia Suchman, Fay Krawchick,
and Larry Scherr*

*To my wife, Eve, and daughters, Anna and Sarah,
who are pleased that Daddy has finished this book.*

To my husband John for his unwavering support in my publication endeavors; to all the professional nurses who maintain practice roles in the context of academic appointments; and to my special physician colleagues who have supported interdisciplinary education, practice and research in the spirit of collaboration, cooperation and respect.

Table of Contents

Foreword

The clinician-patient relationship remains the foundation of the medical care process, but the practice of competent and caring medical care also depends on organization and successful teamwork among many contributors. Changes in medical knowledge and technology, public expectations, and the financial aspects of care all amplify the significance of relational factors and make them essential for the appropriate education of any health care professional. It is curious that such issues of recognized importance in many other sectors have received so little explicit attention in medicine. *Partnerships in Health Care* begins to remedy this shortcoming and brings together examples of thoughtful and practical efforts in improving relations in patient-clinician transactions, teams, and community settings. It is a book consistent with our health care future.

It is not surprising that in a culture as individualistic as ours that even well educated professionals think of professional competence simply in terms of individual values, individual training, individual skills, and individual performance. Yet it doesn't require much reflection to appreciate that most large human problems require mobilization of organizational systems and community relations. The treatment of any serious illness is seldom simply a matter of clinician and patient, but rather depends on specialized supportive processes involving a range of professional input and technical components. No clinician acting alone can accomplish very much. To take an extreme but telling example from far afield, survival in concentration camps depended less on individuals' personal skills and education than on the extent to which they became enmeshed in an informal social structure that could provide information, hide and protect members, obtain food, and support coping in other ways.

Many persistent problems in medical care arise because of failures in communication with patients and their families, ambiguity of authority, breakdowns in communication and coordination among professional caretakers, and poor teamwork. Most negligence and unloving care are not the result of bad people, but rather the product of poor organization, communication and supervision, and the failure of individuals to step forward and assume appropriate responsibility ("it's not my job . . . "). The quality of caring as well as the quality of technical medical work depend on systems of relationships that are organized to give people the information and discretion they need to do their jobs well, to give meaning to each task in relation to the whole, and to give each participant a sense of control over their work. Such control over work and supportive teams insulate against stress, prevent burnout and indirectly contribute to the quality of caring.

Collaboration is essential in medical care, but it must be meaningful collaboration in which each participant can be heard and each is open to new information, corrective feedback, cooperation and change. Studies of linkage in care between inpatient and outpatient settings, and between primary care and specialty care, suggest the importance of true collaboration, not nominal relations. When the outpatient clinician becomes truly involved with inpatient caretakers and the patient, visits with the patient while in the hospital, and maintains relationships with inpatient treatment staff, continuity of care is improved. When specialists become involved with primary care physicians not at arms distance, but in true collaborative relationships, the potentialities for effective, coordinated care and good outcomes are much enhanced.

The reality is that such collaborations are not easy. They take effort and time, and time in modern health care, even more than in the past, is money. Effective collaboration requires a sense of commitment, a willingness to yield some control, and at times a readiness to suffer some frustration and inconvenience. But if successfully and faithfully practiced, collaboration provides opportunities for excellence and a strong basis for trust, for rewarding relationships with colleagues and patients, and a more satisfying sense of self. Collaboration has to be practiced and learned, and we need to draw lessons from the experience and wisdom of others. *Partnerships in Healthcare* provides examples of such experience and wisdom, and will help us move this continuing endeavor forward.

David Mechanic
René Dubos University Professor
of Behavioral Science,
Rutgers University

Preface

If there is a single most important skill in health care today, it must be the ability to form and sustain partnerships. A health care system cannot function well at any level—from the micro level of the patient and the clinician to the macro level of insurers and government—without a commitment to partnership and the knowledge and skills to make it happen. Clinical outcomes are better when patients are active partners in their own care. The increasing complexity of medical care, concerns about service quality and cost, and efforts to address social determinants of health have made interdisciplinary teams a necessity. Unprecedented degrees of partnership among employers, insurers, provider organizations, consumer groups, and government agencies will be necessary to bring coordination, efficiency, and equitable access to our health care system.

Sadly, the actual practice of partnership in healthcare is primitive at best. Patients express dissatisfaction with the quality of patient-clinician relationships. The poor coordination of care between practitioners and across settings is a major source of frustration for patients and families and squanders resources as clinical and administrative work has to be duplicated. The national effort to accomplish health care reform—the need for which is undisputed—is currently stymied.

Why does partnership get so far off track? We believe the problem is one of focus: clinicians and administrators are so preoccupied with outcomes that they fail to pay adequate attention to process—particularly to the values and attitudes that they express through their actions. All too often, and without deliberate intent, their actions communicate disrespect; they disempower and de-motivate those with whom they seek partnership. We believe this happens for three reasons. First, most clinicians and administrators want and expect to be in control—to be able to direct the activity of others. The resulting hierarchical relationships tend to obstruct the free exchange of information and diminish the commitment of the less powerful collaborator(s). There is little of the creativity and enthusiasm that arises from the synergistic interactions of a well functioning, empowered team. Second, only objective experience is taken as real or important. Subjective experience—the feelings and lived experiences of patients and of those involved in their care—is ignored. People are reduced to objects. Third, little explicit attention is paid to relationship building, and especially to understanding and valuing the perspective of the "other." Relationship is the medium of partnership; it requires attention and effort. The control-oriented hierarchical approach produces unnecessary conflict, wasted energy, diminished satisfaction, and impaired creativity and commitment. Not only are the present outcomes thwarted, but future interactions among the participants are put at risk. Thus, by ignoring process in favor of focusing on outcomes, clinicians and administrators get in their own way.

This book is being written to advance the level of partnership skills practiced within the health care system. We propose an alternative approach to working with others, one that is oriented around process and relationship rather than control. Our approach can be reduced to three practical strategies: creating and maintaining respectful relationships that make true dialogue possible, understanding the perspective of one's partner (and one's self), and empowering others through shared decision making.

There is a lot of distance between comprehending these strategies and implementing them. To help fill that gap, readers will find within this volume a number of conceptual frameworks that will enable them to look at the world in new ways and to identify new courses of action to enhance their effectiveness in forming partnerships and in making their work more meaningful. These models offer insights into attitudes towards control (Suchman), negotiation (Botelho), human motivation (Williams), racial bias (Abernethy), education (Westberg), spirituality (Dombeck) and others. Readers will also find descriptions of a diverse assortment of innovative partnerships. The stories of how these partnerships were formed—what barriers had to be overcome, what worked, what didn't—are as instructive as the descriptions of the end results. In reading accounts of how a program to educate complaint-prone physicians was introduced (Blandin Clark), how rural counties worked together to get more effective training for rural physicians (Wheat), how Do-Not-Resuscitate orders are discussed (Brachman), and many other stories, readers will find practical ideas in the specific areas discussed. But more important than this—and this book's unique contribution—they will gain insights into how to improve the quality of partnership process in their own projects and work teams.

The book is divided into five sections. The first section, Perspectives on Partnership Process, contains four conceptual chapters that address issues generic to partnership at all levels. The other four sections address partnership at specific levels of the health care system: Patient-Clinician Partnerships, Partnership in Healthcare Teams, Community Partnerships, and Educational Partnerships. Each section opens with an overview that discusses the core concepts of the book within the context of that particular level and then summarizes the chapters of that section. The section introductions also cross reference related chapters in other sections to help readers with focused interests avoid missing relevant chapters elsewhere in the book.

The chapters have been selected to offer a diversity of professional disciplines, geographical settings and personal backgrounds among the authors. There are nurses, social workers, physicians of various disciplines, psychologists, administrators, ministers, and educators. They are from urban and rural regions; from neighborhood health centers, industry, government, and universities; from the west, south, midwest, and northeastern United States, with a chapter on an international partnership, as well. Through this diversity, there emerges an awareness of common principles that underlie all these perspectives, as well as an appreciation for the range and value of the differences.

By raising the general level of awareness of the process of partnership and the values it implicitly expresses, we hope this book will help practitioners and students of all clinical disciplines and of health care administration to enhance the quality of their partnerships with patients, colleagues, students, and team members. We hope to show that when the process between partners embodies core values of respect, honesty, compassion, and accountability, then the resulting synergistic interactions will unleash creativity, renew personal commitment, enhance satisfaction, and result in better outcomes at all levels of the health care system.

Anthony L. Suchman,
Richard J. Botelho,
Patricia Hinton-Walker
Rochester, New York
December 1997

Acknowledgments

The idea of exploring partnership across the various levels of scale in the health care system originated during the planning of the Symposium on Partnership in Healthcare sponsored by the Program for Biopsychosocial Studies of the University of Rochester. Along with the editors, Dr. O.J. Sahler participated on the symposium planning committee; we are grateful for her contributions to the design of the symposium, which became the structure for the book, and for helping with the selection of presenters, many of whom went on to contribute chapters. We would like to acknowledge all of the symposium participants, whose ideas and enthusiasm stimulated us to proceed with this project. We would also like to thank the Rochester Area Hospitals Corporation (since renamed the Hospital Consortium of Greater Rochester), the Daisy Marquis Jones Foundation, Blue Cross/Blue Shield of the Rochester Area, the Rochester Community Individual Practice Association, the Roerig Division of Pfizer, Inc. and COMSORT for their financial support of the symposium.

We would also like to acknowledge the contribution to the development of the concepts in this book on the part of many colleagues in the Program for Biopsychosocial Studies, the American Academy on Physician and Patient, the Task Force on the Family in Family Medicine of the Society of Teachers of Family Medicine, the American Academy of Nurse Practitioners and the American Academy of Nursing including (but certainly not limited to) Howard Beckman, Rich Frankel, Susan McDaniel, Tom Campbell, Ed Deci, Geoff Williams, Tim Quill, George Engel, Dan Duffy, Penny Williamson, Craig Kaplan and Arlene Katz.

We thank Sean Culhane, our editor at the University of Rochester Press, for his wise counsel on content and production and for his help with the copy editing. We are enormously indebted to Phyllis Joe for her unfailing organization, patience, and good cheer in preparing the manuscript and in communicating with a large and diffuse group of authors.

Finally, we acknowledge with much gratitude the spectrum of support we received from our families that at one time or another ranged from enthusiastic encouragement to forbearance as this project absorbed evenings and weekends, displaced movies, compressed vacations and intruded into transcontinental relocations. Their love and their commitment to our goals, values and visions made this book possible.

A.L.S., R.J.B., P.H.-W.

List of Contributors

ABERNETHY, ALEXIS D.

Alexis D. Abernethy, PhD is a Clinical Associate Professor of Psychiatry at the University of Rochester Medical Center and a research and clinical consultant in cultural competence.

BAGGS, JUDITH GEDNEY

Judith Gedney Baggs, PhD, RN is an Assistant Professor in both the School of Nursing and in the Division of Medical Humanities, School of Medicine and Dentistry at the University of Rochester.

BATURKA, NATALIE L.

Natalie L. Baturka, PhD is an Assistant Professor of Research at the University of Virginia Health Sciences Center and a program evaluator specializing in community-based and substance abuse treatment programs.

BICKLEY, LYNN

Lynn Bickley, MD is an Associate Professor of Medicine and Director of Ambulatory Medicine in the Department of Medicine at the University of Rochester School of Medicine and Dentistry.

BLANDIN-CLARK, MARGARET

Margaret Blandin-Clark, MSW, BCD is a Clinical Assistant Professor of Medicine at the University of Maryland; behavioral consultant to physicians in practice; clinical consultant at the Women's Growth Center, a non-profit organization dedicated to providing psychotherapy services to women and children; and a licensed clinical social worker serving couples, families, and individuals in private practice in Baltimore, Maryland.

BOTELHO, RICHARD J.

Rick Botelho, MD is Associate Professor of Family Medicine and Psychiatry and Director of FellowshipTraining at the University of Rochester School of Medicine and Dentistry.

BRACHMAN, LAURA H.

Laura H. Brachman, MD is an Assistant Attending at The Genesee Hospital and an internist in private practice.

CAMPBELL, THOMAS L.

Thomas L. Campbell, MD is Associate Professor of Family Medicine and Psychiatry at the University of Rochester and is co-editor of the journal *Families, Systems and Health: The Journal of Collaborative Family Healthcare*.

CARDELLE, ALBERTO J.F.

Alberto J.F. Cardelle, MPH is a Research Associate in International Health at the University of Miami's North-South Center. His contribution was written while he served as Director of the International Health Center at the American Medical Student Association Foundation.

COGGIOLA, PETER

Peter Coggiola, RN, MS, NP-C is President of the American Academy of Nurse Practitioners; established Pioneer Health Care, an independent nurse practitioner practice in rural Western New York State; and holds adjunct faculty status at the Universities of Rochester and Buffalo.

CURRY, WILLIAM A.

William A. Curry, MD, FACP practiced Internal Medicine in rural Alabama for seventeen years, and is currently Chairman, Department of Internal Medicine, College of Community Health Sciences, University of Alabama, and Assistant Dean for Rural Medicine, University of Alabama School of Medicine.

DECI, EDWARD L.

Edward L. Deci, PhD is Professor of Psychology and Director of the Human Motivation Program in the College of Arts and Science at the University of Rochester.

DOLAN, JAMES G.

James G. Dolan, MD is an Associate Professor of Medicine at the University of Rochester and Director of the General Medicine Unit at Highland Hospital of Rochester.

DOMBECK, MARY-THERESE B.

Mary-Therese B. Dombeck, PhD, DMin, RN is Associate Professor at the University of Rochester School of Nursing and Associate Counselor at the Samaritan Pastoral Counseling Center.

DWYER, COLLEEN M.

Colleen M. Dwyer, MS, RN, NP-C is an Adult Nurse Practitioner with the Rochester Park Medical Group and an adjunct clerical instructor at the University of Rochester School of Nursing

EDDEY, GARY A.

Gary E. Eddey, MD is a Clinical Assistant Professor of Pediatrics, Columbia University, Associate Medical Director, Matheny School and Hospital, and Director of the Ambulatory Pediatrics Division, Morristown Memorial Hospital, Morristown, NJ.

EPSTEIN, RONALD

Ronald Epstein, MD is an Associate Professor of Family Medicine and Psychiatry at the University of Rochester/Highland Hospital.

EVINGER, JAMES S.

James S. Evinger, MDiv is Instructor at the University of Rochester School of Nursing, and Adjunct Assistant Professor of Medical Humanities at the School of Medicine and Dentistry.

GRAHAM, MARGARET CLARK

Margaret Clark Graham, RN, PhD is an Associate Professor at the School of Nursing and Health, and Director of the Family Nurse Practitioner Program, Wright State University, Dayton, Ohio.

HINTON-WALKER, PATRICIA

Patricia Hinton-Walker, PhD, FAAN, Dean and Professor of the University of Colorado Health Sciences Center School of Nursing, has expertise in building collaborative nurse-managed practices in community-based settings.

JASON, HILLIARD

Hilliard Jason, MD, EdD is a Clinical Professor of Family Medicine at the University of Colorado and Co-Director of the Center for Instructional Support.

MCDANIEL, SUSAN H.

Susan H. McDaniel, PhD is a family psychologist who is a Professor of Psychiatry and Family Medicine and Director of Family Programs at the University of Rochester School of Medicine.

NAUMBURG, ELIZABETH H.

Elizabeth Naumburg, MD is an Associate Professor in the Department of Family Medicine at the University of Rochester and Highland Hospital with academic interests in the effects of diversity on health care and medical education.

QUILL, TIMOTHY E.

Timothy E. Quill, MD is a Professor of Medicine and Psychiatry at the University of Rochester, a primary care internist, and the Co-director of the University of Rochester Program for Biopsychosocial Studies.

ROACH, JULIENNE

Julienne Roach, M.Ed. is Coordinator of the Buckingham Health Education Board and Director of Patient Education at the Central Virginia Community Health Center, New Canton, Virginia

RYAN, RICHARD M.

Richard M. Ryan, Professor of Psychology and Psychiatry at the University of Rochester, is a clinical psychologist who specializes in both basic and applied research on human motivation.

SARKIN, RICHARD T.

Richard T. Sarkin, MD is a Clinical Associate Professor of Pediatrics, and the Coordinator of the Teaching Effectiveness Program at the State University of New York at Buffalo, School of Medicine and Biomedical Sciences.

SCHIEDERMAYER, DAVID L.

David L. Schiedermayer, MD, FACP is an Associate Professor of Medicine at the Medical College of Wisconsin who teaches ambulatory medicine and clinical ethics. He has a special interest in the medical care of the underserved and has worked in Africa and on a Navajo reservation.

SCHORLING, JOHN B.

John B. Schorling, MD, MPH is an Associate Professor of Internal Medicine and Acting Head, Division of General Medicine at the University of Virginia Health Sciences Center, Charlottesville, Virginia.

SCHWARTZ, DIANE G.

Diane G. Schwartz, MLS is Executive Director of the Primary Care Resource Center, Research Associate Professor of Medicine and Family Medicine, School of Medicine and Biomedical Sciences, State University of New York at Buffalo.

SEABURN, DAVID B.

David B. Seaburn, MS is an Assistant Professor of Psychiatry and Family Medicine at the University of Rochester and Director of the Family Therapy Training Program in Psychiatry.

SIEGEL, MARJORIE

Marjorie Siegel, BA is Project Coordinator for the Alliance of Black Churches Health Project at the University of Virginia Health Sciences Center, Charlottesville, Virginia.

SIEGEL, DAVID M.

David M. Siegel, MD, MPH is an Associate Professor of Pediatrics and Medicine at the University of Rochester School of Medicine and Dentistry, and Director of Adolescent Services at Rochester General Hospital.

SUCHMAN, ANTHONY L.

Anthony L. Suchman, MD is an Associate Professor of Medicine and Psychiatry at the University of Rochester and the Executive Director of Highland PO, Inc., a provider network dedicated to accountable, efficient and humane managed care.

SWAIN, GEOFFREY R.

Geoffrey R. Swain, MD is an Assistant Professor of Family and Community Medicine at the Medical College of Wisconsin and Associate Medical Director, City of Milwaukee Health Department.

TALEN, MARY R.

Mary R. Talen, PhD, is an Associate Professor at the School of Professional Psychology and School of Medicine in Multidisciplinary, Primary Family Health Care Programs, Wright State University, Dayton, OH.

WALBROEHL, GORDON

Gordon S. Walbroehl, MD is Professor at the School of Medicine, and Vice Chair for Academic and Community Development, Wright State University, Dayton, Ohio.

WANG-CHENG, REBEKAH M.

Rebekah M. Wang-Cheng, MD, FACP, an Associate Professor of Medicine and Psychiatry at the Medical College of Wisconsin, is active in educating medical students and residents about biopsychosocial patient care. She also writes a bi-monthly "Dear Dr. Becky" column for the Milwaukee Journal.

WEINER, ERIC L.

Eric L. Weiner, MSW, PhD is an Assistant Professor of Family Practice at Michigan State University College of Human Medicine and Assistant Director of Behavioral Medicine Education at McLaren Family Practice Center in Flint, Michigan.

WESTBERG, JANE

Jane Westberg, PhD is Clinical Professor of Family Medicine at the University of Colorado School of Medicine and Co-Director of The Center for Instructional Support.

WHEAT, JOHN R.

John R. Wheat, MD, MPH is an Associate Professor of Behavioral and Community Medicine at the University of Alabama School of Medicine—Tuscaloosa Program, College of Community Health Sciences, and board certified in internal medicine and occupational medicine.

WILLIAMS, GEOFFREY C.

Geoffrey C. Williams, MD, PhD, is Assistant Professor of Medicine and Psychiatry, The Genesee Hospital, and a member of the Motivation Research Group, Department of Clinical and Social Sciences in Psychology, University of Rochester.

PART I

Perspectives on Partnership Process

We begin our exploration of partnership with four chapters setting forth core concepts, strategies, and values as applied across scale, from the individual to the community. *Partnership, Power and Process: An Introduction* presents an overview of the three core concepts of this book. Among the defining characteristics of successful partnerships are the ability to advocate for one's individual needs while also holding in mind the perspective of the system as a whole; the ability to recognize the needs of one's partners; the open-mindedness to recognize diversity of perspectives as a valuable resource rather than a problem; and the ability to communicate and negotiate in ways that enhance trust and mutual commitment between partners. Several perspectives on power are presented. Partnerships must establish ways to distribute power while maintaining accountability for contributing to common goals. Finally, a brief clinical dialogue is used to illustrate the concept of "process" and the way in which values embedded in process are powerful forms of communication.

Tony Suchman suggests that the path to effective partnerships begins with personal reflection—an examination of one's core beliefs, attitudes, and expectations. His chapter describes the orientation towards control that typifies so much individual and institutional behavior in health care and shows how this approach impedes partnership. He then presents an alternative orientation, one based on relation, that predisposes individuals to greater openness and creativity, and therefore, to better partnerships.

The chapter by *Rick Botelho* begins with an overview of the changes currently taking place in the health care system and the resulting imperatives for and impediments to clinical and administrative partnerships. It then presents a comprehensive model of partnership process between individuals, using the patient-clinician relationship as a prototype. Although the chapter primarily employs clinical examples, the model can be easily translated to administrative and educational contexts.

The chapter by *Pat Hinton-Walker* describes the rationale, structures, and processes of community partnerships. She argues that issues such as inadequate access to care and inefficiency in the health care system are beyond the scope of individual practitioners or organizations to achieve; they require community partnerships. She links the events taking place within the health care sector to principles of organizational process drawn from the management literature, and illustrates with a variety of examples from health care administration.

Resonating through all the chapters in this section, and throughout the rest of the book, are the following principles:

1. *Relationship is the medium of partnership. Building and maintaining relationships is an active process that requires attention and effort.* The process of all interactions reflects and influences both the quality and nature of the participants' relationship. Attending to process is a powerful tool; failing to do so is a severe impediment to partnership.

2. *One must strive to understand the perspective of others (and one's own) to be an effective partner.* The process of dialogue by which partners explore each others' perspectives requires a basic stance of respect and openness. When diversity is valued as a resource, dialogue is synergistic; new possibilities will emerge that were unimaginable to any of the partners working in isolation.

3. *Shared decision making is the ultimate hallmark of partnership. It enhances motivation and consolidates mutual commitment.* The term *empowerment* is used to describe a restructuring of decision making in relationships that have a history of hierarchy and unilateral prerogatives.

Relationship, understanding, empowerment—these are the keys to partnership in healthcare.

1 Partnerships, Power, and Process: An Introduction

Patricia Hinton-Walker, Richard J. Botelho, and
Anthony L. Suchman

In this chapter, we will introduce the major concepts of this book: partnership, power, and process. We will briefly define and describe each of these concepts and an approach to integrating them. The rest of the book will provide further elaboration on these themes.

Partnership

The mission of improving the nation's health care system has three interrelated goals: to provide more equitable access to health care; to enhance the quality of health care; and to improve the economic performance of the health care delivery system. Each of these goals has implications for and depends upon partnerships throughout the system.

The potential partners are legion. Partners at the micro level of health care delivery work in the community at ambulatory care settings and/or in hospitals, and they are involved directly in patient care or its local administrative support. These partnerships can involve any combination of patients, their family members, health care providers of many disciplines, office staff, and administrators. Partners at the macro level of health care delivery are involved in working with health care access, system design, ethics, legislation, and policies. Those partnerships involve politicians, academicians, ethicists, economists, provider organizations, third-party payers, health care professional organizations, and governments, among others.

In the United States, the system of partnerships between these various constituencies is fragmented and unstable. Partners often shift their alliances to represent their own interests while also advocating for the goals of access, quality, and cost efficiency; conflict is inevitable. A lack of communication, understanding, and shared mission has created chasms within and between the micro and macro levels. The potential partners at the micro level have focused on delivery of care with particular attention to quality and, to a lesser degree, access. However, they have not generally attended to the causes and consequences of rising health care costs, nor have they initiated solutions. At the macro level, payors, government, provider organizations, and other potential partners have focused on policy and system design with an emphasis on cost and, to a lesser extent, quality. They have not given adequate attention to access and the personal experience of care (from both the patient's and provider's perspectives).

Collaborators at all levels must learn more effective strategies for representing their own needs, looking at the needs of the system as a whole, and addressing conflicts to avoid undermining the success of the overall mission of health care reform. Partners will need to learn how to identify common goals and to explore, respect, and even value differences. They will need to use power in new ways, using win-win approaches that foster collaboration rather than win-lose approaches that create antagonistic competition and mistrust. They must become sensitive to the nuances of language and of process communication (see below) to minimize miscommunication. As they refine their negotiation skills, partners can learn to build upon each others' perspectives and expertise synergistically, achieving better outcomes than they ever could alone.

Power

The second major concept in this book is power. There are many descriptions and names for power in the literature. French and Raven[1] have identified six power bases in their study of power: reward, coercive, legitimate, referent, expert, and informational power. There are also references to personal power, positional power, and charisma. Although power is neither good nor bad, the term frequently evokes negative feelings such as fear and discomfort. These negative perspectives sometimes come from personal and professional experience with the misuse of power to manipulate or coerce. The hierarchical or patriarchal relationships which dominate political and economic institutions in the United States (including health care) are based on a top-down, high-authority, high-control orientation. This type of relationship reinforces both dependence and mistrust in people in subordinate positions—patients, families, and communities, in the case of health care.

Fortunately, in the context of partnership, "power" can take on new meaning. Rusk, a physician, has defined power as "the energy to get things done, and the ability to influence people."[2] He offers a new perspective on power—creative influence without control—which focuses on relationship and team building to increase the energy and commitment of individuals, to create a whole that is greater than the sum of its parts. This pluralistic, participatory approach is well suited to decisions regarding health care, treatment, and care of various populations in the community, decisions which must respect the diversity of feelings and values of the participants. This perspective on power in partnerships acknowledges our interdependence and encourages us to balance control with cooperation and collaboration.

Block[3] identifies four important requirements related to power in partnerships. He advocates that there must be an "exchange of purpose and definition of boundaries" which gets defined through dialogue. Secondly, recognizing that there will always be different levels of authority in decision-making, each partner must retain "the right to say no." Third, there must be "joint accountability," with each assuming some responsibilities for the outcome. Fourth, there must be "absolute honesty." Unfortunately, patriarchal relationships tend to create social distance and vulnerability that sometime cause us to avoid the truth. Similarly, Rusk identifies four core values that guide the use of power to maintain partnership: respect, understanding, caring, and fairness.[2]

Applied in this way, power can be a positive force and have a positive impact on decisions that govern each of our lives, personally and professionally. However, as partners at the micro level and macro levels, we must all learn new behaviors if we are to form partnerships to improve health care delivery. We must learn to choose partnership over patriarchy in roles that have traditionally been dominant, and autonomy instead of dependency in traditionally subordinate roles. Toffler[4] predicted just this type of shift in power in medicine. He saw the demise of the "god in a white coat" and a shift of responsibility for decision-making towards consumers as the latter had increased access to medical knowledge.

As health care professionals attempt to develop effective relationships with patients, other providers, and communities, power must be used more creatively and balanced carefully. Subsequent chapters in the book will provide examples of power use in the context of partnerships to add new energy and unleash creativity in solving health related problems for individuals, families, and communities. Emphasis on autonomy, attention to values, and some new perspectives on balancing power will inform the process of building partnerships.

Process

The third fundamental concept of this book is process. In observing an interaction, we can readily identify its content: the topics which are discussed and the direct, surface level meaning of the words which are spoken. However, there are other important levels of meaning which are expressed through the *process* of the interaction, that is, *how* it takes place: the timing and sequence of statements, the gestures, the inflections, the nuances of the words and imagery, even what is left unsaid.[5] This is how we first discern hidden agendas or recognize unexpressed feelings. Implicit understandings about what topics will or will not be discussed are often reached at a process level without ever being discussed explicitly. Most important, information is communicated about the status of the relationship between the interactants: how is power distributed; do the participants trust or care about one another? Each of these issues can be recognized in the following fragment of a clinical interview:

Dr: So what seems to be the problem?
Pt: Well, I have had a headache for about 3 weeks now and . . .

Dr: (interrupting) Headache, huh? What part of your head is affected?
Pt: Well, it's across my forehead, more often on the right. Sometimes it hurts so much that I really get scared, like I might have a brain tumor or something.
Dr: Does your vision ever get all shimmering, like flashing lights?
Pt: No, but I . . .
Dr: Does it hurt you to bend your neck forward?

The patient in this scenario might come to perceive that it is his role to answer questions and the doctor's role to ask them, thus giving the doctor complete control over the agenda of the visit. He might get the impression that pain is a suitable subject of conversation but fear is not, and thus refrain from speaking about feelings again. He may feel that he is being treated impersonally and conclude (possibly erroneously) that this doctor doesn't care about him. These messages and the responses they engender may not be consciously recognized at all, yet they exert a powerful influence on the expectations, behaviors and reactions of the participants, and thus have a major impact on the ultimate outcome.

Process communication occurs at an organizational level, too. Such details as seating arrangements at a meeting, choices about what does or does not get on the agenda, or the composition of a committee can all convey information about power and relationships within the group which again may or may not be perceived consciously.

More often than not, process is invisible to us; we habitually focus our attention on content and on the outcomes we seek. Thus we can easily find ourselves conveying meanings through our process that we would find unacceptable were we aware of them. The clinician in the interview fragment above meant to be caring. She was conscientiously doing what she had been taught to do—to make a diagnosis. However, by focusing so intently on this one objective, she failed to notice the patient's emotional needs, and she was oblivious to the effects her style of questioning might have on the patient. When we neglect process, we needlessly jeopardize outcomes; we damage relationships and get in our own way.

To remedy this, we can learn to monitor our process, to pay attention to the values which we are expressing through our actions. Using various recording technologies or simply through reflection, we can observe ourselves from the perspective of the other person. What do we do that communicates respect and invites collaboration? How do we try to control others and diminish their autonomy? How well do we listen? Using a set of core values as a reference standard, we can assess the strengths and weaknesses in our communication and relationship skills and work to improve them. This practice of ongoing reflection, so awkward and uncomfortable at first, will soon become second nature. With clear and respectful process, we will truly be able to get out of our own way and harness the full potential of partnerships.

References

1. French J. and Raven B. *The Bases of Social Power*. In Cartwright, D. & Lander, A (eds.): *Studies in Social Power*, Ann Arbor: University of Michigan Institute for Social Research, 1959.

2. Rusk T. *The Power of Ethical Persuasion*. New York: Penguin Books USA, Inc., 1993.

3. Block P. *The Empowered Manager: Positive Political Skills at Work*. San Francisco: Jossey-Bass Publishers, 1987.

4. Toffler A. *Powershift*. New York: Bantam Books, 1990.

5. Schroder AK, Reiser DE. The Interview Process. In Reiser DE, Schroder AK (eds.) *Patient Interviewing: The Human Dimension*. Baltimore: Williams and Wilkins, 1980.

II Control and Relation: Two Foundational Values and Their Consequences

Anthony L. Suchman

The deepest level determinant of partnership—in health care or anywhere else—is our internal predisposition as would-be partners: our attitudes, values, and beliefs. Each of us has a personal philosophy that shapes our perceptions, interpretations and actions. Addressing questions such as why things happen, what is right and what is wrong, and why we are here, our personal philosphies give rise to our own individual ways of being in the world; they also frame our goals and expectations. In adopting a particular personal philosophy, we create the world we inhabit.

Considering its profound importance and implications, the acquisition of our personal philosophy is remarkably uncritical. This process—a gradual assimilation of attitudes and values from our families, culture, education, and life experiences—tends to be so subliminal that we may not recognize our core beliefs as beliefs at all, and instead simply accept them as "reality." We may not recognize that our "truths" are in fact only "assumptions," and that better alternatives might exist.

In this chapter, we will examine one of the most fundamental values in Western culture, one that figures prominently in the personal philosophies of most health professionals: the quest for control. As we explore the control paradigm with regard to its core values, goals, patterns of social relationships, approaches to gathering and using knowledge, and sources of existential security (Table 1), we will see the limitations it imposes on the process of partnership. We will then examine an alternative paradigm, one based on relation, that creates a climate more conducive to creative collaboration and the sharing of responsibility. Although we will use clinicians as exemplars throughout this discussion, the concepts are equally applicable for educators and administrators.

In making an explicit comparison of the control and relation paradigms, we can learn to perceive the values implicit in everyday interactions and to observe their consequences. Then, perhaps for the first time, we can make thoughtful choices about which paradigm to embrace as our own.

The Paradigm of Control

Core Values and Goals

The beliefs, thoughts and behaviors of the control paradigm are organized around a single core value: that the ultimate state to which one can aspire is one of perfect

Table 1: A comparison of critical features of the paradigms of control and relation with regard to core values and goals, patterns of social relationships, approaches to gathering and using knowledge, and sources of existential security.

	Control	Relation
Core values and goals	willfulness; predictability	connectedness; alignment
Primary focus	outcomes	process
Clinician's primary responsibility	fixing patients	being with patients
Social structure	hierarchy	partnership
Use of power	limit choices of others	increase choices of others
Negotiation strategy	win-lose	win-win
Epistemology	reductionistic, analytic	reductionistic and emergent; analytic and experiential
Relevant domain of experience	objective only	objective and subjective
Priority of knowledge	general > particular	general = particular
Implications for clinician's stance	detached observer; affectively neutral	participant observer; fully present
Core affect	fear	trust
Stance	mastery	receptivity
Source of security	self sufficiency	alignment; interdependence
Existential strategy	holding tight	letting go

willfulness and predictability. What one desires happens, with no surprises; all outcomes are intended. For the clinician, the control paradigm is expressed in the questions, "What do I want to happen here?" and "What's wrong and how do I fix it?" The "will to fix" extends well beyond the patient's immediate experience of suffering or dysfunction to encompass intermediate outcomes believed to be associated with future suffering (for example, reduction of blood pressure or glycohemoglobin levels). Thus, the paradigm of control defines the clinician's role as taking deliberate action towards an improvement in the patient's current or future experience. Personal success or failure is judged by the clinical outcome—the extent to which one's intended outcome was realized.

Social Relationships

Interpersonal relationships in the control paradigm are hierarchical. At the level of the physician-patient relationship, for example, the physician, by virtue of expertise and responsibility, makes decisions on behalf of the patient, who is expected to comply. This use of power to limit the choices of others poses no ethical problem for the physician, so long as the intended outcome is the patient's improvement. Relationships on the health care team are hierarchical, as well. Differences in educational and social status, defined areas of responsibility and professional prerogatives create expectations of deference and obedience. In situations of conflict, the control paradigm predisposes the physician to focus on goals and to expect to be dominant, resulting in a win-lose negotiation style (if negotiation takes place at all).

Epistemology

The control paradigm shapes both epistemology—the way we gather and organize knowledge—and clinical reasoning. With control as the principal value, the task of intellectual inquiry is to develop detailed knowledge about causal mechanisms and how to manipulate them. The mode of analysis to produce such knowledge is reductionistic and mechanistic, and tends to move from the particular to the general; that is, it values knowledge of general principles and theories more than knowledge about individual experience (which is dismissed so often as "anecdotal"). When applied to clinical work, this approach leads the clinician to encapsulate an episode of a patient's life experience with a single diagnostic label. Just by this act of classification, the clinician gains a degree of intellectual control and moves from a state of uncertainty towards one of mastery. Treatment is chosen more on the basis of generalized knowledge about the diagnosed disease than about the particularity of the patient's experience. So in both making a diagnosis and planning a treatment, what is unique and particular about an individual patient is less valued and therefore gets less attention than the abstracted characteristics of the "case."

Rational thought, in its orderliness and reproducibility, is most consistent with the philosophy of control, and is therefore valued above all other forms of mental experience. Personal meanings, interpretations and emotions, being "irrational," are deemed inferior, hence less worthy of attention. Objective experience, as the substrate of rational thought, is deemed more relevant—even more real—than subjective experience. In focusing on their own objective experience and suppressing their subjectivity, clinicians take the role of detached observer—distanced from the object of observation and affectively neutral.

Existential Security

The core of the control paradigm—the source of its compelling power—lies deeper than cognition and behavior, at the level of existential security. Questions at this deepest level include: What is the nature of the world? What stance does one take? Where does one find one's grounding? At the heart of the quest for control is fear—a perception of the world as lacking any intrinsic order and of danger lurking in the chaos. One's grounding, or source of security, is in predictability—the order one imposes on the world through collective and individual efforts. Collective efforts include rules of interpersonal engagement and knowledge structures as described above. Individual efforts include staying within prescribed boundaries, obeying rules and resisting the "irrational." Relaxing one's vigilance and losing control of the details of one's life opens the doors to chaos and to bad things happening. And the worst thing of all is something bad happening that could have been prevented but wasn't. The clinician's existential stance in the control paradigm—his basic way of being in the world—is one of mastery: of the instrumental knowledge of medicine and of other people; in metaphorical terms, a stance of holding tight. We may trust others to a point, but we trust nothing and no one as much as we trust ourselves. Therefore, mastery, vigilance and self sufficiency are the ultimate sources of existential security, the best way to keep the fear at bay.

Limitations of the Control Paradigm

That the control paradigm has channeled human energy towards many remarkable technological accomplishments is undeniable. However, as a philosophy for guiding human interactions and for providing existential grounding, it has serious limitations. These are especially significant in medicine, where technology and human experience intersect with particular drama and salience.

The control paradigm interferes in a variety of ways with the formation of relationships, the medium of partnership. The instrumental, reductionistic approach of the control paradigm and the valuation of the abstract general over the particular leads to an inadvertent depersonalization of others. Patients, students and subordinates become objects to be controlled. The hierarchy of the control paradigm blocks the free flow of communication that is necessary to achieve synergistic interaction. It promotes expectations for those lower down in the hierarchy that the people higher up will be able to control things, and leads them to expect little of themselves, thus promoting passivity. There is little opportunity for the kind of mutual goal setting that enhances motivation and commitment. The full power of the team goes untapped.

At the level of thinking and action, the control paradigm encourages an appealing yet utterly unrealistic fantasy of personal control, focusing on the individual as the primary locus of agency. It interferes with the perception of systems and the recognition of emergent phenomena—synergistic processes which are effectively and spontaneously orchestrated through complex interactions among individuals.

The control paradigm accounts for a number of common problems in clinical practice. For example, consider the way doctors become angry with patients who do not take their medicines. The physicians' sense of personal success and mastery depends upon the clinical outcome; the patients' "noncompliance" jeopardizes the outcome and thus threatens the physicians' well being. Consequently, the physicians become angry, as if it was they and not the patients, who had to bear the ultimate consequences. Another common problem is the abandonment of patients with chronic or terminal illness. In these settings, clinicians are predisposed to experience their inability to cure as failure. Not surprisingly, they tend to withdraw, or they may cling to inflated expectations of success and undertake excessive treatment, frequently substituting their own goals for those of their patients.

Ultimately, the control paradigm fails as a source of existential security. It is built on two false premises. First, it is not possible to implement perfectly what one wills, and second, even if it was possible to do so, the outcome would remain unpredictable. We know from chaos theory (enshrined in conventional wisdom as the Law of Unintended Consequences) that predictability pertains only within very narrowly defined situations, and is generally a poor construct for modeling natural and social phenomena.[1] Complexity prevails. Expectations of control are created that cannot be fulfilled, resulting in feelings of disillusionment, anger and personal failure (known also as "burnout"). The control model ultimately must prove to be an inadequate map; at some point its discrepancies with the real world must cause it to create more problems than it solves.

The Paradigm of Relation

Core Values and Goals

An alternative paradigm organized around the core value of relation has the potential to avoid many of the problems of the control paradigm.[2] In the relation paradigm, the most valued state to which one aspires is one of connection and belonging. In this state, one has a feeling of being part of a larger whole—a team, a learning group, a dance troupe, a community, even the world itself. One's individual actions seem spontaneously integrated with those of others to a remarkable degree, contributing to the evolution of a higher order process, i.e., one at a higher system level than that of the individuals of which it is comprised. Words like "serendipity," "flow," "synchronicity," "self transcendence," and "oneness" have been used to describe this ineffable state.

The relation paradigm holds that the best outcomes are realized by maximizing the quality of process, which in turn depends on the quality of relationships and on one's ability to be genuinely present to others and to events as they are unfolding. As often as not, the resulting outcomes will be different from—and better than—those that were originally intended. Being in a state of relation does not imply passivity. Rather, it involves active attentiveness to both inner and outer experience. One asks the question, "What's trying to happen here?" and, according to one's best approximation of an answer, seeks to shape others and the world while also remaining open to being shaped oneself. This balance between control and receptivity puts one in the best possible position to recognize and make use of serendipitous events.

Working within the paradigm of relation, the clinician's role is to "be with" patients: to acknowledge their suffering and identify opportunities to relieve or prevent it, while also respecting their position as principal makers of values, meanings and decisions in their own lives. Thus, the professional is accountable not so much for the outcome as for the quality of the process: the degree to which one understands what the patient is experiencing; communicates that understanding back to the patient; helps the patient formulate goals and priorities; and applies one's expertise and commitment to the creation and implementation of a plan for realizing the patient's goals. These expectations afford the clinician more opportunity to feel successful than do the control paradigm's expectations of perfect clinical outcomes.

Social Relationships

Interpersonal relationships under the relation paradigm are characterized by respect and genuineness. Martin Buber characterized such interactions as "I-Thou," denoting a relationship between two subjects, as contrasted with "I-it," denoting the subject-object relationship of the control paradigm.[3] One acknowledges the subjective experience, personal perspectives, values, and aspirations of others and one's own as well. Self awareness is critically important, as is the ability to act in congruence with one's feelings. One cannot be in a state of relation without being fully present.

The use of power is another key point of differentiation between the relation and control paradigms. The relation paradigm does not negate power or deny the very real differences in power between patients, doctors, nurses, administrators and others. Rather, it maintains that (1) these differences do not imply differences of worth, dignity or the right to have one's perspective respected and (2) to the extent that one has power, one uses it to foster both the power and the accountability of others to the fullest extent possible (within the bounds of the individual's capacity and the external circumstances). In the realm of negotiations, this is expressed as seeking solutions which maximize mutual gain and strengthen mutual commitment (win-win) rather than maximizing one's own gain at the expense of the other (win-lose).

The receptivity of the relation paradigm leads one to value diversity. The unique background, perspective and opinion that each participant brings to an interaction or a team can be a resource to the group (provided it is approached respectfully), just as intra-species diversity contributes to the adaptability and long term survival of a species. However, this attitude of receptivity should not be mistaken for a "do-your-own-thing" kind of uncritical tolerance. One cannot avoid making and acting on judgements about quality if there is to be accountability.

Epistemology

The epistemology of the relation paradigm seeks to join the instrumental, abstract knowledge of the control paradigm with a deep appreciation of context and ecology—a science of stories.[4] This has two key implications. First, one must recognize the existence of "emergent" phenomena, those which arise at higher orders of a system and depend on—yet cannot be explained in terms of—lower order processes. For instance, a cardiology consultation could not take place without normal synaptic function in the brain of the cardiologist, but one cannot understand or even describe the process of making a diagnosis by analyzing neurotransmitters. "Emergentism" can be seen as a style of analysis that moves across the hierarchy of systems in the opposite direction as reductionism, employing very different tools. Both approaches produce useful, distinct and complementary types of knowledge. Second, one must accept data regarding subjective experience as being equally important and relevant as objective data. Rational thought tends to support instrumentality; emotion, spiritual experience and intuition tend to be the substrate of relation.

When translated to clinical reasoning, this epistemology leads us to discover and value what is unique about each patient and his clinical situation, and not to concentrate solely on the abstractable, generalizable features. The accurate understanding of the patient's experience and communication of that understanding back to the patient is a vital therapeutic process,[5] and an essential part of relationship building.[6] Likewise, the clinician must be self-aware to be capable of relationship.[7] So information about the subjective experience of the patient and clinician is important information. Rational analysis and experiential exploration both have their place. The clinician's role, far from one of detachment, is to participate and to observe both her outer (objective) and inner (subjective) experience.[8]

One last aspect of epistemology for us to consider is the relinquishing of answers. Each time one arrives at an answer, one's desire for mastery may be satisfied, but if answers terminate inquiry, further discovery (and the revision of erroneous answers) is precluded. One needs provisional answers to undertake action; medical care, education and administration constantly demand action in the face of uncertainty. Nevertheless, one must avoid holding too tightly to answers and be willing instead to remain always "in the question."[9]

Existential Security

At the core of the relation paradigm is trust. This philosophy accepts that there are sources of order, goodness and meaning beyond one's own creation. This source may be at a collective human level (as exemplified by the consistent performance advantage of well functioning teams over individuals) or on an even higher order: the "Tao," "God," and "Great Spirit" are but a few of the names humans have given in reference to universal sources of order and meaning. Regardless of scale, the basic existential stance in the relation paradigm is one of self transcendence and receptivity, of letting go of control and remaining open. One's source of existential security might best be characterized as alignment: recognizing the interdependence of oneself and others in a larger order of things and learning to be part of the flow of that order.

Further Implications of the Relation Paradigm

It is important to recognize that the relation paradigm does not negate the control paradigm; rather, it promotes balance by adding receptivity to mastery, subjective to objective, emergent to reductionist, participation to autonomy. This last pair—participation and autonomy—has been particularly problematic for practitioners in today's health care environment. The apparently paradoxical convergence of these two opposites can be accomplished using Arthur Koestler's conceptualization of the *holon*.[10] A holon is an entity which is a complete and autonomous whole unto itself and at the same time a part of a larger whole. A holon thus has properties of both agency (corresponding to autonomy, instrumentality and control) and communion (corresponding to participation, context and relation). Until recently, physicians practiced autonomously, making clinical and business decisions from the perspective of their individual patients and their "cottage industry" practices, respectively. They were conscious only of their agency, their completeness in themselves as loci of decision making. Now in this time of integrated health systems, they must maintain their agency, but at the same time think at higher order system level about the care of a population and about the needs of the integrated network. The relation paradigm encourages awareness of multiple system levels and allows for coexistence, rather than mutual exclusivity, of control and relation.

Looking at personal philosophies from a perspective of adult development, we can observe that aspects of the control and relation paradigms correspond to different stages of personal growth as described in two models. In Erikson's model, differentiating oneself from the world and coming to know oneself as an independent locus of agency and control is the culmination of childhood development and arrival into adulthood. However, as one progresses through the adult stages of intimacy vs. isolation (Stage VI), generativy vs. stagnation (Stage VII) and integrity vs. despair (Stage VIII), one increasingly transcends the boundaries of one's self, "[creating] new conditions for, and new avenues of experiencing, enabling a richer empathy, relatedness, and identification with life beyond-the-self.[11 p.201] These themes resonate with willfulness versus alignment, mastery versus receptivity, detachment versus participation and holding tight versus letting go.

Perry's model of adult cognitive and ethical development describes nine stages of growth progressing from Dualism ("Division of meaning into two realms—Good versus Bad, Right versus Wrong, We versus They, All that is not Success is Failure, and the like. Right Answers exist *somewhere* for every problem, and authorities know them") to Commitment in Relativism ("Diversity of opinion, values, and judgement derived from coherent sources, evidence, logics, systems, and patterns allowing for analysis and comparison. Some opinions may be found worthless, while there will remain matters about which reasonable people will reasonably disagree. Knowledge is qualitative, dependent on contexts." "I must be wholehearted while tentative, fight for my values yet respect others, believe my deepest values [to be] right yet be ready to learn.").[12] Here we find resonance with the themes of reductionism and emergence, outcomes and process, and living in the question.

Taken together, these two models describe a process of personal evolution in which one's focus moves from implementing narcissistic willfulness to seeking connectedness, and one's source of grounding shifts from the existence of Authority and external absolute Truths to the experience of self transcendence and connection.

We have been considering the control and relation paradigms in the context of an individual's personal philosophy, but these paradigms are manifest in organizational process and behavior, as well.[13] Given the considerable influence that organizations have on the perceptions, values and behaviors of the individuals who populate them, we might look to the modification of organizational process as a powerful tool for fostering personal transformation (i.e., changing personal paradigms).[14,15]

Conclusions

The control paradigm limits partnership. By focusing our attention on the realization of our will, on outcomes and on objective phenomena, it predisposes us towards dominance and distracts us from making observations and taking actions that might help us cultivate relationships. It creates additional problems by setting unrealistic standards for success and cutting us off from our own subjective experience. The relation paradigm avoids these pitfalls and predisposes us towards partnership. It reminds us to attend to process and to personal experience (our own and that of others), not just to objective data. It encourages us to be perpetually skeptical of answers and their false promise of predictability, and instead to live in the question, remaining receptive even as we undertake specific and decisive courses of action. Once we are able to perceive these two philosophies and their consequences, we are then able to reflect on our own experience and to make thoughtful choices about which paradigms help us live in the most meaningful and effective manner and to move forward in our personal growth.

References

1. Gleick J. *Chaos: Making a New Science*. New York: Penguin Books, 1987.
2. Use of the concept of "relation" as the core of an alternative to the control paradigm

was inspired by arguments in Toulmin S. *The Return to Cosmology*: Postmodern Science and the Theology of Nature. Berkeley, CA: University of California Press, 1982:217–274.

3. Buber M. *I and Thou*. 2nd ed. New York: Scribner's, 1958.

4. Stange KC, Miller WL, Crabtree BF, O'Connor PJ, Zyzanski SJ. Multimethod research: Approaches for integrating qualitative and quantitative methods. J General Internal Med 1994;9:278–281.

5. Suchman AL, Markakis K, Beckman HB and Frankel R. A model of empathic communication in the medical interview. JAMA 1997;277:678–682.

6. Cohen-Cole SA. *The Medical Interview: The Three Function approach*. St. Louis, MO: Mosby, 1991:21–26.

7. Novack DH, Suchman AL, Clark W, Epstein RM, Najberg E, Kaplan C. Calibrating the physician: Personal awareness and effective patient care. JAMA 1997; 278: 502-509.

8. Engel GL. How much longer must medicine's science be bound by a seventeenth century world view? In White KL, ed. *The Task of Medicine. Dialogue at Wickenburg*. Menlo Park, CA: Henry J. Kaiser Family Foundation, 1988:113–36.

9. Goodrich-Dunn B. Walking the critical path. *Common Boundary* 1991 (July-August): 12–21.

10. Quoted in Wilber, K. *A Brief History of Everything*. Boston: Shambala, 1996:20.

11. Acklin MW. Adult maturational processes and the facilitating environment. J Religion and Health 1986;25:198–206.

12. Perry WG. Cognitive and ethical growth: the making of meaning. In Chickering A (ed.), *The Modern American College*. San Francisco: Jossey-Bass, 1981:76–116.

13. Senge P. *The Fifth Discipline*. New York: Doubleday, 1990.

14. Senge P. *The Fifth Discipline Fieldbook*. New York: Doubleday, 1994.

15. Brill PR, Worth R. *The Four Levers of Corporate Change*. New York: AMACOM, 1997.

III Negotiating Partnerships in Healthcare: Contexts and Methods

Richard J. Botelho

The current changes in the US healthcare system have enormous implications for the development of administrative and clinical partnerships. A myriad of factors are creating a demand for unprecedented degrees of collaboration as the current fragmented architecture of "cottage industries" gives way to integrated systems. Leaders, managers, and clinicians need to develop administrative partnerships to design and operate these new systems. Healthcare teams will be called upon to work smarter and to function across a wider portion of the spectrum of care. New levels of partnership between patients and clinicians will be needed to implement population-based healthcare and to resolve its potential conflicts with individually focused care. A variety of obstacles to partnership are also emerging, notably the tension between population-based and individual-focused practice, the corporatization of medicine and the increasing presence of the for-profit sector.

This chapter will address administrative and clinical partnerships. Section I will review in more detail the changes in the healthcare system that constitute the backdrop for partnership formation at all levels: both the imperatives and the obstacles. Section II will describe the detailed behaviors that practitioners can employ to improve the process of partnership and therefore the outcomes of care. Practical examples are provided demonstrating how patients can be engaged as active participants in decision making and in adopting health-promoting behaviors. These techniques can also be applied to promoting self care in chronic diseases.

Developing Administrative Partnerships in a Changing Healthcare System

In the past, our society supported a healthcare delivery approach that rejected limits on healthcare expenditures. All utilization decisions were made within the sacrosanct boundaries of the doctor-patient relationship. This arrangement was culturally consonant with our individualistically oriented society.[1] For the most part, neither third-party payers (employers and government) nor insurers encroached on the decision making of this dyadic relationship. It was in the economic self-interest of insurance companies and physicians alike not to control healthcare costs.

The policy of unlimited healthcare expenditures benefitted physicians' income and helped to expand the growth of the healthcare insurance industry, but it did not help the major payers of healthcare, namely, employers and the government. According to the World Bank, the US accounted for 41% of the world healthcare expenditure ($690 billion, 13% of its gross national product) in 1991.

This economically permissive approach to utilization operated without much public accountability for distributive justice, cost-benefit considerations, and clinical effectiveness. Specialists provided individual patients with high-tech, procedurally oriented care, that contributed to wide practice variations and even some unnecessary care.[2] This approach failed to serve the common good, left large segments of US society with inadequate access to healthcare, and contributed to escalating healthcare costs.[3,4]

The moral and economic crisis in healthcare activated two groups of leaders to address the issues of access, cost, and quality. Political leaders addressed issues related to access to healthcare and cost reduction. Employers revolted against paying for escalating healthcare costs and demanded accountability; managed care companies were called upon to increase the value of their services to employees. (Value is a function of quality divided by costs.) These political and economic leaders raised questions about who is responsible for assuring universal access, who controls healthcare costs, and who has the power to insure that quality services are provided.

The US remains the only developed, democratic country that does not regard healthcare as a human right.[5–7] Consequently, no agency (private or public) has assumed overall responsibility for assuring universal access to healthcare and for controlling healthcare costs.[7] The primary goal of the Clinton administration's health-reform bill was to assure the right of all citizens to healthcare coverage.[8] Initial optimism about passing federal legislative reform dissipated[9–11] as powerful, nonelected interest groups and political opponents within both parties effectively undermined this overly ambitious proposal. Many reasons account for this reform failure: fear of government control, opposition of powerful interest groups to change, lack of bipartisanship and moral leadership, and a complex legislative process—to name a few major factors.[12–20]

The moral principle of "healthcare as a human right" has been ineffective; ironically, the language of human rights has hindered rather than helped to achieve universal access to healthcare. "Healthcare rights" connotes entitlement rather than need. The term "entitlement" is associated with government control and anti-welfare sentiments. Consequently, these associations make healthcare rights an easy target for mobilizing public antipathy against government involvement in healthcare. After the healthcare reform debacle,[21,22] the moral authority and political courage of the Federal government dissipated. Both political parties avoided addressing the access to healthcare problem in the 1996 election year. Given the political leadership vacuum, government and employers turned to managed care.

For-profit, managed companies capitalized on this economic opportunity to benefit shareholders, third-party payers, and chief executive officers, in the name of providing cost-effective, quality care to those who can afford it. These companies took over the economic leadership of healthcare reform, leaving the moral issue of universal access to healthcare on the back burner.[23] They have no responsibilities to address the access problem other than to increase their market share. However, they reorganized an economically out-of-control healthcare system that primarily benefitted the financial self-interests of insurance companies and physicians; it changed the hierarchy of beneficiaries in healthcare: third-party payers, shareholders, chief executive officers, and patients — often in that order.

Healthcare delivery is in a state of rapid change, with a shift toward primary and preventive care, evidence-based approaches to clinical practice, for-profit healthcare delivery systems, continuous quality improvement[24–30] and public accountability in terms of how managed care companies perform. The Health Plan Employer Data and Information Set (HEDIS) reports provide employers with data about the performance of managed care companies, which enables them to make more informed decisions about how to select healthcare coverage for their employees.[31] Competition and innovation drive companies constantly to increase value and quality to customers.[32,33] The economic "savings" from this new-found efficiency benefit shareholders, and thereby reduces the healthcare dollars available for patients.

To its credit, this "economic reform" has advanced population-based approaches to healthcare and developed controls over healthcare expenditure, as opposed to the individually focused, laissez-faire approach. In effect, profits are made from decreasing patient consumption of limited healthcare resources.

The corporatization of healthcare has shifted economic control away from physicians and restructured the delivery of care to reduce costs. Employers, government, insurance and managed care companies have developed a myriad of ways to structure the healthcare delivery system. These changes have heightened the challenge of developing effective partnerships in healthcare in terms of how to collaborate and solve problems together. These factors now affect and impinge on the practitioner-patient relationship, and some of them generate conflicts about how to deliver patient care. These macrolevel changes challenge practitioners, administrators, and patients to develop effective partnerships in healthcare by adopting new roles and negotiating about these conflicts.

Just as nonprofit and for-profit managed care companies are being held accountable to employers and government in terms of reducing costs and increasing quality, practitioners are also being held accountable for the same issues. Administrators, often in collaboration with practitioners, have developed policies (such as restricted drug formularies, practice guidelines[34] and clinical pathways[35]), monitored utilizations (such as practice pattern variations), quality, and costs, and evaluated practitioner performance in a number of specific areas (patient satisfaction, access, and preventive care).

Although practitioners may have some influence over these changes, administrators and managers are the principal organizers. They have increasing responsibility for how practitioners perform in their work. In many instances, practitioners have become the "patients," with administrators and managers becoming the "practitioners" of the healthcare delivery systems.

These administrative partnerships have similarities to clinical partnerships in terms of relational and problem-solving processes, although the tasks are different. Administrators and managers work together with practitioners to address health issues. These partners have different perspectives and different levels of expertise in the clinical, educational, administrative, economic, population-based, political, and legal aspects of healthcare. These differences may give rise to conflicts. For example, administrators emphasize cost-savings, and practitioners advocate for quality, and sometimes these goals conflict with one another. In some situations, practitioners internalize the cost-quality dilemmas when they have incentives to use resources judiciously. For example, patients expect referral to specialists of their choice, and practitioners benefit financially from reducing or restricting appropriate referrals to specialists to reduce their budget expenditures. However, the goals of reducing costs and enhancing quality are not always conflictual. Resources freed up by a reduction in unnecessary care may be redirected toward new effective services, such as evidence-based preventive care, which can reduce (or offset) costs, complications, and demands for unnecessary services and thereby enhance quality of care.

Two dynamic forces (the cultures of business and medicine) shape the opportunities and challenges of developing effective partnerships in an evolving health care system.[36–38] Dialectic tensions do exist between what is good medicine for patients in the long-term and what is good business for investors in the short-term. Table 1 summarizes and juxtaposes these opportunities and challenges to highlight issues in establishing effective partnerships. Practitioners, health administrators, and business managers can learn how to blend the cultures of medicine and business in ways that take a population-based and long-term perspective on improving healthcare outcomes.

Opportunities to Create Effective Partnerships

Stewardship is the pursuit of service over self-interest.[39] This altruistic pursuit resonates with a famous quote by J. F. Kennedy: "Ask not what your country can do for you. Ask what you can do for your country. " Within healthcare organizations, stewardship is the equitable, cost-effective, and appropriate use of limited resources and talents to benefit all members of the community. Leaders and managers in such organizations provide service to practitioners in ways analogous to how practitioners provide patient services to maximize the health of the community. Leaders, managers, and practitioners face the challenging task of balancing individual needs with that of the whole community. The driving force of such organizations is community health improvement; they set up reward systems for leaders, managers, and practitioners that are based on their community performance. The for-profit motive is a factor to the extent that it contributes to and enhances the overall quality of service.

Public accountability of healthcare organizations in terms of meeting patient and community needs is the driving force of stewardship.[39] Such accountable service avoids patriarchal use of control and the short-sightedness of professional eth-

Table 1: Opportunities and Challenges in Developing Effective Partnerships

Factors that create opportunities

1. Use limited resources in cost-effective ways
2. Focus on improving patient outcomes
3. Help organization to transform care
4. Use evidence-based methods to provide care
5. Use continuous quality improvement (CQI) methods to improve performance
6. Work in interdisciplinary ways
7. Use partnership skills to empower patients
8. Manage chronic diseases efficiently
9. Facilitate health behavior change effectively
10. Work collaboratively in a population-based manner

Factors that create challenges

1. Limit use of resources to maximize cost/loss ratio
2. Maximize organizational dominance
3. Help the organization serve investors
4. Use evidence-based methods to deny care
5. Use Quality Assurance (QA) methods to monitor for "bad apples"
6. Use interdisciplinary methods to downsize personnel
7. Lower consumer expectations and demands
8. Use utilization review methods to reduce patient services
9. Avoid addressing specific behavior change issues
10. Selectively recruit healthy patients

nocentrism. In other words, we assume that we know what patients and communities need and how to best meet those needs, but without really involving them in the process. Stewardship in healthcare organizations involves needs assessments: asking patients and communities what are their needs and priorities, and using professional, patient, family and community resources to meet those needs.

> "Stewardship is the choice for service,
> We serve best through partnership,
> Dependency is the antithesis of stewardship and so
> Empowerment becomes essential." (p.6)[39]

Leaders and managers work in autonomy-supportive ways to energize the workforce in an organized pursuit of meeting patient and community needs. This approach fosters intrinsic motivation and collective self-efficacy within the organization in order to achieve a common vision.[40,41]

Within autonomy-supportive organizations, leaders and managers redistribute power to the boundaries of the organization so that practitioners have greater choice about how to serve patients and their communities.[39] Such empowerment decentralizes leadership and management and puts it back where it belongs: where practitioners and patients work together. Effective partnerships help leaders, managers, and practitioners share responsibility and foster reciprocal commitments appropriately among themselves and their patients. The development of such partnerships generates a shared sense of purpose, collective contribution, and accomplishment, fulfilling this lofty goal of stewardship is intrinsically meaningful.

Stewardship in health care sets some parameters within which leaders, managers, and practitioners can develop effective administrative partnerships among themselves, which in turn helps practitioners to develop effective clinical partnerships with patients (see Table 1).

1.Work in cost-effective ways

When practitioners take a population-based approach, they are shifting away from their traditional role of advocating purely for the patient's benefit. This shift increases the complexity of the practitioner's advocacy role in improving the health of both the individual and the population, particularly when attending to the "triple-jump"challenge of reducing overall costs, enhancing quality, and improving appropriate access to healthcare. Cost-effectiveness studies can help practitioners understand better and address these challenges by providing data on the short-term and long-term effects on both individuals and the population as a whole.[42] The initial healthcare reform primarily focused on reducing costs, followed by increasing quality. So much energy is taken up by these two challenges that the US is now losing ground on the third challenge: universal access. These three challenges are not mutually exclusive. The neglect of this access issue may reduce medical costs, but it only increases social costs in other ways. For example, epileptic patients who fail to qualify for Medicaid may receive suboptimal treatment because they cannot afford the cost of their medication. Then, because of uncontrolled seizures, they miss more days from work, which, in turn, increases their financial problems and family stress. As a consequence of poor work performance and inadequate control of seizures, they lose their job and go onto Medicaid.

2. Focus on improving patient outcomes

An important issue to address is what constitutes a need, want, or demand, and what is an outcome.[43] An improved outcome may be:

- medical (e.g., preventable morbidity and mortality, reduced complication rates, and/or reduced pain);
- psychological (e.g., reduced depression, anxiety, and/or enhanced well-being and quality of life);
- social (e.g., improved work performance and/or reduced work absenteeism);
- functional (increased strength and range of motion and/or enhanced level in activities of daily living).

These outcomes are potential needs for patients, but patients may not want some of these needs met. For example, it is estimated that at least 50% of patients with psychiatric morbidity in the community need but do not seek mental health services.

Conversely, patients may appropriately seek mental services but receive ineffective or inappropriate psychological treatments. The distinction between practitioners serving their own professional needs and meeting patients' needs to improve outcomes is often difficult to clarify, particularly when these needs are complementary. Clarification of this issue is important when interventions (for example, some modalities of psychotherapy and complementary medicine) are of questionable or unproven benefit (no evidence that the intervention makes a real difference), and when both practitioners and patients believe that the intervention works. Professional protectionism and patient collusion with unproven practices can impede the appropriate use of limited resources and expertise in the most efficient manner. To avoid practitioners taking inadvertent advantage of patients, practitioners need to combine evidence-based approaches and continuous quality improvement (CQI) methods with cost-effective practices.

3. *Help the organization transform healthcare delivery*

Managed care companies can create a vision, culture, and mission to transform healthcare delivery systems from a reactive to a proactive organization to achieve any of the following goals:

- promote patient self-care of disease;
- develop comprehensive prevention programs, including health behavior change;
- reduce unnecessary healthcare utilizations;
- improve access of services to meet patients' needs.

In collaboration with leaders and managers, practitioners are the key players to help their organization make such a transformation.

4. *Use evidence-based methods to justify clinical practice*

Practitioners can use evidence-based methods to influence which clinical approaches to use or not to use. These methods can guide administrators and patients to use interventions for which there is definite evidence of their effectiveness, and to reject interventions for which there is evidence for their ineffectiveness (for example, screening for prostate cancer).[44,45] Evidence-based methods can guide the development of practice guidelines and clinical pathways.[46]

5. *Use CQI methods to enhance performance*

Practitioners, administrators, and practice managers use continuous quality improvement methods to implement and maintain preventive care, practice guidelines, and clinical pathways in health settings, both in hospital and primary care.[47–52] Practice managers provide feedback to practitioners in ways that can enhance the practitioners performance in working with practice guidelines and patients: for example, reducing smoking cessation rates, lowering hemoglobin A1c in patients with diabetes, enhancing immunization rates in children, and increasing the use of mammography and pap smears for women. In fact, such feedback to practitioners about their performance is more powerful than their knowing the guidelines and discussing review criteria.[53]

6. *Work in interdisciplinary ways*

Many practitioners work as part of teams to provide healthcare services. To maximize the performance of the interdisciplinary team, practitioners need to work in synergistic ways. This approach represents a shift from the solo practitioner to a team approach in which the primary care physician is often responsible for coordinating the delivery of healthcare in an efficient manner.

7. *Use partnership skills to empower patients*

Practitioners can empower patients and families to promote their autonomy, responsibility, and control in addressing health issues and in reducing the need for healthcare services. Managed care systems recognize that self-care is the first level in the healthcare delivery system. Practitioners can play a major role in promoting self-care. They can use their partnership skills to help patients take a greater role in self-care and preventive care and to:

- adapt to acute diseases and adjust to illnesses more effectively;
- assess and manage their own psychosocial problems;
- take greater control over and responsibility for managing their chronic disease;
- improve their functional abilities;
- minimize dependency upon their practitioners and the healthcare system

This is a "need reduction" approach that prevents unnecessary demand on the healthcare system. To work toward this goal, practitioners need to clarify with patients whether they are able to manage their own health issues, or whether they need additional professional assistance, in addition to self-care.

8. *Manage chronic diseases efficiently*

Chronic disease programs assist practitioners to enhance the quality of patient outcomes and to reduce complication rates and healthcare costs. For example, they participate in disease management programs for asthma and diabetes to increase functional abilities and the quality of care while at the same time reducing hospitalization rates and costs.

9. Facilitate health behavior change effectively

According to former Surgeon-General C. Everett Koop (1995): "Diseases are of two types: those we develop inadvertently and those we bring on ourselves by failure to practice preventive measures. Preventable illness makes up approximately 70% of the burden of illness and associated costs."[54] In 1994, medical care consumed 99% of healthcare expenditures, but only 10% of preventable, early deaths is amenable to medical treatment.[55] Yet, only 1% of healthcare expenditure is spent on population-wide, public health initiatives, whereas 70% of preventable, early deaths is amendable to preventive and behavioral approaches. External pressure on the medical care delivery system is beginning to tip the balance from treatment toward prevention.

The HEDIS 3.0 report now includes smoking cessation rates as a criteria for evaluating the performance of managed care companies. Progressive managed care companies, such as Puget Sound, incorporated prevention as part of their mission and developed an organizational culture and delivery system to prevent disease.[56] Managed care companies restructure their organizations and train practitioners in how to enhance the impact of their prevention programs.[57]

10. Work collaboratively in a population-based manner

Practitioners, practice managers, and administrators work in interdisciplinary ways to promote the health of a population. Effective, interdisciplinary partnerships also collaborate to meet patient needs of their population: increase access to effective services, enhance quality, and reduce costs. They recognize and work through the intrinsic conflicts inherent in trying to achieve these three goals, as well as educating patients about these conflicts. This approach attempts to help patients understand why practitioners do not use the most intense, costly, and potent interventions for individual patients. Rather, they will favor the use of necessarily limited resources in the most effective ways so as to improve the overall health of the population.

Challenges in Developing Effective Partnerships

Short-term, for-profit business ventures create barriers to developing effective partnership with patients (see Table 1).

1. Maximize cost/loss ratio

"Cost/loss ratio" is a business term that refers to the total expenditure of providing healthcare (loss) divided by insurance premiums (cost). In other words, how many cents of the dollar do employers and patients "pay" these companies to generate profits that do not benefit the health of the population? Some aggressive, for-profit companies are primarily driven to maximize their cost/loss ratio. The goal of reducing "loss" is to benefit the position of the company in the marketplace, the investment returns to shareholders, and the salaries of chief executive officers, and only secondarily to promote cost-effectiveness for the benefit of patients.

2. *Maximize organizational dominance*

Both for-profit and nonprofit managed care companies struggle to secure a market share (patient population) for survival and to dominate competitors. In similar fashion, sometimes the organizational drive for survival supersedes practitioners' duty to provide effective clinical services. Practitioners maximize business efficiency to maximize profits or professional rewards and thereby reduce potential benefits and resources available to patients.

3. *Help the organization serve investors*

The primary customers in some for-profit companies are investors and, secondarily, patients. They give higher priorities to serving their investors rather than their patients. This priority helps them generate more capital to merge with other companies so that the companies can grow and enhance returns for their investors, which, of course, limits the availability of patient choices in healthcare insurers and providers.

4. *Use evidence-based methods to deny healthcare*

Managed care companies deny services, such as new, expensive procedures, on the basis that there is no evidence that it improves outcome. Practitioners are placed in the challenging situation of explaining to patients why their managed care company will not pay for services that may be of potential good.

5. *Use Quality Assurance (QA) methods to monitor for "bad apples"*

Some managed care companies monitor utilization patterns to dismiss practitioners because of their high-cost profiles, not because of their ability to provide quality care. Such approaches interfere with the process of practitioners developing effective partnerships with their patients.

6. *Use interdisciplinary methods to downsize personnel*

Managed care companies are, in effect, deregulating professional practices by using interdisciplinary methods to downsize personnel, such as the restructuring of nursing services. The primary drive for such an initiative is purportedly to reduce cost and maintain quality, but this process often has the counter-productive effect of demoralizing practitioners so they in fact become less effective in functioning in teams and less resourceful in working with their patients.

7. Lower customer expectations and demands

Managed care companies will set up systems to lower customer expectation and demands for services. They write benefit contracts that limit the utilization of services. They set up a variety of barriers (co-pay, number of visits per year, etc.) to reduce the use of services.

8. Use utilization review methods to reduce patient services

Utilization review methods are used to reduce patient services. For example, the drive to reduce the length of hospital stays may override the personal needs of patients. A recent example of legislation to curtail such practices relates to the length of hospital stays for women after childbirth. Even though there is no evidence that shorter stays cause any problems in low-risk situations, women are now guaranteed a 48–hour hospital stay if they so desire.

9. Avoid addressing specific behavior-change issues

Managed care companies that are in the early stage of their development often have a high turnover of patients who shift from one company to another. Some companies do not invest in preventive program with long-term benefits (e.g., organizing smoking cessation programs) to offset long-term costs because these benefits may not accrue to them. When managed care companies are in a survival mode, they think only about healthcare costs on a yearly basis and are reluctant to invest in long-term benefits, particularly when patient loyalty is in a state of flux, and patients may switch to another company because it has reduced its monthly premium.

10. Selectively recruit healthy patients

Managed care companies would prefer to select the healthiest patients and to work with employers that have healthy employees. This approach helps to avoid an adverse case mix. Some companies also write exclusionary clauses to deny services for preexisting conditions to patients, including exclusion on the basis of genetic markers that suggest susceptibility to certain diseases. Both practices attempt to reduce the losses for managed care companies.

Negotiating Effective Clinical Partnerships

Supports for and barriers to developing effective administrative partnerships in turn affect how practitioners can develop effective clinical partnerships with their patients. This section focuses on how practitioners can enhance the partnership skills with patients. Using the concept of parallel process[58,59] (or isomorphism[60]), they can use the same skills to develop effective administrative (practitioner-manager-leader), professional (specialist-generalist or practitioner-practitioner), and educational (teacher-learner) partnerships. The development of these partnerships involve using similar relational and problem-solving skills.

The term "partnerships in healthcare" has different connotations among administrators, business managers, practitioners, and patients. For some, the term "partnership" implies a sense of equality. However, most practitioner-patient partnerships are far from equal in terms of their respective expertise. The question then becomes how these partners can best use their respective expertise synergistically to address healthcare issues; in other words, how to use each other's strengths, in contrast to focusing on weaknesses and deficits.

Paradoxically, rigid adherence to egalitarianism may restrict how partners can best work together. For example, it may work best when patients take the lead and responsibility for addressing personal issues (such as health behavior change), and practitioners take the lead in addressing clinical issues (choosing a medication for diabetes). Effective partners relate to one another in a flexible manner to address patient care issues. The hallmark of an effective partnership is the extent to which these partners synergistically use their expertise and differences to enhance the process and outcome of healthcare delivery. This process can lead to a reduced reliance on the healthcare delivery system or even a dissolution of the partnership itself because patients are in charge of their self-care or no longer need professional services.

The goal of negotiating partnerships in healthcare is to collaborate and, if necessary, compete in ways that facilitate efficient problem-solving. For some professionals in healthcare, negotiation is associated with adversarial conflicts and differences that lead to win/lose situations. In such competitive situations, one person's gain is another person's loss.[61,62] For example, when practitioners unilaterally set ideal goals for changing health behaviors or make decisions for patients, patients may respond passively, resist their advice, or not participate in decision making. When this occurs, practitioners set themselves up to lose by insisting that patients change or adhere to their advice. Furthermore, they may worsen an already adversarial, win-lose situation by reacting negatively toward their patients. They may project their negative feeling onto patients and blame them. In turn, patients react negatively by feeling frustrated, angry, or guilty. They disengage from their practitioner altogether: in effect, a lose-lose situation.

Practitioner-patient differences do not have to lead to adversarial relationships or conflictual situations. Differences arise from their respective expectations, agendas, concerns, feelings, attributions, beliefs, and values; however, economic, organizational, social, and cultural factors can be viewed as assets rather than sources of conflict. To enhance the prospects of creating win-win situations, practitioners can begin working from "where patients are" in terms of their assumptions, perceptions, expectations, needs, values, and beliefs. Practitioners can discuss openly their differences and constructively negotiate with patients to address healthcare issues.

Proposed here is a model of negotiation designed to help practitioners become more aware of a wide range of negotiation options that they may use explicitly to enhance their partnership skills. This model, which draws on many influences,[5,61–79] provides practitioners with a framework to learn and communicate about how to enhance their partnership skills. It can help to clarify what to address, how to relate to one another, and how to solve problems together.[64] For example, partners can negotiate how to share power, control, and responsibility in working together.

The framework of this model consists of three dimensions:

- content (tasks);
- relational process;
- problem solving.

Although this chapter will focus primarily on clinical partnerships, practitioners can apply this model to professional, administrative, and educational partnerships.

Content in the Clinical Partnership

Practitioners work with patients on their health issues. They have professional expertise about disease, and patients have lay expertise about their illness experiences, risk behaviors, and the context of their lives.[5] For example, practitioners may initially focus on addressing the acute complications caused by risk behaviors, but then use the patients' lay expertise to address issues such as motivating health behavior change.[80]

The content of negotiation arises from patients' and practitioners' agendas. Practitioners and patients may negotiate about disease(s) and/or the impact of health, illness, and disease on patients' personal lives. As patients tell practitioners stories about their health concerns, practitioners can gather more information about the disease processes and/or personal aspects of healthcare: the illness experience, the sick role, and the context of their lives.

Disease-centered domain

Practitioners' attitudes are shaped by their training, which often places a heavy emphasis on the acquisition of clinical knowledge/skills and the use of modern technology. Practitioners are trained to use their technical expertise to diagnose and treat disease. The range of tasks performed by practitioners in addressing diseases is summarized in Table 2.

Person-centered domain

Patients usually come to the practitioner with concerns about disease or the threat of disease. The patients' life experiences shape their concerns about health issues. In addition, patients may have feelings, expectations, and explanations about

Table 2: Tasks of the Medical Domain

- ✔ Diagnose disease
- ✔ Educate patient about the etiology, diagnosis, natural history, treatment, and prognosis of the disease
- ✔ Cure and palliate disease
- ✔ Alleviate and control symptoms of disease
- ✔ Advise about health maintenance and preventive issues
- ✔ Reduce impact of disease upon course of illness and sick role

From Botelho and Skinner.[80] Reproduced with permission of Sage Publications.

the different aspects of their health concerns. The person-centered domain of the clinical partnership deals with the illness experience and the context of patients' lives. Even though patients are the experts on their illness experience, they may not know how best to communicate this experience to their practitioners. Since this partnership deals with the subjective experiences of both patients and practitioners, practitioners' interpersonal skills are crucial for assessing the illness experience and for negotiating a common understanding with patients about their illness experience. They may also address any of the tasks outlined in Table 3 and negotiate with patients about which of these tasks have priority for any given encounter.

Patients rate the personal quality of their partnerships on the basis of how well they relate to practitioners and on the extent to which they feel understood by them. Practitioners also rate the personal quality of their partnership on the basis of the extent to which they like their patients. For technical aspects of acute care, it may not matter whether practitioners and patients like each other. However, when addressing nontechnical issues, the personal aspects of their partnership become more important, particularly when addressing chronic disease, risky health behaviors, and terminal illness.

Relational Process

Relational process has four levels: autocracy, parentalism, egalitarianism, and one-up position (see Figure 1). This figure is an elaboration of Szasz's model for the practitioner-patient relationship but differs from his model in three important respects.[67] First, an additional relationship level, the one-up position, is added to the model to describe the situation in which patients and their families are exerting greater control over and responsibility for healthcare than their practitioner. As the health needs of the community outstrip what the delivery system can provide, patients must take a greater role in reducing the demand on diminished healthcare services by meeting their own needs. In fact, most healthcare is self-care. Second, practitioners can empower patients to enhance their role in self-care of chronic disease and promote health behavior change, thereby reducing the demand on the

Table 3: Tasks of Personal Domain

1. Risk Behaviors
   - ✔ Maintain a therapeutic alliance
   - ✔ Understand patients' assumptions about behavior change
   - ✔ Understand patients' perceptions about benefits and concerns about behavior change
   - ✔ Educate patients about risks and harms of risk behaviors
   - ✔ Advocate for health behavior change
   - ✔ Negotiate a plan for change
   - ✔ Provide care and follow-up support

2. Illness Experience/Behavior
   - ✔ Understand patients' perspectives about their illnesses
   - ✔ Understand the illness context of patients' lives
   - ✔ Educate patients about the expected course of the illness
   - ✔ Develop mutual understanding about illnesses
   - ✔ Reduce impact of illness on patients' lives
   - ✔ Help patients develop strategies to deal with their illness
   - ✔ Collaborate with patients in dealing with psychosocial problems
   - ✔ Understand and alleviate suffering
   - ✔ Facilitate intrinsic healing processes of the patient

3. Contextual Factors
   - ✔ Family—explore family perspectives about health behaviors
   - ✔ Social network and situation—assess supports and barriers
   - ✔ Employment—explore work influences
   - ✔ Culture—identify norms affecting health behaviors
   - ✔ Spirituality and religion—explore health care values

From Botelho and Skinner.[80] Reproduced with permission of Sage Publications.

healthcare delivery system. Third, the relational process is dynamic in terms of how practitioners and patients relate to one another. Within a single encounter, they can change their relationship levels or role at any time, depending on the issues under discussion.

This figure represents a balance of responsibility in the practitioner-patient partnership. At each level, the partners share complementary levels of responsibility for healthcare. Autonomy is defined as the extent to which patients wish to exert responsibility and control over the disease-centered and patient-centered domains of their partnership. The extent to which patients seek autonomy in these domains should influence the level at which practitioners relate to patients. Patients may be uncertain as to what level of autonomy to exert in the practitioner-patient partnership. Practitioners may need to explore with patients which level is appropriate for a given situation or decision.

For any given health issue, these partners may relate to one another at different

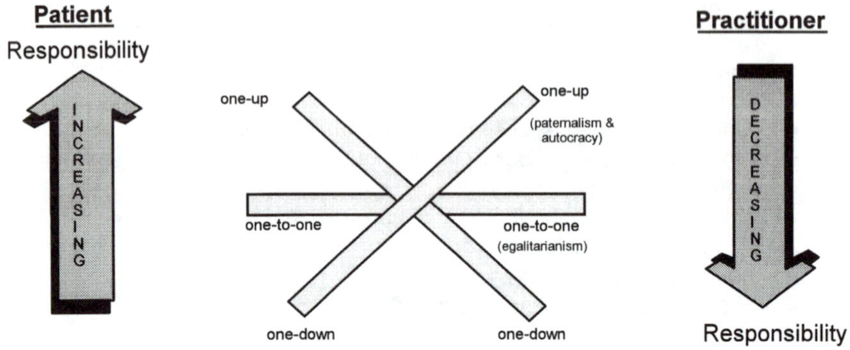

Figure 1. A Balance of Responsibility.

levels in the personal and medical aspects of care. In addressing the medical aspects of care, a patient may not want to know all the nuances of treatment options, but simply want to follow what practitioners consider to be the best treatment for their disease. But, at a personal level, patients tell their practitioners how they are going to deal with the effects of the disease on their lives.

In interpersonal terms, power refers to the capacity of partners in healthcare to influence others, and control refers to the extent to which partners regulate and restrain others. Power and control can affect how practitioners and patients relate to one another in their partnership. Many factors affect the issues of control, power, and responsibility for both practitioner and patient. These factors shape how patients and practitioners relate to one another in the disease-centered and personal domains. For example, patients whose past experiences make them doubt the competence of practitioners may attempt to control the practitioners in their relationship, and may reject appropriate clinical advice because of their distrust of practitioners. Practitioners who understand the reasons for such behavior may relate better with these patients and help them become more willing to accept clinical advice.

Relationship levels

The one-up position describes the relationship level at which patients assume more responsibility than the practitioners for some or all aspects of their healthcare. In effect, patients and their families determines the course of the healthcare, with practitioners acting as consultant experts who defer to the directives and preferences of patients and their families. Requests for tests or second opinions, disease/illness-induced self-determinism, medical consumerism, and practitioner-shopping are all examples of patients taking the one-up position with their practitioner.

Patients are justified in requesting second opinions or tests when practitioners are uncertain about the diagnosis or treatment options. However, when patients make inappropriate requests, practitioners do not have relate to patients and their families from the one-down position. To avoid unproductive power and control struggles with patients and their families, practitioners need to understand the patients' reasons, concerns, and expectations with regard to their request. This process may help practitioners establish a more egalitarian relationship before educating patients about the reasons why the request may be inappropriate.

Emotional reactions to disease and illness can induce patients and their families to take the one-up position. This one-up position may be inappropriate when patients act in ways detrimental to their own health: for example, patients who refuse hospitalization or treatments for a life-threatening disease. Practitioners who understand and address the emotional reasons (fear, terror, and anxiety) underlying this behavior are more likely to change the relationship level in ways that enable them to work more effectively with patients and their families.

Medical consumerism is a reflection of economic mores; patients view disease and health as a commodity. Patient are empowered by the consumer-rights movement and tell their practitioners what they want. They control the process of medical care, and practitioners defer to their requests. Patients are able to have their demands met if they have the economic resources.

"Practitioner-shopping" is a form of medical consumerism running amok: continually seeking attentions of different practitioners over time, whether traditional or complementary ones. Complex motivations often underlie such behavior. These patients have had traumatic or unsuccessful interaction with the healthcare system, or have personality, interpersonal, psychological and/or psychiatric problems. Practitioners may contribute to this phenomenon by failing to accommodate patients' explanatory model of illness. Taking a history about past healthcare encounters may alert practitioners to this kind of patient. Practitioners may need additional time to develop a more trusting relationship with such patients before dealing with healthcare issues. At best, practitioners may be able to engage this kind of patient only in a relationship that functions episodically, interspersed with periods when patients and their families may revert to practitioner-shopping. Nonetheless, practitioners can view such a patient response as an improvement when compared to previous patterns of behavior.

Egalitarianism in practitioner-patient partnerships represents a needed departure from the traditional forms of paternalism in healthcare. However, these partners may still be on an unequal footing: practitioners have professional expertise about disease, and patients have lay "expertise" about their illness experiences and the context of their lives. The ideal of egalitarianism works to redress the balance of power and control that favored practitioners in their relationships to patients. As these partners work together, their task is to use their respective expertises synergistically in ways that efficiently achieve outcomes that are satisfactory to both of them.

Practitioners encourage patients and their families to enhance their responsibility in decision making and their influence over healthcare issues. However, severe illnesses and life-threatening situations can diminish patient autonomy to the point that egalitarianism is inappropriate, particularly when patients and their families implicitly or explicitly request that their practitioner take charge of the medical care. Rigid adherence to egalitarianism is inappropriate when practitioners expect patients to share control and responsibility equitably in these kinds of situations. Patients and their families may suffer more when practitioners decline to take a leading role in the decision-making process. Egalitarian relationships may inappropriately allow practitioners to reduce their level of responsibilities in providing care. Consequently, patients may feel abandoned by their practitioners.

Thus, practitioners need to develop a flexible approach in relating to patients. For example, a patient may relate to their practitioner at the level of the one-up position in the personal domain and at the level of the one-down position in the disease-centered domain. Overall, this partnership may be described as egalitarian, but such a label conceals the nuances of sharing responsibility, power, and control.

Paternalism may exist between two extremes along a continuum, (from a practitioner-centered to patient-centered style). A description of these styles will help to clarify when practitioners can appropriately use paternalism in health care.[81–84]

With traditional or practitioner-centered paternalism, practitioners act in a controlling and authoritarian way to treat disease for patients and their families; their practice style is disease-focused, cure-oriented, and biomedically fixated. They make decisions that reflect their values rather than the patient's values. Ardent patient-rights opponents have rejected this traditional format for being immoral under any circumstances because practitioners are unilaterally imposing their values without regard to the patient's wishes and needs, even though practitioners try to act in the "best" interest of patients and their families. In some circumstances, this kind of parentalism is ethically justifiable when patients explicitly state that they want their practitioner to act in this manner.

Patient-centered parentalism is characterized by authoritative and autonomy-supportive practitioners who take into consideration the values and needs of patients and their families in the problem-solving process. Practitioners may act in response to patients' preferences or their diminished autonomy due to the effects of an illness. In these circumstances, they initially support the implicit or explicit request from patients and their families to assume more responsibility and control for healthcare issues, but they attempt to activate patient involvement in healthcare over time, particularly as patients recover from their illness and regain full, active, and healthy lifestyles. Practitioners may use relational strategies (described later) to activate patients' involvement in healthcare.

Autocracy is justified when practitioners need to take charge of medical, surgical, and psychiatric emergencies. When patients and their families' autonomy are impaired in life-threatening situations, practitioners must take control to save or to protect life. However, this relationship level has only limited applications in nonemergency situations.

Problem-Solving Process

The problem-solving phases of this negotiation model are as follows:[68]

- building relationships;
- setting an agenda;
- assessing health, behavior, illness, and disease;
- enhancing mutual understanding;
- implementing a plan;
- following through.

Ideally, practitioners work through these phases in sequence, but, in reality, they implicitly move back and forth between these phases depending on the issues that need attention. Sometimes, they may explicitly negotiate with patients to shift from one phase to another.

Practitioners can use partnership skills through the problem-solving phases to activate patients into taking greater control and responsibility for their healthcare, such as the self-care of chronic disease. Practical suggestions are described that can help practitioners develop effective partnerships with patients in dealing with decision making and behavior change.

1. *Building the relationship*

Practitioners can use a variety of communication skills throughout the encounter to understand and validate the patient's experience in ways that build empathic relationships.[85–87] Empathic relationships help practitioners understand patients' perspectives. However, such relationships alone may be insufficient to motivate patients to change their behavior or activate them to make healthcare decisions. Practitioners can use relational strategies to help patients assume more control and responsibility for their own healthcare; for example, practitioners taking the one-down position and putting patients in the one-up position. When uncertain about patients' capacity to assume greater responsibility for change, practitioners may use these strategies tentatively, observe the impact of this approach, and make adjustments accordingly. When practitioners perceive patients as capable of assuming greater responsibility for their health, they can use these strategies in an affirmative manner. Practitioners can use either of these strategies, tentatively or affirmatively, by changing the way they present questions to patients.

Decision making. Practitioners take the one-down position by telling patients that they:

- Do not know how they can best address a decision: "I don't know how you can best make this decision. Perhaps you can talk with your family and let them help you decide what to do. I will certainly help you. "

- Are uncertain when patients should make decision: "I'm not sure when you should decide to put your wife in a nursing home. It's a complicated decision."
- Cannot make decisions for patients: "I don't know what it would take for you to be sufficiently concerned about your health and seriously consider changing your behavior."
- Do not know whether patients feel up to making a decision: "I don't know whether you feel up to making a decision now about whether to have a lumpectomy or a mastectomy. Maybe you need a few days to think about it."

Practitioners can put patients in the one-up position by:

- Encouraging them to make their own decisions: "What would it take for you to feel more confident in making this decision?"
- Reminding them that they have to decide what to do: "I'll certainly share with you what I think, but it is really for important for you to decide what to do in your situation."

Health behavior change. Practitioners may take the one-down position by assuming a lower relationship level than that of their patient. This approach may be appropriate when practitioners are uncertain about patients' capabilities and/or interest in assuming more responsibility for healthcare. They may take the one-down position by telling patients that they:

- Do not know whether to address an issue: "I'm not sure whether to address this issue, particularly if you don't think you'll change. What do you think?"
- Are uncertain when patients should address their risk behavior: "I don't know when would be a good time for you to set a quit date."
- Cannot make patients change: "I don't know what it would take for you to be sufficiently concerned about your health and consider seriously changing your behavior."
- Do not know what is the best decision for patients: "I don't know what it would take for you to decide that you need to change your behavior (stop smoking, reduce alcohol consumption, assume self-care of chronic disease)."
- Do not know whether patients have the self-efficacy to change: "I don't know whether you feel that you're capable of changing if you do decide to change."

This strategy can help patients explore whether they are capable of assuming greater responsibility for changing their behavior. Practitioners can put patients in the one-up position by:

- Deferring to their opinion: "If your health were damaged by alcohol, would you consider changing your drinking habit?"

- Encouraging them to make their own decisions: "It's up to you whether you want to change and feel better."
- Reminding them that they know how to change: "You gave up drugs because it caused health problems. You know how to quit. But what would it take for you to quit smoking?"
- Deferring to patients' self-efficacy about changing: "You seem the sort of person that if you decide to change, you would do it. What would it take for you to change?"
- Deferring to their expertise in how to change: "You know how to control your blood glucose levels because you've done it before. What would help you decide to do it again?"

By putting patients in the one-up position, practitioners acknowledge that patients have the capacity to take charge of the situation. Such approaches may further enhance patients' sense of confidence and help them find their own ways to take charge.

Misuse of relational strategies. Practitioners may inappropriately use these strategies. For example, they may take the "one-down" position when patients are unable to exert greater autonomy in dealing with their disease and/or illness. In such circumstances, patients and their families may feel pressured, uncomfortable, and/or abandoned. Sometimes, practitioners misuse these strategies as a way of abdicating responsibility to patients and their families, especially those patients who create difficulties for the practitioners. However, the threat or effects of ill health can have a number of consequences for patients and their families: diminished self-efficacy and self-determination, the emotional sequelae from anticipated or actual losses, unhelpful perceptions, and family dysfunction. Practitioners may need to consider these consequences before trying to activate patients and their families to assume greater involvement in healthcare.

Setting an Agenda

Agenda-setting may occur any time during the practitioner-patient encounter. To manage time efficiently, this task is best accomplished early in the encounter. This issue is particularly important when practitioners and patients have multiple, competing agenda items to address in brief encounters. To work more effectively together, practitioners can ask patients to think in advance what they want to address and set priorities for the encounter.

Decision making. Practitioners and patients may need to clarify their respective agendas for the encounter and then negotiate about their differences in priorities to decide what to address at this encounter or at subsequent follow-up appointments. Resolving their differences helps to preempt unproductive conflicts and ineffective interactions. Together, they can negotiate to develop a common agenda by addressing the patient's presenting complaints as well as issues raised by their practitioner, such as health behavior change.[88]

Health behavior change. Practitioners often give higher priorities to addressing certain issues, such as health behavior change, than do their patients. Consequently, patients are less ready to discuss and address this agenda item.[89–92] For example, a practitioner may say,

> Let's decide what we can deal with today and what we may need to leave for the next appointment. We will certainly deal with your concern about high blood pressure , but I would also like to work out whether your alcohol is causing your high blood pressure. Can I explain why this may be so?

When practitioners need to address an issue that is not on a patient's agenda, they may work more effectively if they also inform the patient about its importance, appropriateness, and relevance. They can provide a brief rationale for raising an issue and then obtain the patient's consent to discuss this issue further. Such an approach may prevent conflicts that arise from differences in priorities, perspectives, and/or understandings about healthcare issues.

Assessing Diseases, Psychosocial Issues, Risk Behaviors, and Decision Making

Practitioners and patients exchange information in order to assess medical problems and psychosocial issues. A medical assessment consists of assessing diseases only. To treat psychological problems, promote health, and prevent diseases, practitioners may need to assess psychosocial factors, preventive issues, and health behaviors. Most practitioners are well trained to assess diseases but are much less well prepared to address patients' motivation to make their own decisions, participate in prevention procedures, or change their risk behaviors. Most practitioners have not been adequately trained to motivate patients to change their risk behaviors and/or to assess patients' willingness to participate in decision making. These two issues will be briefly used as examples for assessment. Practitioners can assess how patients perceive the advantages and disadvantages, or benefits and concerns in comparing one option vs. another.

Decision making. Practitioners and patients can make healthcare decisions in three ways: practitioner or patient makes the decision independently of the other, or they share this responsibility to varying degrees. For a decision about a particular health issue, a practitioner may say,

> How would you like me to help you with this decision? Would you like me to give you all the information that you want so you can make the decision by yourself ? Or, would you like us to work together in deciding what you think is best for you? Or would you prefer that I made the decision for you?

With the last suggestion, practitioners do not necessarily have to accept patients' expressed initial preference. Practitioners may need to clarify if patients do not want to make the decision, and then ask what it would take for them to feel more confident: if they jointly participated in this process or if they made the decisions by themselves. Such an approach may help patients take a more active role. However, the decision-making process is dynamic. Patients may change their level of involvement in this process depending on the particular health issue and on whether they are physically and/or psychologically ill.

Health behavior change. When practitioners ask about patients' readiness to change their risk behaviors, patients may fall into one of four categories:

- Not considering change;
- Thinking about change;
- Preparing to change;
- Taking action.

Practitioners can give different explanations for using a decision balance with patients, depending on their readiness to change (see table 4). A decision balance is part of a motivational assessment that helps to clarify how they differ with patients in terms of their perceptions about health behavior change. They first gather information about how patients perceive the benefits and risks of their behaviors, and their concerns about and benefits of adopting healthy behaviors.

When patients are not thinking about change or are not ready to change, practitioners can seek to learn more about why they do not want to change. When patients are thinking about change, practitioners can explain to patients that they want to learn more about why they are ambivalent about change. When patients state they are not thinking about change or are not ready to change, practitioners can explain to patients that they want to learn more about why they do not want to change by using a decision balance.

The decision balance can help practitioners understand patients' reasons not to change (resistance or the cons of change) and reasons to change (motivation or the pros of change).

A motivational assessment prepares practitioners to work with patients in the next problem-solving phase (enhancing mutual understanding). This process prepares patients to "confront" and advise themselves about the need for change so that they decide whether, why, what, and how much to change their behavior.

Enhancing Mutual Understanding

Opportunities to develop skills at enhancing mutual understanding are suboptimal in many professional training programs of practitioners, particularly in medical education.[93] Consequently, many practitioners are not skillful in enhancing mutual

Table 4: A Decision Balance for Assessing Patient Motivation for Behavior Change

Behavior Change	Reasons not to change (resistance)	Reasons to change (motivation to change)
Risk Behavior	*1. Benefits* • "What do you like about . . . ? And anything else . . ." • "What do you like most about . . . ?"	*2. Concerns* • "Do you have any concerns about . . . ?" • "What concerns do your family have about your . . . ?"
	3. Concerns • "What concerns would you have if you were to change?" • "In what way would your life be different if you were to change?"	*4. Benefits* • "In what ways would your health be better if you were to change?" • "In what ways would other aspects of your life be better if you were to change?"

understanding with their patients about their investment in health behavior change and decision making. Sometimes mutual understanding about a risk behavior or the decision-making process can be achieved, and a plan for change implemented with minimal discussion.

Decision making. Practitioners can work with patients to evaluate the extent to which they want to participate in the decision-making process. Whatever the patients' expressed preferences, practitioners can help them reassess their options either to enhance their role in decision making ("I think that you can return back to work even though your back is still hurting") or encourage them to relinquish full control over the decision-making process ("I know how much you want to go back to work, but maybe you should consider taking a few days off work to help you cope better with what is going on").

Health behavior change. When patients and practitioners have different perceptions about the benefits and concerns associated with health behavior change, implicit and explicit disagreements can impede the process of implementing effective plans. Practitioners can use four strategies to enhance mutual understanding with patients: educate about need to change risk behaviors; use nondirect interventions to reduce patient resistance; use direct interventions to enhance patient motivation; and clarify patients' perception about their self-efficacy. Practitioners can use the decision balance to discuss how their own perceptions of benefits and concerns of health behavior change differ from the perceptions of patients in order to clarify areas of agreement and disagreement, enhance areas of agreement, and address areas of disagreements: in other words, to understand their differences concerning the reasons to change and not to change.

When patient resistance is high, patients usually lack motivation to change. Conversely, when resistance is low, patient motivation is more likely to be high. Understanding this dynamic between resistance and motivation helps practitioners work with patients in more deliberate and effective ways to enhance the prospects of health behavior change. First, they can use nondirect intervention to lower patient resistance to change. For example, they may ask patients how they could obtain the same list of benefits in alternative ways:

> How could you relax in other ways rather than (smoke cigarettes, drink alcohol, use drugs, or eat excessively)? Could you make a list of other ways of relaxing other than using alcohol between now and the next appointment?

Practitioners can ask these questions and write down patients' responses on a sheet of paper. They can give patients this sheet of paper with the following explanation:

> I think that reasons to change are more important than the reasons not to change, but I think that you probably see it the other way around.

If patient agrees with this statement, practitioners can ask if patients would like to change. If patients disagree with this statement, practitioners can point out a discrepancy between what patients say and do.

Implementing a Plan

After enhancing mutual understanding about the decision-making process and health behavior change, practitioners can work with patients to develop and implement a plan.

Decision making. Practitioners can clarify with patients the extent to which they would like to make decisions. Patients may want to make their decisions independently or with varying degrees of input from their family and their practitioner. Practitioners defer to patients' autonomy about their level of involvement in the decision-making process. Patients may decide to let their practitioner make some decisions for them: for example, the combination of chemotherapy agents needed to treat leukemia. Or they may decide to use complementary therapies of unproven effectiveness, without the support of their practitioner.

Health behavior change. The extent to which practitioners reach a common understanding with patients about the need for change should guide how they develop a plan for change together. Many other factors (e.g., patients' depression) also affect how patients may develop a workable plan and set realistic goals. These factors may influence the extent to which practitioners and patients are able to set goals for change: think more about it, prepare for it, and take incremental steps toward change or accept the ideal goals for change (short-term vs. long-term change, and pragmatic vs. ideal). Practitioners can suggest goals compatible with patients' stage-of-change. Then they can negotiate together to set appropriate goals.

Practitioners may help patients by evaluating and enhancing their commitment to change. Patients may or may not need our assistance in deciding when and how to implement a plan for change. Before finalizing the plan, practitioners may check on how patients perceive their competence, confidence, and ability to achieve these goals. For patients who are ready to change their behavior, practitioners can help patients develop contingency plans to deal with possible lapses and relapses.

Following Through

After developing a plan, practitioners and patients need to address the issue of follow-through arrangements.

Decision making. Practitioners and patients may need to make a number of decisions about follow-up arrangements (timing, duration, and frequency of follow-up contacts—appointments, telephone calls). Traditionally, practitioners make arrangements for the dates for follow-up. Alternatively, practitioners can provide a rationale and explain the purpose for follow-up to patients, inquire about the patient's understanding or need for follow-up, and then ask when it would be convenient to make an appointment. If the practitioner has concerns about the patient's response, they can negotiate about the timing of the next appointment, telephone contact, etc. Some patients are surprised by this approach and expect practitioners to tell them when to return. Practitioners can tell patients that they will share their opinions with them, but they are also interested in what the patients think. This approach may enhance patient motivation to keep their follow-up appointments.

Health behavior change. The purpose of a follow-up visit is to assess the extent to which patients are maintaining their goals for change; to discuss the successes, difficulties, lapses, and relapses that patients may have experienced in working toward their goals; and to help patients evaluate how their contingency plans helped to deal with lapses or relapses. Practitioners can use a variety of approaches to help patients sustain their motivation to maintain change and for preventing lapses and relapses.

Conclusion

Section I addressed the factors that support and hinder the prospects of leaders, managers, and practitioners in developing effective administrative partnerships in work together in managed care organizations. In turn, these factors affect the ability of practitioners to develop effective clinical partnerships with their patients. With the shift toward population-based approaches to healthcare delivery, managed care organizations and practitioners are now called upon to work in ways that maximize the health of the population, such as:

- Promoting patient self-care of risk behaviors and diseases;
- Developing comprehensive prevention programs;
- Reducing unnecessary healthcare utilizations.

Such an approach helps to reduce the need for services within the healthcare delivery system while serving the community's goal of improving health. Stewardship is a guiding force that promotes the development of effective administrative and clinical partnerships in healthcare.

Section II describes a negotiation model that provides practitioners with a framework to enhance their partnership skills. Practitioners can use these skills to:

- Encourage patients and families to assume more responsibility and control for healthcare;
- Enhance patient participation in decision making;
- Motivate patients to change their risk behaviors.

Practitioners can also use these skills to enhance patient self-care in acute and chronic diseases and to reduce unnecessary patient demands on the healthcare delivery system. Overall, they can use these skills to enhance the process and outcome of healthcare delivery.

Merely understanding this model does not mean that practitioners will have the skills to work effectively with patients. But, by using the framework of this model, they can assess where negotiation was particularly effective and where it broke down. They can then use a variety of learning strategies (review of videotaped practitioner-patient encounters, role-plays, etc.) to understand these communication problems and to practice how to improve their partnership skills.

Administrative and professional partnerships have similarities to clinical partnerships in terms of relational and problem-solving processes even though the specific tasks are different. Practitioners can reflect about how to apply these clinical partnership skills to administrative, healthcare team, and educational contexts.

References

1. Botelho RJ. Overcoming the prejudice against establishing a national health care system. Archives of Internal Medicine 1991; 151:863–869.

2. Franks P, Clancy C, Nutting PA. Gatekeeping revisited—Protecting patients from overtreatment. New England Journal of Medicine 1992; 327:424–429.

3. Schroeder SA. The medically uninsured–Will they always be with us? The New England Journal of Medicine 1996; 334:1130–1133.

4. Blumberg L, Liska DW. The uninsured in the United States: A status report. 1996; (abstract)

5. Kleinman A. Patients and healers in the context of culture. Berkeley: University of California Press, 1979.

6. Curran WJ. Law-Medicine Notes: The Constitutional Right to Health Care. New England Journal of Medicine 1989; 330:788–789.

7. Levey S, Hill J. National health insurance: The triumph of equivocation. New England Journal of Medicine 1989; 321:1750–1754.

8. Brock DW, Daniels N. Ethical foundations of the Clinton Administration's proposed health care system. JAMA 1994; 271(15):1189–1196.

9. Lundberg GD. National health care reform: The aura of inevitability intensifies. JAMA 1991; 265:2566–2567.

10. Lundberg GD. National health care reform: The aura of inevitability Intensifies. JAMA 1992; 267:2521–2524.

11. Lundberg GD. American health care system management objectives: The aura of inevitability becomes incarnate. JAMA 1993; 269:2554–2555.

12. Morone JA. Nativism, hollow corporations, and managed competition: Why the Clinton Health Care Reform failed. Journal of Health Politics, Policy and Law 1995; 20:391–398.

13. Brodie M, Blendon RJ. The public's contribution to Congressional gridlock on health care reform. Journal of Health Politics, Policy and Law 1995; 20:403–410.

14. Peterson MA. The health care debate: All heat and no light. Journal of Health Politics, Policy and Law 1995; 20:425–430.

15. Baumgartner FR, Talbert JC. From setting a national agenda on health care to making decisions in Congress. Journal of Health Politics, Policy and Law 1995; 20:437–445.

16. Brady DW, Buckley KM. Health care reform in the 103d Congress: A predictable failure. Journal of Health Politics, Policy and Law 1995; 20:447–454.

17. Steinmo S, Watts J. It's the institutions, stupid! Why comprehensive national health insurance always fails in America. Journal of Health Politics, Policy and Law 1995; 20:329–372.

18. Yankelovich D. The debate that wasn't: The public and the Clinton Plan. Health Affairs 1995; 14:7–23.

19. Skocpol T. The rise and resounding demise of the Clinton Plan. Health Affairs 1995; 14:66–85.

20. Heclo H. The Clinton Health Plan: Historical perspective. Health Affairs 1995; 14:86–98.

21. Skocpol T. Boomerang: Clinton's health security effort and the turn against government in US politics. New York: W.W. Norton, 1996.

22. Johnson H, Broder DS. The system: The American way of politics at the breaking point. 1st ed. Boston, MA: Little, Brown & Co. 1996.

23. Barer ML, Marmor TR, Morrison EM. Health care reform in the United States: On the road to nowhere (again)? [editorial]. Social Science & Medicine 1995; 41:453–460.

24. Blumenthal D. Part 1: Quality of care—what is it? [editorial]. New England Journal of Medicine 1996; 335:891–894.

25. Chassin MR. Quality of health care. Part 3: Improving the quality of care. New England Journal of Medicine 1996; 335:1060–1063.

26. Brook RH, McGlynn EA, Cleary PD. Quality of health care. Part 2: measuring quality of care [editorial]. New England Journal of Medicine 1996; 335:966–970.

27. Berwick DM. Harvesting knowledge from improvement [editorial; comment]. JAMA 1996; 275:877–878.

28. Berwick DM. Eleven worthy aims for clinical leadership of health system reform. JAMA 1994; 272:797–802.

29. Berwick DM. A primer on leading the improvement of systems. BMJ 1996; 312:619–622.

30. Yank G. Quality improvement in health care organizations: A general systems perspective. [Review]. Behavioral Science 1995; 40:85–103.

31. National Committee for Quality Assurance. Health plan employer data and information set 3.0. 1997; (abstract)

32. Teisberg EO, Porter ME, Brown GB. Making competition in health care work. Harvard Business Review 1994; 72:131–141.

33. Solberg LI, Isham G, Kottke TE, et al. Competing HMOs collaborate to improve preventive services. Joint Commission Journal on Quality Improvement 1995; 21:600–610.

34. Leape LL. Practice guidelines and standards: an overview. QRB Quality Review Bulletin 1990; 16:42–49.

35. Pearson SD, Goulart-Fisher D, Lee TH. Critical pathways as a strategy for improving care: problems and potential [see comments]. Annals of Internal Medicine 1995; 123:941–948.

36. McArthur JH, Moore FD. The two cultures and the health care revolution. JAMA 1997; 277:985–989.

37. Emanuel L. Bringing market medicine to professional account. JAMA 1997; 277:1004–1005.

38. Showstack J, Lurie N, Leatherman S, Fisher E, Inui T. Health of the public: The private-sector challenge. JAMA 1996; 276:1071–1074.

39. Block P. Stewardship: choosing service over self interest. 1st ed. San Francisco: Berrett-Koehler, 1996.

40. Bandura A. Self-Efficacy: The exercise of control. New York: W.H. Freeman, 1997.

41. Deci EL, Ryan RM. Intrinsic motivation and self determination in human behavior. New York: Plenum Press, 1985.

42. Russell LB, Gold MR, Siegel JE, Daniels N, Weinstein MC. The role of cost-effectiveness analysis in health and medicine. JAMA 1996; 276:1172–1177.

43. Gillam SJ, Murray SA. Needs assessment in general practice. 1996; Occasional Paper 73:(abstract)

44. Walsh PC, Brooks JD. The Swedish prostate cancer paradox. JAMA 1997; 277:497–498.

45. Johansson J, Holmberg L, Johansson S, Bergstrom R, Adami H. Fifteen-year survival in prostate cancer. JAMA 1997; 277:467–471.

46. Clancy CM, Kamerow DB. Evidence-based medicine meets cost-effectiveness analysis. JAMA 1996; 276:329–330.

47. Solberg L, Kottke T, Brekke M. Continuous quality improvement (CQI) as a method to improve prevention services in primary care clinics [abstract]. AHSR & FHSR Annual Meeting Abstract Book 1994; 11:85

48. Young MJ, Ward R, McCarthy B. Continuously improving primary care. Joint Commission Journal on Quality Improvement 1994; 20:120–126.

49. Jackson RS, Leininger LS, Harris RP, Kaluzny AD. Implementing continuous quality

improvement in primary care: Implications for preventive services. Journal of Ambulatory Care Management 1994; 17:8–14.

50. Palmer RH, Hargraves JL. Quality improvement among primary care practitioners: An overall appraisal of results of the Ambulatory Care Medical Audit Demonstration Project. Medical Care 1996; 34:SS102–13.

51. Brook RH, Kamberg CJ, McGlynn EA. Health system reform and quality. JAMA 1996; 276:476–480.

52. Howie JGR, Heaney DJ, Maxwell M. Measuring quality in general practice. 1997; Occasional Paper 75:(abstract)

53. Palmer RH, Louis TA, Peterson HF, Rothrock JK, Strain R, Wright EA. What makes quality assurance effective? Results from a randomized, controlled trial in 16 primary care group practices. Medical Care 1996; 34:SS29–39.

54. Koop CE. Editorial: A personal role in health care reform. American Journal of Public Health 1995; 85:759–760.

55. US Public Health Service. A paper on population-based core functions. The Core Function Project. US Public Health Service, 1993.

56. Thompson RS, Taplin SH, McAfee TA, Mandelson MT, Smith AE. Primary and secondary prevention services in clinical practice: Twenty years' experience in development, implementation, and evaluation. JAMA 1995; 273:1130–1135.

57. Skinner H, Botelho RJ. Beyond advice: 1. Changing organizations to promote healthy behaviors. Thousand Oaks, CA: Sage Publications, Inc. 1998

58. Shapiro SH. On parallel processes in therapy and teaching. Psychoanalytic Quarterly 1976; 45:394–415.

59. Shapiro J. Parallel process in family medicine system: Issues and challenges for resident training. Family Medicine 1990; 22:312–319.

60. Botelho RJ, Seaburn DB, Harp J. Isomorphism in resident consultations: implications for education and patient care. Family Systems Medicine 1990; 9:137–149.

61. Lazare A, Eisenhart S. Outpatient psychiatry. A negotiated approach to the clinic encounter: Conflict and negotiation. Baltimore, MD: Williams & Wilkins, 1979:157–170.

62. Fisher R, Ury W. Getting to yes: Negotiating agreement without giving in. Boston: Houghton-Muffin, 1981.

63. Rollnick S, Kinnersley P, Stott N. Methods of helping patients with behaviour change. British Medical Journal 1993; 307(6897):188–190.

64. Botelho RJ. A negotiation model for the doctor-patient relationship. Family Practice 1992; 9(2):210–218.

65. Siegler M. Searching for moral certainty in medicine: A proposal for a new model of the doctor-patient encounter. Paul Turner, New York. Academy of Medicine 1981; 57(1):56–69.

66. McWhinney IR. Are we on the brink of a major transformation of clinical method? Canadian Medical Association 1986; 135:873–878.

67. Szasz TS, Hollenender MC. A contribution to the philosophy of medicine. The basic models of doctor-patient relationship. Archives of Internal Medicine 1956; 97:585–592.

68. Byrne PS, Long BE. Doctors talking to patients. London: HMSO, Royal College of General Practitioners, 1976.

69. Berne E. Games People Play. Penguin Books, 1978.

70. Cohen-Cole SA. Medical Interview: The Three-Function Approach. Mosby Year Book, Inc. 1991.

71. Pendleton D, Schofield T, Tate P, Havelock P. The consultation: An approach to learning and teaching. 6th ed. Oxford General Practice Series, 1984.

72. Lipkin, Jr. M. The medical interview and related skills. In: Branch WT, ed. The office practice of medicine. Philadelphia: WB Saunders Co. 1987:1287–1306.

73. Levenstein JH, Brown JB, Weston WW, Stewart M, McCracken EC, McWhinney I. Patient-centered clinical interviewing. In: Stewart M, Roter D, eds. Communicating with medical patients. London: Sage Publications, 1989:107–120.

74. Working party of the Royal College of General Practitioners. The future general practitioner: Learning and teaching. London: British Medical Association, 1972.

75. Scott NCH, Davis RH. The exceptional potential in each primary care consultation. Journal of the Royal College of General Practitioners 1979; 29:201–205.

76. Casell EJ. The function of medicine: Restoring autonomy to the patient. The Hastings Centre, 1977:16–19.

77. Putnam LL, Roloff ME. Communication and negotiation. Newbury Park, CA: Sage Publications, 1992.

78. Emanuel EJ, Emanuel LL. Four models of the physician-patient relationship. JAMA 1992; 267:2221–2226.

79. Lazare A. The interview as clinical negotiation. In: Lipkin, Jr., Putnam SM, Lazare A, eds. The medical interview. New York: Springer-Verlag, 1995:50–64.

80. Botelho RJ, Skinner H. *Beyond advice: 2. Becoming a motivational practitioner*. Thousand Oaks, CA: Sage Publications, Inc. 1998 (in press).

81. Brewin TB. Truth, trust and paternalism. Lancet 1985; 1:490–492.

82. Lomas HD. Paternalism; medical or otherwise. Social Science & Medicine 1981; 15:103–106.

83. McKinstry B. Paternalism and the doctor-patient relationship in general practice. British Journal of General Practice 1992; 42:340–342.

84. Thomasma DC. Beyond medical paternalism and patient autonomy: A model of physician conscience for the physician-patient relationship. Annals of Internal Medicine 1983; 98:243–248.

85. Lipkin Jr. M, Putnam SM, Lazare A. The medical interview: Clinical care, education, and research. Springer-Verlag, 1995.

86. Spiro H, McCrea M, Perschel E, St. James D. Empathy and the Practice of Medicine. Yale University, 1993.

87. More ES, Milligan MA. The empathic practitioner: Empathy, gender, and medicine. New Brunswick, NJ: Rutgers University Press, 1994.

88. Butler NM, Campion PD, Cox AD. Exploration of doctor and patient agendas in general practice consultations. Social Science & Medicine 1992; 35:1145–1155.

89. Prochaska JO, DiClemente CC. Toward a comprehensive model of change. In: Miller WR, Heather N, eds. Treating addictive behaviors: Processes of change. New York: Plenum Press, 1986:3–27.

90. Prochaska JO, DiClemente CC. Stages of change in the modification of problem behaviors. In: Hersen M, Eisler RM, Miller PM, eds. Progress in behavior modification. Sycamore, IL: Sycamore Press, 1992:184–214.

91. Prochaska JO, DiClemente CC, Norcross JC. In search of how people change: Applications to addictive behaviors. American Psychologist 1992; 47(9):1102–1114.

92. Prochaska JO, DiClemente CC, Velicer WF, Rossi JS. Standardized, individualized, interactive and personalized self-help programs for smoking cessation. Health Psychology 1993; 12(4):399–405.

93. Brown JB, Weston WW, Stewart MA. Patient centered interviewing Part II: Finding common ground. Canadian Family Physician 1989; 35:153–157.

IV Community Partnerships: Building Collaborative Models for Care

Patricia Hinton-Walker

In 1994, concern over rising costs, poor health outcomes for certain populations, and documented overuse of the medical care delivery system caused great political debate, but there was no consensus about what approach to adopt (from a legislative perspective) to solve these health care problems. Although no national "health care reform" policy was legislated, market-driven forces began to change the health care delivery system at an unprecedented rate. In response to these changes, business and state governments have been struggling to find ways to control health care costs, yet improve the health of their communities. Consequently, managed care is emerging as one of the primary tools for change. The term "managed care" describes many different types of emerging approaches to health care delivery, all of which involved various forms of partnership between and among "provider communities" such as hospitals, home health agencies, groups of primary care providers, and "consumer communities," the recipients and purchasers of care. Common to all the "community partnership" models for managing care are the goals of controlling costs, increasing access to providers, and improving outcomes and the health status of covered populations.

To meet these goals, the newly created partnerships of provider and consumer communities must learn to coordinate care across the full continuum of needs and services, improve performance in prevention and health promotion, attend to the role of the family and community in health and disease, involve patients as active partners in their own care, and create interdisciplinary practices designed around the needs of patients rather than the needs of providers and institutions.[1] As health care delivery shifts to the community and as many provider organizations struggle with the development of new structures of partnership, the values and skills described in the previous two chapters will be essential: the process of partnership development at the organizational and community level cannot be accomplished without focusing on relationships. The challenge of developing "community" partnerships to address these changes in the health care delivery system is the focus of this chapter.

Context for Partnership Development

One of the ways that businesses have tried to contain costs on a community-wide basis and to control cost-shifting from employer to employer is through the development of partnerships and/or coalitions.[2] Although this approach met with marginal success, the use and development of many forms of partnerships to address many of the health care problems of today has continued. Now we are seeing many new forms of partnerships springing up including alliances and affiliations of providers, hospitals, and other health care delivery agencies; networks developed to address fragmented care of specific populations; consortia for quality improvement and outcomes research; and community, business, and government partnerships designed to enhance population-based, community-sensitive care for underserved groups.

The literature is full of references to various forms of partnerships and uses different terms to label them. A thesaurus will list alliance, network, affiliation, consortium, and coalition in connection with the use of the word "partnership." *Partnership* usually refers to a legal relationship involving a close cooperation between parties with specified and joint rights and responsibilities; an *alliance* is an association that furthers the common interests of its members; and a *consortium* is a combination of members that is formed to undertake an enterprise beyond the resources of any one member.[1] For the purposes of this chapter, the various synonyms for partnership will be used interchangeably, depending on the particular reference; the principles related to the development, management, and functions of partnerships apply regardless of the particularities of form or composition.

Is an evolving *partnership* a new type of organization? Can we learn how to develop, work within them and manage them in the same ways that we study traditional organizations? Can we better understand how to build successful community partnerships and to contribute meaningfully to this latest evolution of organizational change in response to shifts toward community-based care? According to Tichy, we can apply the insights of organizational models to study the evolving organizational structures that result from partnership-development.[3] He describes a "model" as a set of assumptions or beliefs that guide actions, assist in diagnosing organizational problems, and focus the efforts of people working in organizations. He the offers three models of organizations: classical/mechanistic, human resource/organic, and political.

The classical/mechanistic model focuses on structure, span of control, and specialized task functions. This model describes a typical bureaucratic organization with detailed job descriptions, a rigid chain of command, and decisions made along vertical lines with minimum flexibility. The human resources/organic model is more concerned about the human side of organizational behavior and involves more participatory decision-making with lots of interaction. Leaders and administrators of such systems are more likely to function as motivators and facilitators, and to communicate through expectations rather than through strict job descriptions. The political model views organizations as political arenas in which multiple partners may vie for control through ongoing processes of bargaining and negotiation. Of the three models, the political model may be most helpful in understanding the dynamics of community partnerships, with their full complement of internal (frequently full-time employees) and external partners (alliance members, and/or networked groups).[4] Community partnership strategies are frequently based on making political adjustments, understanding and possibly realigning power and influence, and mastering the skills of negotiation, bargaining and collaboration. The political model is the most flexible, adaptive and responsive of the three models previously discussed. This flexibility is essential in addressing the difficult problems that arise from differences in production (structure, processes, outcomes) and culture between the different organizations or communities in a partnership.[3(p.49)]

Reasons for Development of Community Partnerships

The evolution of alliances, mergers and partnerships is a global phenomenon. Many companies in a variety of industries are choosing partnerships, alliances and coalitions as a vehicle for change in a tight market economy; this phenomenon is occurring on a worldwide basis. Competition is the name of the game. In a timely book, *Competing for the Future*, the authors write that "competition for the future is competition to create and dominate emerging opportunities–to stake out new competitive space."[5] Further, the authors say, organizations must not only compete within the boundaries of existing industries (such as hospitals, home health, insurance industries, etc.), but they must shape the structure of industries that will be able to compete and dominate the market in the future. This means that future competition and collaboration will take place within and between partnerships of companies and communities, not just between individual businesses.[5]

With many managed care organizations still in early stages of development and many state and county governments still exploring new ways of providing care to their communities, no particular structure has emerged as the "best way" to provide cost-effective care. Consequently, regardless of the type of partnership on either the provider or consumer side of the equation, it is important for all health care providers across disciplines to be active players shaping future partnerships and to contribute meaningfully to new ways of doing things. We might reasonably predict that partnerships fashioned after the political model will be the dominant structures for health care delivery organizations at this time. If we understand the nature of this type of organizational change in which we are all participating (either by choice, chance, or requirement), perhaps we can facilitate collaboration and the development of *healthy community partnerships*. It is clear that health care providers who are willing to unlearn old paradigms of control, hierarchical relationships, and patriarchy and to embrace new ways to assist organizations and communities in partnership-building will be more likely to survive and be valued in a changing health care environment. If health care providers understand that many companies are trying to "shape the structures" for managed care through development of different kinds of alliances and partnerships, then they can position themselves within these new evolving structures to have more influence on quality of care and customer satisfaction. They can also participate in positive ways in the evolution of partnerships into integrated delivery systems (IDSs), physician/hospital organizations (PHOs), Independent Provider Associations (IPAs), or a variety of other network structures which enhance community participation.

One key to developing successful partnerships for change is to understand the process of partnership building. One principal reason that partnerships develop is that no one organization has all the required resources, skills, knowledge, and expertise to produce "the product." In health care, the product may be cost-effective quality care across the continuum, or accessible care, or perhaps improvement of the health status of populations in communities. (The chapters in the Community Partnerships section of this book describe partnerships that were formed for similar reasons.) Another reason that partnerships are formed is to achieve some type of potential political advantage. Partnership may be a good way to co-opt potential future competitors, to keep crucial resources out of the hands of a competitor, or to provide access to new markets. A third reason for partnership development is to share risks and costs. "Alliances allow participation in highly volatile industries, where knowledge spreads rapidly, at substantially lower investment and risk than would be the case for a single organization."[6]

Types and Formation of Partnerships in Health Care

There are two general types of alliances or partnerships in health care, lateral and integrative. Similar types of organizations come together to achieve economies of scale by pooling resources (e.g., sharing information and human resources, group purchasing) and increasing collective power. When this occurs, it is called a lateral alliance or partnership.[6] Examples of this include the merging or networking of hospitals based on type of service, geographic distribution, and/or religious preferences, e.g., rural hospital networks and community hospital partnerships. On the other hand, integrative alliances focus on marketing, strategic positioning, and competitive advantage. Kanter uses the term "stakeholder alliance" to describe such linkages among buyers, suppliers, and customers. These partnerships feature both vertical and horizontal integration including clinical, administrative, financial, and delivery components of the organization.[7] Examples of this type of partnership in health care are the emerging corporate linkages between providers (physicians), suppliers (medical supply companies or pharmacies), provider organizations (hospitals), and consumers (insurance companies and/or industry purchasers of health care). In the literature this may be described as a Health Maintenance Organization (HMO), physician-hospital organization (PHO), or an integrated delivery system (IDS).

Other types of partnerships are emerging that affect health care, although they are not formalized as "business partnerships." With the shift from hospital to community-based care, there will be increasing numbers of partnerships that bring together health care businesses, volunteer-run community programs, and in some cases state and local governmental groups to change patterns of access, lower cost, improve outcomes, and influence health care policy. The chapters in the Community Partnerships section of this book describe three such examples: a partnership between churches and a health care organization to improve hypertension treatment, a partnership between a school district and a department of pediatrics to improve diagnosis and coordination of care for "at-risk" students, and an alliance of business and healthcare leaders to improve rural health. Another illustrative case study involves a nurse-provider in an African-American community who mobilized a racial-ethnic minority community to impact state-level child care policies. She created a partnership involving mothers linked to a child care advocacy group, a welfare watch group, and a welfare reform advisory committee to the State Welfare Department.[8] Rooney described the development and evolution of business partnerships among employers who were concerned not only about rising costs, but also about quality of care. Approximately 140 employers from nine states formed the Midwest Business Group on Health. This partnership developed educational programs, designed quality measures, and conducted research designed to assist purchasers of health care.[2]

Two additional examples of community-wide partnerships demonstrate the bridging of providers, local communities, and state government to address consumer, provider, and political issues related to population-based care. In one case, the fragmentation of care provided to technology-dependent children and their families stimulated the development of a community partnership developed to coordinate services for this at-risk population. This partnership was an example of an interagency, professional, and community collaboration at the state, regional, and local levels in New York, and was established through local agencies—a hospital, a pediatric pulmonary center, and a health systems agency—in which providers, parents, and community-based agencies worked together to achieve common goals.[9] A rural community-wide partnership was formed to promote farm health and safety. The Occupational Health Nurses in Agricultural Communities (OHNAC) Project is an example of a partnership developed by a professional organization, the state department of health and local public health agencies to design intervention strategies using education, engineering, and enforcement in an attempt to reduce farm accidents and influence the health of agricultural families and workers.[10] These two examples illustrate the development of partnerships in order to improve population-based care.

New Paradigms, Collaboration, and Energy: Keys for Successful Partnerships

Many partnerships in today's health care environment may seem to be linkages formed for political reasons, as well as the need to acquire resources, skills, and knowledge.[1] In many cases, partnership-building may not follow any distinctive pattern. However, several key principles of collaboration, and new ideas and paradigms must be considered when building community partnerships.

Collaboration offers a viable strategy for partnership development among businesses, communities, and governments. According to Gray, complex societal issues affecting our global future make it difficult for individual organizations to act unilaterally to solve problems without creating unwanted consequences or encountering constraints. Because of interdependence, collaboration offers a process whereby parties who see different aspects of a problem can search for solutions that go beyond their own limited vision, resources, knowledge and skills.[11]

Characteristics of collaborative partners that are key to the successful development of partnering relationships are highlighted in Weaver and Farrell's TARGET approach.[12] These key characteristics are: *T*ruth, *A*ccountability, *R*espect, *G*rowth, *E*mpowerment, and *T*rust. Without *truth*, it is difficult to attain respect and trust. Mutual *accountability* is required for all members to accept responsibility for the partnership's failure or success. By being willing to *grow* together, partners can learn to complete more difficult tasks, work more efficiently, and be more creative together to solve problems. *Respect* is usually based on members experiencing others as acting with integrity and being respectful themselves. A critical change begins to occur as members move from adversarial to partnering behaviors, when members feel *empowered* enough to act on their own initiative to improve the success of the partnership. Although interdependent action is clearly required, empowerment to act independently in the partnership's best interest is an important sign of growth. Lastly, but not least important is the characteristic of *trust*. Partners in successful relationships have learned to trust each other and count on each other to uphold their parts of the process and to honor decisions made. Groups interested in the development of partnerships can use the TARGET approach to measure the growth and success of the partners at different points in the development of the relationship.

Successful collaboration usually results in the development of successful partnerships. However, collaboration will not always work—particularly when one party has unchallenged power to influence a decision, or when some of the stakeholders are unwilling to engage each other in meaningful and participative relationships. Sometimes a skilled third party is needed to facilitate collaboration, particularly for complex multiparty situations. Gray reinforces the use of negotiation, consensus building and mediation as tools used to facilitate conversational interactions in order to define problems, agree on recommendations, design action steps, and/or to help structure a dialog where parties can work out differences.[11] All of these skills are important to the process of collaboration and are needed in the development of complex community partnerships.

Power is another factor that influences the success and/or failure of community partnership development. Frequently the company with the most relevant competencies and resources for a given product is the strongest partner and has the most influence and power. Therefore, it is not unusual to see traditional hospital-driven and provider-based systems enjoying relatively strong positions in this changing health care market. However, with the shift in emphasis from traditional hospital and physician-based services to community-based care and with the increasing intensity of competition, the voice of the consumer and/or the community is growing, creating an opportunity for a more balanced sharing of power between providers and communities. Block emphasizes just this kind of paradigm shift as being important to the development of partnerships, shifting away from hierarchical, patriarchal power with its top-down, high control orientation to shared power and authority; and from dependency to interdependence in relationships.[13]

Attention to relationships is necessary but not sufficient for partnership development. The sign of an effective partnership between organizations lies not in whether the people involved feel good about their relationship, but in the value of the specific operational links and activities of the organizations involved.[6] Clear and explicit agreements on structure and procedures are an essential foundation for inter-organizational or inter-community partnerships. One of the reasons that many partnerships fail is that they are built mainly on personal relationships, and lack formalized structures, decision-making mechanisms, and tracking systems to measure success and identify problems.[6] Once community and/or organizational partnerships are formed, the management of these new entities is another challenge. Kaluzny et al describe two types of administrative challenges involved in sustaining these inter-organizational or inter-community partnerships: 1) management of the inherently fragile relationships among the partners; and 2) managing the changes internal to each partner's organization or community.[6]

The administration of partnerships on a day-to-day basis requires new ways of thinking and new energy. Gadman, in his new book entitled *Power Partnering*, describes the need to shift from "automatic energy" which keeps things going without any act of attention to higher levels called "sensitive energy," "creative energy," and "unitive energy."[14] Sensitive energy requires a higher level of awareness which brings meaning to the sensing of what is needed in what he calls day-to-day power partnering. He also emphasizes the need for the use of "creative energy" which is associated with spontaneity and requires creative thinking, communication, and problem solving. One of the more important higher levels of energy mentioned by Gadman is "unitive" energy which is the power to engage the collective will of many people and unite them around a common cause."

Clearly, the development and sustaining of community partnerships requires the use of new energies, new paradigms, and renewed investment in collaboration. In addition to learning these new skills of community building, health care providers must also increase their clinical knowledge, skills, and competencies in health promotion and prevention, care across the continuum, cost-effectiveness, and attention to customer satisfaction and community-level outcomes.

Case Study: Community Nursing Center and School-Based Health Center Partnership

The School of Nursing at the University of Rochester created a Community Nursing Center (CNC) to facilitate faculty practice in the community. A number of community partnerships were developed over the past seven years (from 1990 to 1997) for urban and rural community-based care. Advanced practice nurses in the context of interdisciplinary care, provided a variety of services through community settings such as rural community hospitals, a rural hospital consortium, group homes for troubled teens, primary care offices, county health departments, and in a rural county jail. With a reputation and commitment to development of true community partnerships, an opportunity was presented to compete for a school-based health center grant sponsored by the New York State Department of Health (NYSDOH).

A major requirement for the proposal was the development of partnerships among physician providers for referral and interdisciplinary consultation by advanced practice nurses; other provider organizations (such as a hospital) for 24–hour care; the county health department which provided some school health services; the high school where the school-based center would be housed (including students, parents, teachers, and administrators); and the city school system. Since the State of New York was using these grants to stimulate development of capitated care arrangements for school age and adolescent populations in the context of managed care, the participation of insurance companies in these partnerships for care was also expected. Partnership development in this case study was consistent with the political model described previously.

The CNC's mission and values were consistent with partnership development, but it was important to CNC leadership that the spirit of partnership with the community not be violated when other providers and health care organizations were added to this new evolving structure. The sharing of risk was one of the purposes mentioned for the development of partnerships. In the context of managed care in this school-based health care partnership, financial risk was primarily shared between the CNC and the Strong Memorial Hospital (SMH). The opportunity to penetrate the market, develop positive community-based care partnerships, and participate in future Medicaid managed-care contracts for adolescent care initiated by the State of New York was the hospital's purpose in agreeing to share the risk and participate in this partnership. For the CNC, the partnership with the hospital provided access to 24–hour care, emergency room access, and inpatient care for enrollees. To keep costs down, the CNC was also able to use the administrative and financial services of the hospital such as billing, purchase of supplies, pharmaceuticals, lab and X-ray services.

Partnership relationships with the Monroe County Health Department (MCHD), the Rochester City School System, East High School (EHS), and the Adolescent Medicine Department in the University of Rochester School of Medicine were developed for the purposes of coordinating activities among provider and consumer groups. These partnerships were examples of public-private partnership arrangements that were developed based on a shared vision of school-based care for underserved adolescents. As in many situations, these partnerships are less formal in arrangement than traditional business partnerships. However, they required significant nurturing and communication to keep the partnerships alive and well in a managed care environment.

Key stakeholders and players from each of the partnership groups were identified early in the process and ongoing meetings were held to involve the players in the development, implementation and evaluation of the school health center. Additionally, expectations, authority, financial risk sharing, and roles/responsibilies of the different groups in the partnership were clarified. This was accomplished through a combination of letters of agreement, memoranda of understanding, and contracts. An advisory committee structure was designed to facilitate involvement of the consumers in the community (students, parents, faculty, and administrators), contracted physicians, and stakeholders from the health department, the school system, and the School of Nursing. The advisory group has remained very active in policy development and review, marketing, quality assurance, and problem solving especially related to student and parent issues. Advanced practice nurses provide the leadership at the clinic site; nurture the relationships in this alliance for care of adolescents; negotiate financial arrangements and contracts; and assume full responsibility for the care of this underserved, at-risk population. The CNC is at the center of this partnership and has the strongest power and influence because the student population needs the skills, knowledge and competencies of nurses (nurse practitioners and psych-mental health clinical nurse specialists). The community expressed a need for a preventive model of care with a psych-mental health component, and the CNC had practitioners with expertise in adolescent care available for the care of this population, along with the business knowledge and expertise to put this partnership together.

Conclusion

Concluding thoughts for this chapter on community partnerships come from three sources: Kaluzny, Zuckerman, and Ricketts' book *Partners for the Dance: Forming Strategic Alliances in Health* Care,[6] Gadman's book, *Power Partnering*[14] and Gray's book *Collaborating: Finding Common Ground for Multiparty Problems*.[11] Each of the authors writing is so clear and elucidating that I have chosen to present their thoughts in their own words–from their respective books. Gadman writes, "In our Western culture, our loathing for ambiguity has given way to a craving for truth and certainty . . . institutions have been founded on a clock logic of certainty, predictability, and control . . . and our 'clock workplaces' have become 'cultures of devaluing and distrust.'"[14 (p. 16)] Partnerships do not exist or thrive in an environment of distrust, certainty and control. Gadman encourages us to reverse this clock logic and explore the benefits of power partnering relationships. This approach demands that we set aside the time to share and mutually value one another's perspective. This valuing means "you pay attention to them as a real person existing out there in the real world–and, more importantly, as a reflection of your inner world as you see yourself mirrored in them. . . if you want to understand your environment you must first begin by understanding yourself."[14 (p. 17)]

To develop lasting partnerships and alliances, according to Kaluzny, Zuckerman, and Ricketts, "the message is simple. . . Be friendly, develop trust, and aim toward good working relationships with the 'other folks', but trust isn't' the only basis for an effective relationship. . . It is not enough to simple create an excellent structure, or to train people. . . or to revise linkages . . . or to communicate the purposes of a new idea within and without the organization . . . all are necessary, and all must be carried out in concert."[6 (p .28)]

Barbara Gray's comments provide an appropriate summary for this chapter. She also highlights keys for success when she writes that collaboration "[must] overcome the barriers associated with different institutional cultures. . . be accomplished laterally without the hierarchical authority to which most managers are unaccustomed. . . involves a process of joint decision making among key stakeholders. . . and establishes a give and take among the stakeholders that is designed to produce solutions that none of them working independently could achieve."[11(p 9)] Finally, she makes a very important point that is consistent with the title of this book, "collaboration is a essentially an emergent process rather than a prescribed state of organization. By viewing collaboration as a process, it becomes possible to describe its origins and development as well as how its organization changes over time."[11 (p. 15)]

References

1. Walker PH. Toward a comprehensive health care system: Business and provider coalitions. In J McCloskey and HK Grace, eds., *Current Issues in Nursing*. Fifth Edition. St. Louis: Mosby-Year Book, Inc., 1997:427–433.

2. Rooney E. Business coalitions on health care: An evolution from cost containment to quality improvement. American Association of Occupational Health Nurses, 1992:40;342–351.

3. Tichy NM. *Managing Strategic Change: Technical, Political and Cultural Dynamics*. New York: John Wiley and Sons, 1983:43–55.

4. Mintzberg H. Policy as a field of management theory. *Academy of Management Review*, 1977:2;88–103.

5. Hamel G & Prahalad CK. *Competing for the Future: Breakthrough Strategies for Seizing Control of Your Industry and Creating Markets of Tomorrow*. Boston, MA: Harvard Business Press, 1994:23.

6. Kaluzny AD, Zuckerman HS, & Ricketts III TC. *Partners for the Dance: Forming Strategic Alliances in Health Care*, 1995.

7. Kanter RM. Becoming PALs: Pooling, allying, linking across companies. *Academy of Management Executives*. 1989:3;183–193.

8. Russell K. Strengthening Black and minority community coalitions for health policy action. Journal of National Black Nurses 1992:6;42–47.

9. Lobosco A, Eron N, Bobo T, Kni L, and Chalanick K. Local coalitions for coordinating services to children dependent on technology and their families. *Childrens Health Care* 1991:20;75–86.

10. Lexau C, Kingsbury L, Lenz B, Nelson C, and Voehl S. Building coalitions: A community wide approach for promoting farming health and safety. American Association of Mental Health Nurses 1993:41;440–463.

11. Gray B. *Collaborating: Finding Common Ground for Multiparty Problems*. San Francisco: Jossey-Bass Inc., 1989.

12. Weaver RG, and Farrell JD. *Managers as Facilitators*. San Francisco: Barrett-Koehler, Inc., 1997:80–81.

13. Block P. *The Empowered Manager: Positive Political Skills at Work*. San Francisco: Jossey-Bass, Inc., 1987:23–24.

14. Gadman S. *Power Partnering*. Boston: Butterworth-Heinemann, 1997.

PART 2

Patient-Clinician Partnerships

The patient-clinician relationship is where partnership in healthcare begins. It takes historical precedence (when the first patient sought attention from the first healer, health system issues were probably not very complicated); it is the front line of health care delivery; and it is the most thoroughly researched level of partnership, and thus paradigmatic for the study of all healthcare partnerships. The conceptual models emerging from this research—collectively known as the Biopsychosocial model,[1] Medical Family Systems Theory,[2] and Relationship-Centered Care[3]—are the sources of the core principles of partnership that we propose in this volume:

> (1) respectful relationships (including the clinician's genuine presence as a participant observer, not merely a detached observer); (2) the recognition by the clinician of the patient's (and family's) perspective as well as her own, resulting in true dialogue (in which the clinician is open to being changed and not just bent on changing the patient); and (3) shared, flexible decision making in which the patient is encouraged to be an active partner.

These principles are "tried-and-true" in the realm of patient-clinician relationships, but their application remains infinitely varied as new modes of health care delivery, new types of treatment, and new decision support tools are created. And, as in any area of intellectual endeavor, new questions arise as quickly as the old ones are answered, leaving us always "in the question." The chapters in this section touch on all of these areas.

Geoffrey Williams, *Ed Deci*, and *Richard Ryan* anchor the link between good partnership process and improved outcomes in an empirically tested model of human motivation known as Self Determination Theory. They review their own considerable body of research and that of others demonstrating that patients' motivation to undertake and maintain behavior change (and subsequent achievement of improved health outcomes) is strongly associated with the use of autonomy-supportive approaches and the communication of personal interest and commitment on the part of their clinicians. Self Determination Theory has also been shown to predict motivation and behavior change in the workplace. The themes of this chapter resonate with those of the Westberg and Hill chapter that contrasts authoritarian and collaborative approaches to patient care and professional education and can be seen at work in vignettes of the chapters by Brachman (later in this section) and Blandin-Clark (in the Educational Partnerships Section).

Thomas Campbell, *Susan McDaniel*, and *David Seaburn* focus on the need for mental health professionals and primary care practitioners to collaborate more effectively in providing mental health services in a population-based manner. They also address the need to incorporate family members in the provision of healthcare services. The principles which they offer to guide this process of establishing and providing collaborative family care include the core principles of partnership:

1. Heal the mind-body split.
2. Respect culture differences.
3. There's no substitute for a good relationship.
4. Respect each other's paradigm.
5. Promote a "community of caring."
6. Adopt a population-based perspective.
7. Make families part of the health care team.
8. Develop a common mission.

Mary Dombeck and *James Evinger* take the exploration of partnership to a deeper level. They use the novella *The Death of Ivan Illyich* by Leo Tolstoy and a case example from psychotherapy to illustrate the spiritual aspects of partnership in clinical practice. Such partnerships involve close cooperation between parties having specified rights and responsibilities toward each other. What distinguishes the spiritual partnership is its covenantal nature that is rooted in the personhood of both the clinician and the patient. Spiritual partnerships involve aspects of persons (values, meanings, purpose, and connectedness to a higher power) that are crucially important and unique to them. The scope and depth of a patient's spiritual needs, when heard by the clinician, will draw out the clinician as a person to share in what is universally human, and, at the same time, transcends each individual human life. Note the strong resonance here with Suchman's description of the relation paradigm—especially the clinician's role of being with the patient. Dombeck and Evinger address some of the barriers that hinder the process of developing spiritual partnership with patients; these become particularly salient in dealing with issues of death and dying. The reunification of spirituality and health care, after a three-century separation, is a leading-edge topic in the study of clinician-patient relationships.

Laura Brachman and *Tim Quill* address the importance of an effective working relationship between the doctor, patient and family in order to address "Do-Not-Resuscitate" decisions. They describe the limitations of providing resuscitation in particular circumstances, explore the issue of clinical uncertainty in determining medical futility, and address the challenge of presenting survival data to patients. Two cases—one routine and one complex—illustrate how physicians can develop partnerships with patients and their families in dealing with this emotionally charged issue. Achieving a thorough understanding of the "other," in this case the patient or family, flexibility and engagement of all principal participants in decision making are critical factors to success.

Rebekah Wang-Cheng and *David Schiedermayer* explore an interesting, common, and seldom discussed aspect of partnership. Some of the factors that promote partnership building—mutual exploration, understanding, genuine presence and respect—may also contribute to the development of friendship. What happens to professional/personal boundaries when friends become patients and when patients become friends? Clinicians, administrators, and educators who pursue good partnership process will need to be prepared for this situation. Wang-Cheng and Schiedermayer provide some suggestions about how to deal with the challenges of this situation. They also discuss cases in which friendship interfered with development of effective clinical partnerships through loss of objectivity, inability to discuss personal details or perform appropriate examinations, physician countertransference, and excessive guilt and anger about poor outcomes.

Alexis Abernethy, Betsy Naumburg, and *David Siegel*, in an outstanding example of parallel process, describe the strategies they used to explore and work through their own biases in preparing a workshop on cultural competence (the ability to form effective, transcultural partnerships with patients) for pediatric residents. Their workshop addressed the need to understand each other's assumptions, world views, values, and biases and the need to acquire skills to openly and respectfully talk about differences; these were the very same tools that allowed them to come together as a team to put on the workshop. Although this workshop focused on racial bias, the principles can be applied to other types of biases in healthcare partnerships.

Jim Dolan describes how a microcomputer-supported decision making method called the Analytical Hierarchial Process (AHP) can help practitioners develop an efficient and effective decision making partnership. This method is an amplification of and formalized semiquantitative approach to the principle of understanding the patient's values and perspective. Dolan uses the example of counseling a woman about the benefits and risks of hormone replacement therapy: through the AHP process, the clinician learns the relative values (positive or negative) that the patient places on heart disease, endometrial cancer, breast cancer, resumption of vaginal bleeding, hormonal side effects, and the need for close monitoring for endometrial cancer. The patient's weighted preferences then drive the ranking of three options: estrogen with progesterone, estrogen without progesterone, and no hormone replacement therapy. With a small degree of irony (born of a perception of computers as impersonal), we note that computer programs such as this one can improve patient-clinician communication and improve the quality and quantity of patients' participation in decision-making.

References

1. Lipkin Jr M, Putnam SM, Lazare A (eds.). *The Medical Interview: Clinical Care, Education, Research*. New York: Springer-Verlag, 1995.

2. McDaniel S, Campbell TL, Seaburn DB. *Family-Oriented Primary Care*. New York: Springer-Verlag, 1990.

3. Tresolini CB and the Pew Fetzer Task Force. *Health Professions Education and Relationship-centered Care*. San Francisco: Pew Health Professions Commission, 1994.

V Building Healthcare Partnerships by Supporting Autonomy: Promoting Maintained Behavior Change and Positive Health Outcomes

Geoffrey C. Williams, Edward L. Deci, and Richard M. Ryan

In healthcare partnerships, providers and patients develop a collaborative responsibility for the patients' health and actively exchange information about how to achieve the preferred outcomes. There is substantial humanistic appeal to the sharing of power in such relationships, and on those grounds alone, one could provide a strong argument in favor of healthcare partnerships. Such an argument would not be wholly satisfactory for an empiricist, however, who would argue that one should specify clearly the outcomes one expects to achieve from healthcare partnerships and then explore scientifically whether partnerships actually contribute to those specified outcomes.

Everyone agrees that patients' physical health is an important healthcare outcome. As advocates of the biopsychosocial orientation toward medicine,[1] which emphasizes that every level of organization (including not only the biological, but also the psychological and social) affects every other level, we believe that patients' functional status and psychological health are also important healthcare outcomes. Accordingly, we maintain that it is important to investigate empirically whether healthcare partnerships do in fact contribute to the outcomes of physical health, effective life functioning, and psychological well-being.

Our research on these issues, which is reviewed in this chapter, has been guided by the psychological concepts and paradigms of self-determination theory.[2,3] At the core of the theory is the distinction between autonomous and controlled behavior. To be *autonomous* means to act with a full sense of volition and choice, whereas to be *controlled* means to act with a sense of pressure and demand. Beginning with this theoretical distinction, we have focused on two critical questions: (1) Does the extent to which patients' motivation is autonomous versus controlled affect the healthcare outcomes we have specified? and (2) Do healthcare partnerships—or, more precisely, does the providers' orientation toward supporting patients' autonomy, which we argue is the central element of healthcare partnerships—influence patients' autonomous versus controlled motivation and, in turn, their health outcomes? We will address these two questions in turn by reviewing relevant research.

Studies on the physical and psychological health consequences of individuals' autonomous motivation, and on the factors that promote autonomous motivation have been conducted in several life domains, including parenting,[4] education,[5] work,[6] and religion,[7] in addition to the healthcare domain.[8] The research in all these domains is relevant not only because it highlights the fact that psychological and social factors do affect individuals' well-being, as the biopsychosocial model suggests, but also because it has helped to clarify which psychological and social variables in healthcare settings are important for promoting health and well-being.

Our research within the healthcare system, which will be the central focus of this chapter, is concerned primarily with patients' maintained change of behaviors that are unequivocally linked to health outcomes—for example, stopping smoking, increasing exercise, adhering to medication prescriptions, dieting, and controlling glucose levels. As psychologists, the issue of changing health-compromising behaviors interests us greatly because we view motivation as the key to effective behavior change and understand it as a mediator of health and well-being.

The Relation of Behavior to Health

Each year about two million Americans die. The official causes for the vast majority of these deaths are diseases familiar to us all—heart disease, cancer, strokes, pneumonia, and diabetes. Although most people think about these diseases in terms of their biochemical causes and cures; the diseases can be more fully understood by using the biopsychosocial approach. Increasingly, it has become clear that the psychological and social factors which underlie people's behavior (in addition to the biological factors) play a significant role in the onset and progression of these illnesses, and thus in people's morbidity and mortality. According to McGinnis and Foege,[9] for example, diet and activity patterns, alcohol dependence and abuse, and tobacco use account for approximately eight hundred thousand deaths annually.

It is indeed amazing that nearly half of all American deaths are directly attributable to people's own actions. By engaging in unhealthy behaviors, people bring on disease and death, and yet they show great resistance to changing those unhealthy behaviors. The success rates for behavior-change programs and physician-supervised interventions for weight loss and smoking cessation are startlingly low. Large percentages of people drop out of behavior-change programs,[10] and very few of the people who do complete them demonstrate maintained behavior change over long periods. Indeed, Kramer, Jeffery, Forster, and Snell[11] reported that only three percent of people who had participated in weight-loss programs had maintained their losses four years later. Direct clinical interventions by physicians have also yielded discouraging results. For example, Schwartz[12] reported that success rates for smoking cessation typically range between only five and 20% at one-year follow-up.

A related problem concerning patients' behavior that also affects their health and troubles providers is the problem of adherence to medical regimens. As Phillips and Jones[13] put it, physicians prescribe treatments and count on their authority to motivate patients to adhere to these prescriptions, but according to DiMatteo and DiNicola[14] the rate of adherence is very low. Many patients do not reliably take their medications, carry out prescribed dietary modifications or exercise regimens, return for scheduled appointments, nor engage in preventive activities such as undergoing mammograms.[15]

Clarifying how to facilitate maintained behavior change and medication adherence is clearly of utmost urgency. The research reviewed in this chapter represents a step toward that clarification.

Motivating Change through Healthcare Partnerships

The biopsychosocial approach to medicine[1] emphasizes the importance of providers' awareness of their own feelings and reactions to patients because these factors are believed to have a meaningful impact on patients' responses to medical treatment. In other words, providers' behavior toward patients is considered a social factor that affects the patients' motivation and behavior, and thus their health and well-being. The concept of healthcare partnerships emerged out of the biopsychosocial model as a way of describing the type of patient-provider relationship that is expected to promote the most positive health outcomes and the type of provider behaviors that are expected to facilitate an optimal relationship.

One of the important elements of self-determination theory[2] is a set of dimensions that characterize interpersonal environments. Perhaps the most critical of these dimensions is the extent to which an interpersonal environment supports autonomy versus controls behavior. *Autonomy support* in healthcare is defined as the extent to which providers listen fully to patients, take the patients' perspective, provide patients with meaningful information about their condition, and offer patients choice about how to deal with their health. In contrast, control refers to providers' pressuring patients to behave in particular ways.

We argue that providers' being autonomy-supportive is implicit in the biopsychosocial approach to healthcare and is essential for establishing healthcare partnerships. One of the important advantages of formulating partnerships in terms of providers being autonomy-supportive is that there is a history of empirical work on the effects of autonomy support that can inform an exploration of the effects of healthcare partnerships on healthcare outcomes.

Accordingly, we have investigated whether providers' being experienced as autonomy-supportive by their patients will lead the patients to be more autonomous in regulating their health-relevant behaviors and whether that in turn will lead to greater maintenance of behavior change and improvements in health, well-being, and functional status. In presenting that work, we begin by describing more fully what is meant by being autonomous. We then review studies linking autonomy to healthcare outcomes, and finally we consider whether partnerships (as operationalized by patients' describing their providers as autonomy supportive) promote autonomous motivation and healthy behaving.

The Regulation of Behavior

Self-determination theory posits that autonomous and controlled behaviors have different underlying regulatory processes that are reflected in different reasons for behaving and are accompanied by different experiences.

Autonomous behaviors are regulated by the process of choice, which is reflected in people's experiencing a full endorsement of or sense of willingness to engage in a behavior. Their perceived locus of initiation and regulation for such behaviors is within themselves. When acting autonomously, people behave because they find the activity either interesting or personally important. It emanates from the self and is thus considered self-determined.[3]

Controlled behaviors, in contrast, are regulated by the process of compliance and are reflected in people's feeling as if they have no choice. Instead, they feel pressured to behave by interpersonal or intrapsychic forces that are either coercive or enticing. When controlled, people say they are behaving because of demands, threatened punishments, or proffered rewards from an external agent such as a health-care practitioner, or alternatively because of a rigid belief that they *should* do the behavior—that they *have to* do it to feel worthy.

If a woman were to take her diabetes medication because she fears her husband would be angry at her if she did not, the behavior would be controlled. Evident in this woman's reason are a lack of personal investment in the behavior and the fact that the behavior is initiated by factors outside herself. Alternatively, if she were to take it because she believes the medication is important for her own well-being and she has made a personal commitment to maintain the best possible health, the behavior would be autonomous. In contrast to the controlled reason, this reason contains both personal investment and self-initiation.

The distinction between autonomous and controlled motivation actually represents a continuum rather than just a dichotomy, so any action can be described in terms of the degree to which it is autonomous versus controlled. Thus, our research has explored the extent to which people experience a given behavior such as dieting or smoking cessation as being freely chosen (i.e., autonomous) versus being pressured by some interpersonal or intrapsychic force (i.e., controlled).

Distinguishing between these two types of motivation or regulation represents an important step beyond previous health-relevant motivation theories such as the Health Belief Model,[16,17] self-efficacy theory,[18] and the health locus of control model[19] because each of those other theories has treated motivation as a unitary concept. Stated differently, they have focused on what leads people to have more versus less motivation without considering which type of motivation it is. As such, they have been unable to predict various quantitative and qualitative aspects of patients' behavior and experience that are associated with the different types of motivation.

Becoming More Autonomous: The Integration of Regulations

The reason healthcare professionals find it necessary to encourage patients to change their behaviors is that many patients do not do it on their own. These new behaviors (e.g., dieting or abstaining from tobacco use) are neither fun nor easy. Indeed, they are difficult and require considerable determination. Furthermore, the old behaviors that must be abandoned (the overeating or smoking) have served a purpose for the patients, such as reducing anxiety or providing needed stimulation. Consequently, for successful behavior change, patients must develop self-regulatory capacities that are sufficiently effective not only to overcome inertia and prompt the new behavior but also to manage powerful forces such as anxiety that maintain the old behavior.

Patients develop greater self-regulation by internalizing regulations. *Internalization* entails transforming external regulatory processes into internal regulatory processes.[20] Through internalization, regulations that initially reside in the urgings of a healthcare professional can be taken in by the patient. However, there are two types of internalization, only one of which results in patients' becoming autonomous.[3] The less effective type is referred to as *introjection*, in which people take in a regulation but never accept it as their own. They then use the internalized regulation to coerce themselves into acting. Such regulation is still considered controlled rather than autonomous even though the regulation is within the person.

The type of internalization through which regulations become truly autonomous is referred to as integration. *Integration* involves identifying with a regulation and bringing it into harmony or coherence with other aspects of oneself.[21] When a regulation has been integrated, the person will act with a full sense of choice.

The phenomenon of initially external regulations becoming internalized occurs in healthcare all the time. Frequently, when patients are diagnosed with a common illness such as hypertension, they demonstrate little interest in performing appropriate behaviors like monitoring their blood pressure, reducing their salt intake, or taking their medication. Often, health educators must initially serve the function of prompting these patient behaviors. But for successful healthcare, the patients must assume responsibility for these regulatory functions. Internalization is the general process through which that happens, and integration is the optimal form of internalization.

To recapitulate, self-determination theory distinguishes between controlled behaviors and autonomous behaviors. Controlled behaviors encompass both external regulation and introjected regulation. External regulation describes behaviors that are dependent upon rewards, demands, or evaluations overtly external to the individual, whereas introjected regulation refers to behaviors that are dependent on internal demands and internally administered consequences. With introjected regulation, it is as if one part of a person were pushing the rest around. An example would be a woman who prods and pressures herself with the threat that she will feel worthless if she does not stick to her diet.

The primary problem with external regulation is that when the external demands and contingencies are not present, the behavior will not be maintained. The problem with introjected regulations is more subtle, because introjected regulations are within the person so there is a greater likelihood that they will persist when the external agent is no longer prompting them. But our research has shown that introjected regulation is associated with poorer persistence than integrated regulation, and it also has a variety of negative consequences.

In contrast to controlled behaviors, autonomous behaviors result when people have fully accepted the personal importance or value of a behavior and have thus integrated its regulation into their coherent sense of self. People do not behave simply because they feel they should, but rather because they have identified with the underlying value of a behavior and have genuinely accepted the responsibility for doing it. When a regulation has been integrated, people will do the behavior with a full sense of willingness and choice. An example would be a man who exercises because he understands how important exercise is for his health and has decided that he is willing to expend the effort necessary to do it on a regular basis.

To the extent that patients are successful in integrating behavioral regulation, the behavior will be autonomous, whereas to the extent that they are less successful, the subsequent behavior will occur only if controlled by external agents or their introjected counterparts.

Autonomous Regulation and Healthy Behavior

In the realm of health-relevant behavior change, we have tested the hypothesis that when patients behave for autonomous (rather than controlled) reasons, they will be more successful in long-term maintenance of behavior change, which, in turn, will have positive health consequences. This is shown schematically in Figure 1. The studies testing this hypothesis have used the Treatment Self-Regulation Questionnaire (TSRQ) to assess whether patients' reasons for participation in health-relevant treatments are more autonomous (i.e., more integrated) versus more controlled (i.e., more external and introjected).

In one study, Ryan, Plant, and O'Malley[22] used the TSRQ with one hundred outpatients who entered an alcohol treatment program. Outcomes were assessed eight weeks later in the form of attendance records and clinicians' ratings of patient involvement in the treatment program, and results confirmed that those people whose reasons for entering the program were more autonomous (i.e., more integrated) attended more regularly and stayed in the program longer than people with more

Figure 1. The hypothesized effects of patients' autonomous motivation on behavior and health.

controlled reasons. There was also a significant correlation between patients' autonomous motivation and clinician ratings of patients' degree of involvement in the program. Thus, there is evidence that patients with more autonomous reasons for participation will value the treatment more highly and be more likely to remain in treatment programs, relative to patients who are more controlled in their regulation.

Another study involved severely obese patients participating in a medically supervised, very-low-calorie-liquid-diet, weight-loss program.[8] In addition to following the diet, patients attended weekly meetings at which they consulted a doctor or nurse and met in a group counseling session to support each other and discuss issues that came up for them. Nutritionists and exercise physiologists also consulted with the patients at those meetings. A few weeks into the program, patients completed the Treatment Self-Regulation Questionnaire, indicating the reasons why they were following the program guidelines. From their responses, the extent to which their motivation was autonomous versus controlled was calculated. Then, at the end of the six-month program their attendance at the weekly program meetings and their weight loss over the period were tabulated. Finally, at a two-year follow-up, their long-term weight loss as well as the amount of their regular exercising were assessed.

Results of the study revealed that patients who reported more autonomous reasons for program participation attended more regularly and, in turn, lost more weight. Even more importantly, the extent to which patients' reasons for participation were autonomous was a positive predictor both of the amount they exercised two years later and of their maintained weight loss over that same period. It appears that when patients are autonomous in their motivation for participating in a weight-loss program, the outcomes are likely to be substantially more positive than when they are controlled.

Williams, Rodin, Ryan, Grolnick, and Deci[23] did a study of medication adherence that also supported self-determination theory. One hundred and twenty-six patients who had been prescribed long-term medication participated in the study, and adherence was assessed with both self-reports and pill counts. Regression analyses revealed that patients' autonomous regulation—that is, their integration of the value and regulation of medication taking—was a strong, positive predictor of adherence, whereas controlled regulation was unrelated to medication taking.

In this Williams, Rodin et al. study, physicians were asked to predict which patients would adhere to their prescriptions, and analyses revealed that physician predictions were unrelated to actual adherence. This point is an interesting and important one because it suggests that people can look just as motivated whether their motivation is autonomous or controlled, and the physicians were probably using overt signs of any motivation to make their predictions. However, these and other results suggest that it is only when that motivation is autonomous that positive consequences (such as maintained adherence) are likely to be manifest. To detect the difference between autonomous and controlled motivation, physicians would need to listen more fully to patients and be able to take their perspective—in other words, the physicians would need to be more autonomy-supportive.

A recent study of diabetic patients[24] explored the relation between autonomous regulation and a physiological indicator of health. Maintained glucose control (i.e., low HgbAlc) is widely acknowledged to indicate a positive prognosis for diabetics.[25] In the Williams, Freedman, and Deci study, 136 patients completed the TSRQ, and their HgbAlc scores were obtained from blood samples. Analyses revealed that autonomous regulation of the treatment regimen (viz., maintaining a proper diet, exercising, and taking medication) was a positive predictor of glucose control.

Williams and Deci[26] performed a study in which physicians counseled patients to quit smoking. They assessed patients' reasons for trying to quit, using the TSRQ, and results confirmed that patients whose motivation for quitting was more autonomous evidenced greater cessation six months later.

When taken together, these studies indicate that patients' motivation for behavior change can be either autonomous or controlled, but it is their autonomous (rather than controlled) motivation that positively predicts adherence to prescriptions, attendance in programs, display of positive behavior change, and maintenance of those changes over time.

As indicated earlier, we also consider sound mental health and functional status to be important outcomes for healthcare. Several studies have found strong relations between autonomous motivation and both psychological adjustment and effective functioning, although these studies have been primarily in domains other than healthcare.

In a study of school children, Ryan and Connell[27] found that children's controlled reasons for doing their school work was correlated strongly with anxiety and maladaptive coping, whereas the more autonomous style was correlated strongly with enjoyment and proactive coping. Ryan, Rigby, and King[7] found that controlled regulation of religious behaviors was negatively correlated with a variety of mental health indicators whereas autonomous regulation was positively correlated with those indicators.

Other studies have related controlled versus autonomous regulation to various types of effective functioning. For example, Benware and Deci[28] and Grolnick and Ryan[29] found autonomous regulation to be associated with enhanced conceptual understanding. Grolnick, Ryan, and Deci[30] found positive correlations between autonomy and school achievement; Koestner, Ryan, Bernieri, and Holt[31] found positive relations between autonomy and creativity; Blais, Sabourin, Boucher, and Vallerand[32] found positive relations between autonomy and marital satisfaction; and Sheldon and Elliot[33] found strong relations between autonomous motivation and persistence at goal pursuits.

Taken together, all these studies provide confirmation of the hypothesis presented in Figure 1, namely, that autonomous regulation leads to maintained behavior change and to enhanced physical and mental health. Thus, because there are clear advantages to being autonomous, understanding the antecedents of people's autonomous regulation for health-relevant behaviors is of considerable import. As psychologists, we look to two general types of antecedent variables—individual differences in patients' personality, and differences in the interpersonal environment within which they are behaving.

As mentioned previously, we consider the providers' interpersonal style (viz., the degree to which it is autonomy-supportive) to be the central aspect of the interpersonal healthcare environment for patients. We put greater emphasis on the relation between this interpersonal-environment variable and patients' autonomous motivation for healthy behaving than on the relation between patients' personality and their autonomous motivation, because the interpersonal environment created by providers' style is more directly relevant to the issues of partnerships and to the promotion of healthy behaving by healthcare professionals. However, before we address those issues we briefly consider the effects of individual differences in patients' personalities.

Individual Differences: Causality Orientations

The concept of general causality orientations[34] refers to general individual differences in the way people relate to their environment with respect to initiating and regulating their behavior. The *autonomy* orientation describes the general tendency to be self-initiating and to orient toward contextual factors that promote choice; whereas the *control* orientation describes the general tendency to be controlled and to orient toward contextual factors that pressure or coerce people to think, feel, or behave in particular ways. Thus far we have been talking about autonomous versus controlled motivation to engage in a particular behavior, such as losing weight or adhering to a medication prescription. The concept of causality orientations addresses the autonomy-control distinction at a much broader level of general personality differences. Causality orientations are therefore viewed as cutting across domains and being relatively stable aspects of peoples' personality.

We hypothesize that people who are high on the general autonomy orientation (as assessed with the General Causality Orientations Scale, or GCOS) will tend to be more autonomous in regulating specific health-relevant behaviors. This hypothesis is shown graphically as the lower path in Figure 2. We now review research relevant to that hypothesis.

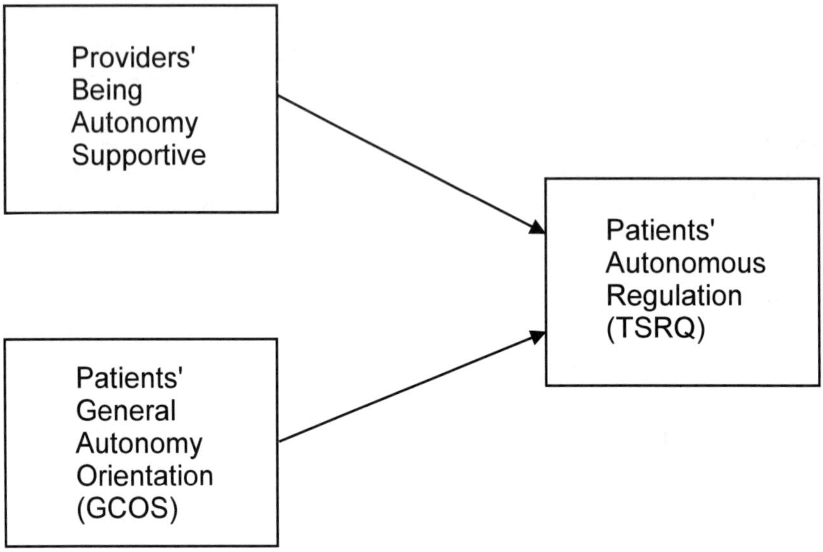

Figure 2. The hypothesized antecedents of autonomous motivation for healthy behavior.

Past research revealed that the autonomy orientation is positively related to self-esteem, self-actualization, and ego-development,[34] as well as to greater integration in personality.[35] The autonomy orientation has also predicted cardiac surgery patients' viewing their surgery as a challenge rather than a threat and having more positive post-operative attitudes.[36] The control orientation in contrast has been correlated with less positive characteristics such as the Type-A coronary-prone behavior pattern and public self-consciousness.[34]

These findings relating the general autonomy orientation (assessed with the GCOS) to positive outcomes are complementary to the earlier-reported findings that related autonomous reasons for specific behaviors (e.g., stopping smoking, doing school work, or staying in a marital relationship) to positive outcomes. Both indicate that there are advantages to greater autonomy. However, there is an important difference between the two sets of findings because they refer to different levels of generality in human functioning. The autonomy orientation is a general personality orientation that has positive relations to outcomes by contributing to people's being autonomously regulated with respect to particular behaviors. In other words, the effects of the general personality orientation are theorized to be mediated through the autonomous regulation for particular behaviors. People's general personality is, however, only one of the two important contributors to autonomous regulation; the other is their social environment. Thus, although we hypothesize that being high on the autonomy orientation will predict greater autonomous regulation and more positive outcomes, this relation will be attenuated in highly controlling environments. For example, a man who is high on the autonomy orientation may not be high on autonomous regulation for taking medication if his doctor is controlling and authoritarian.

Having a full understanding of the motivation for behavior change and positive health consequences therefore requires a consideration of both the general autonomy orientation (GCOS) and the social environment as they impact specific autonomous regulation (TSRQ). We begin with research testing the hypothesis that there is a relation between one's general autonomy orientation (GCOS) and behaving autonomously with respect to a particular health-relevant behavior (TSRQ).

In our longitudinal weight-loss study[8] patients' general causality orientations were assessed before they entered the program, and we found that patients who, in general, were more autonomy oriented tended to have more autonomous reasons for participation in the diet program, which, in turn, led to maintained weight loss. Thus, there is initial support for our hypothesis that personality plays a role in facilitating people's being more autonomous in the regulation of particular health-relevant behaviors. Still, from the point of view of healthcare practitioners, the more important question concerns whether it is possible for providers to influence the extent to which people are autonomous in such behaviors.

Supporting Autonomous Regulation and Health Outcomes

Much of our research is directly relevant to how healthcare professionals can promote autonomous regulation of patients' behaviors. Several studies have explored whether interpersonal conditions influence people's success at integrating behavioral regulations. Because physicians and other practitioners create the interpersonal conditions in which patients learn about their health status and about the behaviors that could improve it, we have explored whether the interpersonal style of the healthcare practitioner influences the extent to which patients will be more autonomous.

As noted earlier, we focus on whether a healthcare context tends to be autonomy-supportive versus controlling. *Autonomy-supportive* refers to listening fully, providing choice, and encouraging self-initiation, whereas *controlling* refers to pressuring one to think, feel, or behave in specific ways. Our hypothesis, which is shown as the upper path in Figure 2, is that providers' autonomy support will predict patients' autonomous regulation. We have typically assessed this dimension either by asking the patients their perception of the context—using an instrument called the Health Care Climate Questionnaire (HCCQ), or by making observations and using a rating system based on the HCCQ to characterize the healthcare climate along the autonomy-supporting to controlling dimension.

In the Williams, Grow et al.[8] weight-loss study, we investigated the patients' perceptions of the extent to which the providers were autonomy-supportive, using patients' reports on the HCCQ, and analyses revealed that these perceptions positively predicted the degree to which patients were autonomous in their participation in the program (assessed with the TSRQ). As mentioned already, this in turn resulted in better maintenance of behavior change.

The Williams, Rodin et al.[23] medication-adherence study, yielded complementary results. Patients who reported that their primary physician was more autonomy-supportive also reported more autonomous reasons for adhering to their prescriptions and they evidenced the best adherence. Similarly, in the diabetes study,[24] patients also completed the HCCQ and results confirmed that their perceptions of the providers' autonomy support predicted autonomous regulation, which in turn predicted HgbA1c.

Finally, in the smoking cessation study[26] involving physicians using the National Cancer Institute guidelines (NCI)[37] to counsel patients, the patients who perceived their physicians as more autonomy-supportive were more autonomous and evidenced better six month smoking cessation (biochemically verified) than patients who experienced more controlling physicians.

The patient-physician interactions were tape recorded in this smoking cessation study, and observers rated the physicians' autonomy support. These independent ratings of autonomy support were significantly related to patients' autonomous regulation assessed with the TSRQ, just as the patients' perceptions of physician autonomy support had been related to the patients' autonomous regulation. This is important because it means that the frequently-replicated relation between patients' perceived autonomy support of the physician and patients' autonomous regulation is not just a result of the autonomous patients perceiving their doctors to be more autonomy-supportive.

Still, it is possible that when patients are more autonomous, physicians will actually be more autonomy-supportive of them, thus indicating that the causal direction goes from patient to physician. Accordingly, Williams and Deci[26] had each doctor use the NCI guidelines in an autonomy-supportive way with some smokers and in a controlling way with others (randomly assigned). Results revealed that in the condition where doctors were supposed to be autonomy-supportive, observers rated them as significantly more autonomy-supportive than in the condition where doctors were supposed to be controlling. In turn, these ratings were significantly predictive of patients' autonomous regulation and their six-month cessation rates, thus confirming that physicians' behavior does have an effect on patients' motivation and behavior.

Although it is possible that the relation between physicians' interpersonal style and patient's motivation is bi-directional, these findings emphasize, consistent with the partnerships approach, that physicians should take the lead and be autonomy-supportive with all patients. That way, physicians will facilitate patients becoming more autonomous.

Together this set of studies provides support for the hypothesis shown in Figure 2—that providers' autonomy-support (assessed with the HCCQ) as well as patients' general autonomy orientation (assessed with the GCOS)—will predict patients' being autonomous for a particular health-relevant behavior (assessed with the TSRQ).

We turn now to studies that revealed direct relations between autonomy-supportive versus controlling interpersonal environments and both psychological adjustment and effective functioning. We assume these relations occur because the autonomy-supportive environments promote people's autonomous motivation, which in turn promotes positive outcomes. A study by Grow and Ryan,[38] conducted in a home for the aged, revealed that the residents' perceptions of the staff's being autonomy-supportive was positively predictive of the residents' vitality, perceived well-being, and life satisfaction. Ryan and Grolnick[39] found that perceived autonomy support in the classroom was associated with autonomous motivation and with several indicators of adjustment, including self-esteem, perceived competence, and less aggressive ideation. Additionally, Deci, Connell, and Ryan[6] found that when managers in a major corporation were perceived as more autonomy-supportive by their subordinates, the subordinates were more trusting and expressed higher job satisfaction.

Elements of autonomy support

A laboratory experiment by Deci, Eghrari, Patrick, and Leone[40] explored three specific aspects of interpersonal contexts that constitute autonomy support and thus facilitate integration: namely, a meaningful rationale so individuals will understand the personal importance of the activity; an acknowledgement of people's feelings and perspective, so they will feel understood; and an interpersonal style that emphasizes choice and minimizes control. Results of the experiment indicated that these three facilitating factors did lead to greater integration. Furthermore, results indicated that any internalization that occurred in the conditions low on these three factors (i.e., when the context was controlling) was only *introjected*. It thus seems that when healthcare practitioners create controlling contexts, they decrease the likelihood that patients will integrate regulation of the necessary life changes, and instead ensure that the regulation will either remain external or only be introjected.

Other experiments have found complementary results. For example, various studies have consistently indicated that when people such as healthcare providers, who are in a position to motivate others, use language replete with words such as "should," "must," and "have to," those others become less autonomous.[41,42] Additionally, experiments have confirmed that material rewards[43] and threats of punishment[44] also tend to be experienced as controlling and thus tend to undermine personal autonomy. These are familiar means of attempting to promote change, and it seems increasingly clear that these methods are less likely to produce maintained change than is autonomy support.

There is considerable convergence between the results of these laboratory experiments and the conclusions made by Bein, Miller, and Tonigan[45] who conducted a review of thirty-two studies of brief interventions for problem drinkers to determine the common elements of successful interventions. Among the most important elements they identified for promoting change in patients' drinking behaviors were giving nonevaluative feedback, providing choice, encouraging patients' taking responsibility, and being empathic. Bein et al. refer to providers who behave in these ways as being patient-centered, yet these are also central features of autonomy support. Thus, these elements describe the ways providers behave to establish healthcare partnerships.

Autonomy-supportive programs

Studies from other laboratories have explored the relative effectiveness of healthcare programs that tend to be more autonomy-supportive versus more controlling and have also yielded results that are complementary to ours. For example, Harackiewicz, Sansone, Blair, Epstein, and Manderlink[46] studied smokers who were randomly assigned to various self-help treatment conditions, including an intrinsic (i.e., autonomy-oriented) self-help condition in which the importance of one's own initiations and efforts was highlighted. Results showed that subjects in this autonomy-supporting self-help condition remained abstinent significantly longer than those in the two more controlling conditions or in the comparison condition. In another study,[47] participants were assigned to one of four groups: an intrinsic, personalized-feedback (i.e., autonomy-supportive) group; an extrinsic, financial-incentive (i.e., controlling) group; an intrinsic-plus-extrinsic group; and a comparison group. Results of the study indicated that the autonomy-supportive condition yielded the best maintained abstinence record (biochemically confirmed) at 3– and 12–month follow-ups.

Other relevant studies of behavior change have explored the effects of providers' style of counseling patients. For example, Ockene et al.[48] had physicians' vary their style of counseling patients for smoking cessation and found that using a more patient-centered (i.e., autonomy-supportive) style resulted in more positive attitudes toward the physicians and greater abstinence at a six-month follow-up than did a more directive or controlling style. Similarly, Miller, Benefield, and Tonigan[49] found that when providers were patient-centered (rather than controlling and confrontational) in counseling problem drinkers, their patients were less resistant and drank less alcohol at a one-year follow-up.

Finally, Langer and Rodin[50] found that supporting the autonomy of institutionalized elderly patients by providing them control over aspects of their lives and encouraging them to be more self-initiating and accepting of responsibility led to enhanced physical and psychological well-being and lowered mortality.

The Self-Determination Model

The general healthcare model that has emerged from all these studies confirming our hypotheses is shown in Figure 3. It suggests that (1) the degree of autonomy

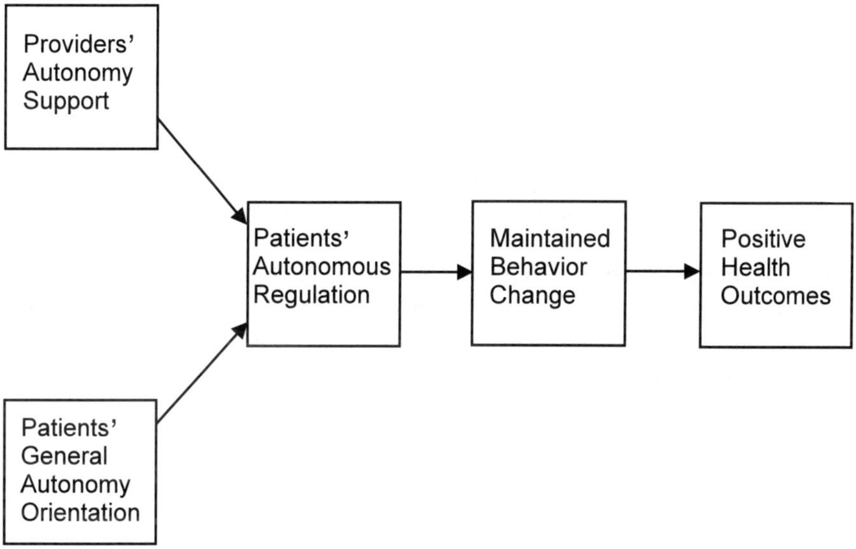

Figure 3. The self-determination model of health care.

supportiveness of healthcare providers, and (2) the strength of a patient's general orientation toward autonomy will predict the patient's being more autonomously motivated to participate in medical treatment. This autonomous participation will in turn enhance the patient's maintained behavior change and positive health status.

The Responsibility for Change

Supporting a patient's autonomy starts with taking the patient's perspective and thus understanding and accepting the patient's reasons for engaging in unhealthy behaviors or failing to engage in healthy ones. Patients have reasons for smoking and overeating, for example, and patients have the right to smoke or overeat if they decide to (as long as their behaviors do not harm others). We argue that it is not the providers' responsibility to change patients' behaviors. The patient's behavior is the patient's responsibility.

Providers do however have several responsibilities, namely: giving patients the information they need to make informed decisions; making recommendations based on medical knowledge; encouraging patients to be responsible in making health-relevant decisions; and supporting patients' attaining healthy outcomes by providing a nurturing relationship. We maintain that being autonomy-supportive is one of the important factors that accounts for providers' effectiveness in facilitating patients' responsibility. It is truly fascinating that when providers are genuinely supportive of patients for making their own choices, rather than pressuring them to make the "right" choice, the patients are more likely to behave in healthy ways.

Supporting patients' autonomy is a very active process. It does not mean leaving patients to take care of themselves. Instead, it means being "in there" with the patients; it means listening to and understanding their perspective, providing information when appropriate, accepting their needs and anxieties, and allowing them to make their own choices, without criticizing or judging them. Autonomy support involves caring and being involved with the patients in a very active way—not a controlling or pressuring way, but still an active way. Indeed, it is often harder work to be autonomy-supportive than to be controlling. Autonomy support begins with respect for the patients, it includes an understanding of psychological and social, as well as biological, contributors to illness, and it requires a set of listening skills and a communication style that may be difficult but are nonetheless essential for the patient's health and well-being.

Promoting Autonomy Support in Health Care Professionals

Because research results have repeatedly confirmed that providers' being autonomy- supportive has positive behavior-change and health consequences, we have also been exploring how to teach providers to be more autonomy-supportive. This work has been conducted in medical schools and has extended previous work we had done in other education settings. For example, we had reliably found that teachers' being autonomy-supportive was associated with greater internalization and autonomy in students.[5]

Williams and Deci[51] used interviewing courses for second-year medical students at two university medical centers as the setting to extend these findings to medical education. We hypothesized that autonomy support by instructors would lead to greater autonomy on the part of students; that student autonomy would increase the likelihood that they would internalize the biopsychosocial values that were extant in the course; and that internalization of biopsychosocial values would lead these doctors-in-training to be more autonomy-supportive in their subsequent interactions with patients.

At each university, several instructors taught the course, and students were asked to rate the degree to which their instructor was autonomy-supportive, using a Learning Climate Questionnaire (patterned after the Health Care Climate Questionnaire). The students' reasons for attending the course and their beliefs about psychological and social contributors to illness and health were measured at the beginning of the course and again at the end. Results of the study indicate that, relative to students who had perceived their instructors as controlling, those students who had perceived their instructors as more autonomy-supportive attended the course more regularly, became more autonomous in the reasons they gave for why they attended class, and, in turn, developed a stronger belief in the importance of psychological and social contributors to illness and health. Further, these differences were still in evidence at a two-year follow-up.

This study also confirmed that the students who had developed stronger psychosocial beliefs during their second-year interviewing course used a more autonomy-supportive (i.e., patient-centered) approach in their third year when they counseled simulated patients about changing cardiovascular risk behaviors. It seems, therefore, that medical school instructors' being autonomy-supportive in their style of teaching medical interviewing leads medical students to develop a stronger biopsychosocial orientation and to relate to patients in more autonomy-supportive ways. That, in turn, as we have argued throughout, leads patients to behave in more autonomous and healthy ways.

In a related study of medical students, Williams, Weiner, Markakis, Reeve, and Deci[52] found that students who perceived their internal medicine instructors as autonomy-supportive were more likely to select a residency in internal medicine than were students who perceived their internal medicine instructors as controlling. The autonomy-supportive instructors facilitated students developing greater interest in internal medicine and integrating the value of that field of medicine. A study by Williams, Saizow, Ross, and Williams[53] replicated these findings and also extended them to the field of surgery.

Conclusions

The closely related issues of adherence and maintained behavior change represent a major challenge to the healthcare field with respect to prevention, well-being, health outcomes, and cost. We have argued that understanding the motivational basis of adherence and maintained behavior change requires distinguishing between autonomous and controlled motivation, and we reviewed a variety of studies indicating that patients will be more likely to adhere and maintain healthy behavior if their motivation is more autonomous, resulting from their having integrated the regulation of relevant behaviors.

The idea of autonomous regulation in health requires a rethinking of healthcare delivery, and it even emphasizes the importance of talking about the issues differently. For example, the term *compliance* is very widely used in healthcare circles, with behavior change often being spoken of in terms of achieving compliance. This is problematic, however, because the term "compliance" itself implies *control* rather than *autonomy*. One complies with a control—with the demands of an authority. But complying with a demand is quite different from making a personal choice to behave in healthy ways. Accordingly, we speak about supporting patients who behave in healthy ways, rather than trying to make patients comply.

Considerable research now indicates that patients do tend to be more autonomous, and, in turn, do behave in healthier ways when providers are more autonomy-supportive in interacting with them. In terms of healthcare partnerships this is a very important point. As we have argued, being autonomy-supportive is a crucial element for building partnerships, so the results of our research suggest that partnerships in healthcare, which result from providers' being autonomy-supportive, will lead patients to be more self-determined and to display more positive and enduring behavior change.

We have emphasized that the term "autonomy support" does not mean that providers should abandon patients or leave them to their own devices. Quite the contrary. Patients need to feel related to and actively supported by their providers. They need to feel that they can rely on their providers for information and encouragement. It is clear that behavior change is difficult, and our studies have now consistently shown that patients will be more successful in their attempts to change when they feel understood and cared about by their providers. Stated differently, patients tend to accept greater responsibility for their own health when they perceive their healthcare providers to be encouraging their autonomy or volition.

Studies reviewed in this chapter have also shown that when medical school instructors are more autonomy-supportive in their teaching, doctors-in-training become more biopsychosocially oriented and are in turn more autonomy-supportive of their patients. Taken together, the parallel results from the healthcare and health education studies are very interesting in their suggestion that there are wide-ranging advantages to being autonomy-supportive. When medical educators and healthcare providers are autonomy-supportive, their students and patients are likely to become more autonomous, which, in turn, has a variety of positive consequences. In education, autonomous self-regulation has been associated with enhanced learning and better adjustment, and in medical care, it has been associated with maintained behavior change, more positive health status, and better mental health. By impacting the orientation of providers-in-training, it seems possible to have the kind of impact on health delivery that will result in patients' behaving more responsibly and evidencing more positive health status.

References

1. Engel GL. The need for a new medical model: a challenge for biomedicine. Science. 1977;*196*:129–136.

2. Deci EL, Ryan RM. *Intrinsic Motivation and Self-determination in Human Behavior*. New York: Plenum; 1985.

3. Deci EL, Ryan RM. A motivational approach to self: integration in personality. In Dienstbier R, ed. *Nebraska Symposium on Motivation: Vol. 38. Perspectives on Motivation*. Lincoln, NE: University of Nebraska Press; 1991:237–288.

4. Grolnick WS, Ryan RM. Parent styles associated with children's self-regulation and competence in school. J Educational Psychology. 1989;*81*:143–154.

5. Deci EL, Schwartz AJ, Sheinman L, Ryan RM. An instrument to assess adults' orien-

tations toward control versus autonomy with children: reflections on intrinsic motivation and perceived competence. J of Educational Psych. 1981;*73*:642– 650.

6. Deci EL, Connell J., Ryan RM. Self-determination in a work organization. J Applied Psychology. 1989;*74*:580–590.

7. Ryan RM, Rigby S, King K. Two types of religious internalization and their relations to religious orientations and mental health. J Personality and Social Psychology. 1993;*65*:586–596.

8. Williams GC, Grow VM, Freedman Z, Ryan RM, Deci EL. Motivational predictors of weight loss and weight-loss maintenance. J Personality and Social Psychology. 1996;*70*:115–126.

9. McGinnis MJ, Foege WH. Actual causes of death in the United States. JAMA. 1993;*270*(18):2207–2212.

10. Brownell KD, Kramer FM. Behavioral management of obesity. Med Clin North Am, 1989;*73*:185–201.

11. Kramer FM, Jeffery RW, Forster JL, Snell MK. Long-term follow-up of behavioral treatment for obesity: patterns of weight regain among men and women. International J Obesity, 1989;*13*(2):123–136.

12. Schwartz JL. *Review and Evaluation of Smoking Cessation Methods: The United States and Canada, 1978–1985*. Bethesda, MD: National Cancer Institute; 1987.

13. Phillips GM, Jones JA. Medical compliance. American Behavioral Scientist. 1991;*34*:756–767.

14. DiMatteo MR, DiNicola DD. *Achieving Patient Compliance: The Psychology of the Medical Practitioner's Role*. New York: Pergamon Press; 1982.

15. Sackett DL, Snow JC. The magnitude of compliance and noncompliance. In Haynes RB, Taylor DW, Sackett DL, eds., *Compliance in Health Care*. Baltimore, MD: Johns Hopkins University Press; 1979:11–22.

16. Janz NK, Becker MH. The health belief model: a decade later. Health Education Quarterly. 1984;*11*:1–47.

17. Rosenstock IM. The health belief model and preventive health behavior. Health Education Monographs. 1974;*1*(4):354–386.

18. Bandura A. *Social Learning Theory*. Englewood Cliffs, NJ: Prentice-Hall; 1977.

19. Wallston KA, Wallston BS. Development of the multidimensional health locus of control (MHLC) scales. Health Education Monographs. 1978;6:160–170.

20. Schafer R. *Aspects of Internalization*. New York: International Universities Press; 1968.

21. Ryan RM. Agency and organization: intrinsic motivation, autonomy and the self in psychological development. In Jacobs J, ed., *Nebraska Symposium on Motivation: Developmental Perspectives on Motivation*. Lincoln, NE: University of Nebraska Press; 1993;*40*:1–56.

22. Ryan RM, Plant RW, O'Malley S. Initial motivations for alcohol treatment: relations with patient characteristics, treatment involvement and dropout. Addictive Behaviors. 1995;*20*:279–297.

23. Williams GC, Rodin GC, Ryan RM, Grolnick WS, Deci EL. Compliance or autonomous regulation: new insights about adherence to medical regimens. Unpublished manuscript, University of Rochester; 1996.

24. Williams GC, Freedman Z, Deci EL. Promoting motivation for diabetics' self-regulation of HgbAlc. *Diabetes*. 1996;*S45*(2):13.

25. Diabetes Control and Complications Trial Research Group (DCCT). The effect of intensive treatment of diabetes on the development and progression of long-term complications in insulin-dependent diabetes mellitus. N Engl J Med. 1993;*329*:977–986.

26. Williams GC, Deci EL. The national cancer institute guidelines for smoking cessation: do they motivate quitting? J General Internal Medicine. 1996;S1(11):138.

27. Ryan RM, Connell JP. Perceived locus of causality and internalization: examining reasons for acting in two domains. J Personality and Social Psychology. 1989;*57*: 749–761.

28. Benware C., Deci EL. Quality of learning with an active versus passive motivational set. American Educational Research Journal. 1984;*21*:755–765.

29. Grolnick WS, Ryan RM. Autonomy in children's learning: an experimental and individual difference investigation. J Personality and Social Psychology. 1987;*52*:890–898.

30. Grolnick WS, Ryan RM, Deci EL. The inner resources for school achievement: motivational mediators of children's perceptions of their parents. J Educational Psychology. 1991;*83*:508–517.

31. Koestner R, Ryan RM, Bernieri F, Holt K. Setting limits on children's behavior: the differential effects of controlling versus informational styles on intrinsic motivation and creativity. J Pers. 1984;*52*:233–248.

32. Blais MR, Sabourin S, Boucher C, Vallerand RJ. Toward a motivational model of couple happiness. J Personality and Social Psychology. 1990;59:1021–1031.

33. Sheldon KM, Elliot AJ. Not all personal goals are personal: comparing autonomous and controlled reasons for goals as predictors of effort and attainment. Unpublished manuscript, University of Rochester; 1996

34. Deci EL, Ryan RM. The general causality orientations scale: self-determination in personality. J Research in *Personality*, 1985;*19*:109–134.

35. Koestner R, Bernieri F, Zuckerman M. Self-determination and consistency between attitudes, traits, and behaviors. Personality and Social Psychology Bulletin. 1992;*18*: 52–59.

36. King KB. *Coping with Cardiac Surgery*. Rochester, NY: University of Rochester; 1984. Thesis.

37. Glynn TJ, Manley MW, Pechacek TF. Physician-initiated smoking cessation program: the national cancer institute trials. In Engstrom PF et al., eds., *Advances in Cancer Control: Screening and Prevention research.* New York: Wiley-Liss; 1990;*339*.

38. Grow VM, Ryan RM. Autonomy and relatedness as predictors of health, vitality, and psychological well-being for elderly individuals in a nursing home facility. Unpublished manuscript, University of Rochester; 1996.

39. Ryan RM, Grolnick WS. Origins and pawns in the classroom: self-report and projective assessments of individual differences in children's perceptions. J Personality and *Social Psychology*. 1986;50:550–558.

40. Deci EL, Eghrari H, Patrick BC, Leone DR. Facilitating internalization: the self-determination theory perspective. J Pers. 1994;*62*:119–142.

41. Deci EL, Spiegel NH, Ryan RM, Koestner R, Kauffman M. The effects of performance standards on teaching styles: the behavior of controlling teachers. J Educational *Psychology*. 1982;*74*:852–859.

42. Ryan RM. Control and information in the intrapersonal sphere: an extension of cognitive evaluation theory. J of Pers and Social Psych. 1982;*43*:450–461.

43. Deci EL. Effects of externally mediated rewards on intrinsic motivation. J Personality and Social Psychology. 1971;*18*:105–115.

44. Deci EL, Cascio WF. Changes in intrinsic motivation as a function of negative feedback and threats. Presented at the meeting of the Eastern Psychological Association; April,1972; Boston, MA.

45. Bein T, Miller W, Tonigan J. Brief interventions for alcohol problems: a review. Addiction. 1993;88:315–336.

46. Harackiewicz JM, Sansone C, Blair LW, Epstein JA, Manderlink G. Attributional processes in behavior change and maintenance: smoking cessation and continued abstinence. J Consulting and Clinical Psychology. 1987;55:372–378.

47. Curry S, Wagner EH, Grothaus LC. Evaluation of intrinsic and extrinsic motivation interventions with a self-help cessation program. J Consulting and Clinical Psychology. 1991;59:318–324.

48. Ockene JK, Kristeller J, Goldberg R, et al. Increasing the efficacy of physician delivered smoking interventions: a randomized clinical trial. J General Internal Medicine. 1991;6:1–8.

49. Miller W, Benefield G, Tonigan J. Enhancing motivation for change in problem drinking: a controlled comparison of two therapist styles. J Consulting and Clinical Psychology. 1993;*61*:455–461.

50. Langer EJ. Rodin J. The effects of choice and personal responsibility for the aged: a field experiment in an institutional setting. J Personality and Social Psychology, 1976;*34*:191–198.

51. Williams GC, Deci EL. Internalization of biopsychosocial values by medical students: a test of self-determination theory. J Pers and Social Psych. 1996;*70*:767–79.

52 Williams GC, Weiner MW, Markakis KM, Reeve J, Deci EL. Medical students' motivation for internal medicine. J General Internal Medicine. 1994;9:327–333.

53. Williams GC, Saizow R, Ross L, Williams SA. Motivation for choosing internal medicine and surgery. J General Internal Medicine. 1995;*S10*(4):96.

VI Principles of Collaborative Family Healthcare[1]

Thomas L. Campbell, Susan H. McDaniel, and David B. Seaburn

We are in the midst of a corporate revolution in healthcare that is changing the lives of everyone involved. The current changes are spreading rapidly across the country and represent the greatest upheaval since the Flexner reforms of eighty years ago. Those reforms established medicine as a science, and relocated training and research from private clinics to medical schools and universities. The current corporate revolution is transforming health care from a cottage industry of small group practices into a big business dominated by highly competitive regional and national corporations.

This new health care system will be comprised of large, vertically integrated, regional health care systems that are fully capitated. Increasingly, providers are being paid for managing patients' overall health, not for providing specific services; they are being held accountable, as they have never been before, first for cost and later for quality.

As a result of this corporatization of health care, there will be:

1) a modest increase in primary care services provided by family practitioners, pediatricians, general internists, nurse practitioners, and physician assistants. Primary care providers are the centerpiece of this new health care system—which is the way it has always been in most other countries.

2) a dramatic reduction in hospital beds and services, with as many as one third of all hospitals closing or merging.

3) a sharp reduction in specialty services and numbers of specialists. Mental health will be no exception. For example, one analysis estimated that under managed care, there will be a need for one third the current number of psychiatrists.[1]

The managed care revolution is creating an imperative for radical redesign of delivery systems. Although these changes will put tremendous strains on health care professionals and institutions, managed care and capitation do create an environment in which there is adaptive value to the development of new partnership-based delivery systems that provide integrated, biopsychosocial, and family centered health care.

It would be difficult to argue that the delivery of mental health services is not in need of change. The current mental health care system, based in private offices, clinics, and mental health centers, has failed to serve the majority of individuals and families with mental health problems. Numerous studies have shown that over 70% of all patients with diagnosed mental health disorders are never seen by a mental health professional, and are treated only in primary care settings.[2] Primary care has been called the *de facto* mental health care system in this country.[3] Furthermore, up to one quarter of primary care patients have a mental health disorder, but fewer than half of these are ever detected and even fewer are adequately treated.[4]

Primary care patients with mental health disorders differ from patients seen in the mental health sector.[5] They are likely to be older, poorer, and non-white and often have multiple chronic physical and mental health disorders. They utilize much more medical care and have higher disability, illness, and mortality than other patients. For example, the Medical Outcomes Study demonstrated that depressed primary care patients have the same degree of functional disability and morbidity as patients with serious chronic medical disorders such as diabetes, heart disease, and cancer.[6] Mental health patients in primary care settings tend to be resistant to accepting a mental health diagnosis or treatment, and are more likely to drop out of any kind of treatment, either medication or counseling.

Attempts to apply standard psychiatric interventions in primary care settings have largely been unsuccessful. Most of psychology and psychiatry, including the Diagnostic and Statistical Manual, is based upon the more selected population of patients of the mental health system and is not easily transported to primary care.[7] Mental health problems in primary care usually present with physical symptoms (such as fatigue, chronic pain, and insomnia) rather than psychological symptoms. These disorders are often less severe than those seen in mental health settings—they would be classified as subthreshold in the DSM system—and have higher rates of spontaneous remission.

Collaborative Family Healthcare

Closer collaboration between primary care and mental health practitioners is essential to improve the quality of care for the large population of patients receiving their mental health care in primary care settings. An integrated, biopsychosocial approach might yield better patient and family outcomes, be cost-effective, and therefore be accepted and promoted by managed care. Collaborative family health care is one such model of integrated care that can be beneficial to primary care providers, mental health professionals, managed care systems, and, most importantly, to patients and their families.

There are two distinguishing characteristics of collaborative care. The first is close and proactive collaboration and communication between medical and mental health professionals *on location* at the primary care practice site; this is considered essential for quality care. Second, the family members–construed broadly to mean anyone connected to the patient by legal, blood, or emotional bonds—are actively encouraged to participate in the health care process as equal partners.

Neither type of collaboration comes easily. For health care providers to work in teams and collaborate successfully in primary care, there must be a shared understanding of what primary care is, how it differs from specialty care, and how mental health disorders in primary care differ from the problems seen in the mental health sector. For patients and family members, bridges must be built to span such differences as cultures, social status, terminology, and belief systems.

From the successes and failures of ten years' experience working with families in a primary care setting and teaching family-oriented care to family medicine residents, we have distilled eight principles of collaboration. Some apply across all types of collaboration; some pertain to interdisciplinary collaboration among professionals, and some pertain to working with families.

Heal the Mind-Body Split

For a collaborative primary care practice to function successfully, the participating clinicians must share a common medical model. Traditionally, medicine and the mental health disciplines have been rooted in the Cartesian view that mind and body are separate entities, and so those components of the health care system devoted to the mind and the body have been kept separate and distinct. The wholeness of the patient's experience is compartmentalized and partitioned; responsibility for diagnosis and treatment is parsed out to different parts of the health care system which work in relative isolation from one another. This is an unsatisfactory approach for the large number of primary care patients who present with an undifferentiated mix of biomedical and psychosocial problems that are woven together to form the tapestry of their experience.[8] The compartmentalized model is inadequate for guiding care in such common problems as somatization, pain, and adherence to treatment plans in chronic illness. (The inadequacy of the model may explain why many clinicians find these problems to be so vexing.)

The collaborative family health care model recognizes the fundamental indivisibility of the patient's experience. It integrates the biopsychosocial model with systems theory, emphasizing the interdependency and interplay among different levels, from the molecular through tissue and organs to the person, the family, the community, and the culture.[9] This approach assumes that *every psychological issue has some biological component and every biological event has psychosocial consequences*. From this point of view, understanding a patient is something like trying to understand a planet in the solar system: we need to know a lot about the individual characteristics of that planet, but we also need to understand the movements and gravitational influence of other planets to predict the behavior of that one planet. The challenge for the biopsychosocial practitioner or team is to evaluate simultaneously the biomedical and psychosocial aspects of the problem and to decide at which level or levels one should intervene.

The integrated systems perspective of collaborative family health care can also be seen in its emphasis on encouraging both agency and communion for the patient, the family, and the health care team.[8] Agency is a sense that one can make personal choices in dealing with illness, the health care system, and one's colleagues, family, and friends. Communion refers to strengthening emotional and spiritual bonds that can be frayed by illness, disability, and contact with the health care system, whether one is a patient or a provider. Taken together, agency and communion describe individual autonomy in a relational context. It's another kind of yin-yang, or balance, that is important to health outcomes.

Operationalizing a biopsychosocial systems approach that encourages agency and communion is difficult—in part because this is not the dominant paradigm, and in part because the methodology is still to be developed. A biopsychosocial approach is more than just adding psychosocial data to biomedical data. The biopsychosocial systems approach emphasizes the interdependency and interplay among different levels of the system—whether it is the cardiovascular system, the individual, family members, or the community.

We are in the process of defining what collaborative family health care means, and there may be disagreements among us on the fine points. For example, a competition exists between an individually-oriented integrated approach and a systemic, collaborative approach to biopsychosocial care. Pitting integration against collaboration is a misguided enterprise. It is systems theory that distinguishes one way of working from another, not the number of patients or the number of providers. Many physicians and therapists work with individual patients. Sometimes biopsychosocial care requires only one provider. Health care problems such as self-limited illness, prenatal care, child health care, older adult health care, and uncomplicated chronic or terminal illness typically are managed by one provider, collaborating with the patient and the family. The same is true for uncomplicated mental health problems, where only Level One or Two collaboration between professionals is necessary.[10] Of course, the difficult problems in primary care benefit from an approach that involves close collaboration with the patient, the family, and a multidisciplinary health care team. Whether working directly with an individual patient as an individual provider, or in a health care team with a patient and family, we believe a biopsychosocial systems approach is extremely useful and offers some correction to the current weaknesses in our health care system.

Respect Culture Differences

In workshops that we have presented on collaboration, we often ask physicians and therapists to list the adjectives that best describe their stereotypes of each others' professions. The results are very revealing. In describing the stereotypic physician, therapists respond with "cold, insensitive, rigid, controlling, egotistical, reductionistic, obsessive-compulsive, pressed for time, technician, counterdependant, and somatically fixated." On the other hand, physicians describe the stereotypic mental health professional as "too cerebral, impractical, touchy-feely, wishy-washy, impotent, neurotic, weird, flaky, not a *real* doc, and psychosocially fixated." These stereotypes seriously interfere with our ability to collaborate, and at times make us behave in ways that actually elicit the stereotypical behavior in the other professional.

To collaborate successfully, both professions need to understand that the stereotypes originate from differences in professional cultures rather than obnoxious personal characteristics. As in all cross cultural endeavors, behaviors that seem unusual become more understandable and functional when one grasps the other's culture.

We have identified several characteristics that distinguish medical and mental health cultures:

1. *Physicians and therapists have different value systems.* The mandate for physicians is to cure physical ailments (sometimes at almost any cost). Profound socialization processes throughout medical training inculcate a focus on objective biomedical outcomes—diseases must be cured, laboratory values must improve. Our medical-legal system reinforces this approach. We know of no physician who has been sued for not diagnosing or treating depression, but missing cancer is one of the most common reasons for malpractice suits. In contrast, the culture of mental health professionals gives more emphasis to the patient's (or client's) subjective experience and emphasizes such outcomes as mood and capacity to resume one's usual family, occupational and other social roles.

2. *Physicians and therapists have different practice styles.* Doctors often use the approach, "When in doubt, do something." They are taught to be like mechanics, plumbers, or carpenters whose job it is to fix the body when it is not working. As a result, they are quick to give advice. In contrast, mental health professionals are more inclined to observe, facilitate, and follow the patient's lead; they are less focused on achieving immediate results.

3. *Physicians and therapists have very different orientation to time.* Primary care physicians have 1500–2000 active patients. They see patients every 12–15 minutes who each have an average of 2–3 complaints each. Mental health professionals typically schedule one patient per hour, and carry a correspondingly lower caseload. There tends to be one consistent problem or set of problems that is the main focus of multiple visits. Time for reflection on each case is often built into their schedules.

4. *Physicians and therapists have their own special languages.* For successful communication to take place, each must become familiar with the other's language.

5. *Physicians and therapists differ in their approaches to confidentiality.* This difference can sometimes lead to conflict. Because physicians generally work as part of a medical team, information about somatic disease processes (which may be just as sensitive as psychosocial information) is often shared with other members of the team (specialists, nurses, allied health professionals, and others) without explicit consent being obtained. Mental health professional usually work alone, and disclose clinical information less routinely.

6. *Responsibility*. Physicians tend to feel very responsible for their patients' health outcomes. This may work adequately in the context of acute medical illness, but for chronic medical and psychosocial problems (which vastly outnumber acute illnesses), this approach creates a situation of physician overresponsibility and patient underresponsibility. Therapists are less apt to feel personally responsible for patients' clinical outcomes, recognizing the fundamental importance of the patient's goals and motivation.

Awareness of these differences in professional cultures can help would-be collaborators identify sources of disagreement. In discussing these differences, they can offer each other a wider perspective and greater range of approaches than is available within their own disciplines alone. If handled with respectful dialog, interprofessional diversity becomes a resource rather than a problem.

There's no Substitute for a Good Relationship

The single most important ingredient for effective collaboration is relationship. The quality of the collaboration between a mental health professional and a physician, physician assistant, nurse, or nurse practitioner depends upon the quality of the relationship between them. Sharing the same work space, the same paradigm, and the same patients means very little if the professionals cannot work together. It is remarkable how often the behaviors most of us learned as "good manners" are breeched. Offering simple politeness and consideration; demonstrating respect for each other's expertise, skill, and perspective; and recognizing one's own limitations are all behaviors that invite relationship.

When medical and mental health professionals have not previously worked together, starting a new collaborative relationship can be anxiety-provoking; aspects of each one's culture and core beliefs in some way threatens the other, raising concerns about whether it is possible to blend these two cultures at all. They may worry (or hear from peers) that they do not really understand what they are getting into. Or they may be afraid that they will not live up to the expectations of their new collaborative colleagues. These anxieties are similar to those of starting any new relationship. Often the first step is the hardest, especially for the mental health professionals, who are venturing out onto the home turf of the medical professionals. So we have a few suggestions to keep in mind when getting started.

1. *Have lunch with your health care colleague*. We used to advise mental health providers to "take a physician to lunch," but considering the time demands of a medical practice, it may be more reasonable to take a lunch to a physician or nurse. Physicians, nurses, nurse practitioners, and other health professionals can do the same—invite a mental health professional to lunch in your setting.

2. *Identify potential collaborators in your community*. Mental health professionals can contact the directors of residency programs in pediatrics, internal medicine, and

family medicine in their areas to discuss opportunities for working with patients, families, and residents. Primary care physicians who have graduated from residency programs in the last 10–15 years are more likely to be trained in the psychosocial aspects of clinical practice and to be more interested in collaboration. Therapists can also talk with their own primary care providers about physicians and other health care professionals in the community who may be interested in collaborative care.

In a similar fashion, primary care providers can contact departments of psychiatry, family therapy training programs, and other training institutes. An increasing number of family therapists are being trained in collaborative models. The Collaborative Family Health Care Coalition sponsors local chapters to further the implementation of collaborative care.

3. *Get permission from patients to contact their medical or mental health providers.* Mental health professionals should routinely get the name of their patients' primary care provider as part of the health history, even if the primary care provider is not the referring person. They can then contact the primary care provider to introduce themselves and to discuss their availability for referral and collaboration. Primary care professionals can do the same thing. When they learn that a patient is seeing a therapist, they can talk with that patient about the value of having confidential communication with the patient's therapist as part of the overall health care plan.

4. *Spend a little time seeing patients together.* The primary care practitioner and mental health professional might arrange to spend a few hours seeing patients together in the primary care practice to learn each other's perspectives and styles. In this way, mental health professionals can quickly learn the medical culture and understand what medical practice is like. It also conveys to the medical practitioner an attitude of interest, respect, and openness to learning about her or his world. The primary care professionals can learn about the helpful perspective which the mental health professional can bring to the care of many otherwise perplexing and vexing patients.

Respect Each Other's Paradigm

In the 1950s the "ugly American" emerged as a description of the American tourist who entered other countries with a suitcase full of cultural stereotypes and behavior that was disrespectful. To add insult to injury, they often approached their host country as a missionary might approach so-called heathen, introducing them to Coca-Cola, Mickey Mouse, and other advantages of our more "advanced" culture.

Systems oriented mental health professionals can make similar mistakes when entering the culture of medicine and health care. Seeking to create a common mission, they may project a missionary zeal that alienates their medical colleagues, insisting that a paradigmatic exorcism and rebirth is necessary before collaboration can begin.

Although it is true that a systems paradigm provides an effective framework for collaboration, a strategy of forced conversion seldom works. Most medical professionals are not particularly interested in discussing new theoretical paradigms; they tend to have a more concrete and practical focus. They are far more likely to adopt a systems approach through observation of its implementation in the mental health professional's clinical reasoning and behavior, and the resulting outcomes. This is comparable to the way a therapist would start working with a family. Rather than suggesting that the way they look at the world is wrong and telling them how they need to think, a skilled therapist would try to enter their world much as an ethnographer might enter a new culture—with curiosity, respect, inclusiveness, and a desire first and foremost to understand the territory.

The missionary zeal must also be understood for what it is: a response to the uncertainty and insecurity that a mental health professional often feels entering the foreign land of medicine and primary care. It can be intimidating, to say the least. A normal reaction is to somehow demonstrate that, "I deserve to be here." Parading one's paradigm is often an attempt to legitimize oneself, to show "I know something," or even "I know something you don't know." Unfortunately, this seldom works.

There are some simple things mental health professionals can keep in mind in order to avoid turning off their health care colleagues.

1. *Monitor anxiety*. Watch for telltale signs of missionary zeal; it is most often a sign of feeling intimidated or stereotyped.

2. *Take initiative*. Don't wait for medical professionals to approach and then wonder why they don't; approach them not to advocate a point of view but to learn from and about them, their work, their needs, and their goals.

3. *Ask questions before giving answers*. This would apply to almost everything, but especially to questions of how the system is organized and structured, who people are, what they do, what the history of the setting is, and what the mission is.

4. *Be available*. Think of yourself as a resource to the context, try to maintain a flexible schedule, and notify others of the easiest way to contact you.

These reminders will make it easier to travel and be accepted in the medical culture.

Promote a "Community of Caring"

Differing goals and perspectives often get in the way of promoting a sense of community among health care professionals. In the days of the Clinton Health Care Plan, Hillary Clinton called on the nation to keep a human face on medicine and to participate in what she called a "community of caring." Collaborative family health care offers that opportunity. It can promote communion for patients, families, and professionals through the facilitation of communication and understanding, and by decreasing blame.

Medical and mental health professionals can all get so busy or so burnt-out that they lose a sense of agency or communion. A team of colleagues can function like a family does in private life, to share frustrations and victories both formally and informally, and to provide feedback and support when job responsibilities seem overwhelming.

> My husband is the pediatrician for the Hemophilia Center in Rochester where staff were devastated by a series of AIDS deaths. As you may know, most hemophilia patients must receive blood concentrate to survive, so they were infected by HIV in the 1970s and early 1980s before the blood supply was appropriately screened. This means the very treatment prescribed and administered by these dedicated health care workers resulted in patients' deaths. As they died one after another, the staff became more and more upset. The Executive Director approached my husband to ask for help, so we worked to establish a support group. Since my husband was part of the group, I couldn't lead it and asked [a colleague] to do it. Everyone attended, from the Executive Director to the physicians, nurses, and secretaries. The benefits were tremendous. At a recent retirement dinner for someone at the Hemophilia Center, several people made jokes about their "group therapy," but it was clear that it had made an enormous difference for them in making it through a very difficult time.

Therapists can also benefit from the support and consultation of physician and nursing colleagues, whether through formal activities like Balint groups with health and mental health care providers or by more informal means such as hallway conversations.[8] They would be unwise to move into the new health care arena without carefully attending to their own needs; they need an opportunity to deal with their feelings so their patients don't have to.

In building a community of caring, the family is an apt metaphor for the primary health care team. If the family is held together by marriage, ideally, it is not a marriage of necessity, but rather one of mutual desire. It is a family characterized by respect and shared values, which enriches all its members. It is a family in which members are secure enough on their own to enjoy the interdependency of close collaboration. The physician or nurse in this relationship may function as the instrumental, action-oriented partner who ideally is also sensitive to his or her own needs and emotional cues. The therapist may be expressive and concerned with process while hopefully also able to get the job done practically and efficiently. While each could be independent and effective on her or his own, the team is greater than the sum of its parts.

Adopt a Population Based Perspective

Medical and mental health providers alike need to redefine their goals in this new age of accountable health care systems. Until now, the goal has been "Providing medical or psychotherapy services to motivated individuals and families who come to our offices." A new more inclusive goal might be "To work collaboratively to insure better physical and mental health outcomes for a defined population of patients in a cost-effective manner."

Mental health care for primary care patients must be entirely reinvented. Mental health professionals will need to be more flexible and creative to meet the needs of these patients. This means seeing different types of patients, including difficult medical patients with whom primary care professionals have the most difficulty: for example, patients with somatization syndromes, chronic pain, chronic and terminal illnesses, and patterns of high risk behaviors or minimal cooperation with treatment plans.

It also means showing greater flexibility in the way that patients and their families are treated: seeing patients in different locations rather than waiting for them to come to one's office; seeing patients and families jointly with the primary care physician, perhaps right there in the primary care exam room; making use of shorter, more frequent contacts as part of primary care office visits; meeting patients in the hospital, in the nursing home or at home. A wide range of new interventions will be needed, from multi-family psychoeducational groups for chronically ill patients and their families to cognitive-behavioral approaches for chronic pain. Primary care and mental health professionals need to work together to develop innovative, cost-effective approaches to treat a wide range of health problems.

> Over the past five years, the Mental Health Service at our Family Medicine Center has developed several new approaches to helping our patients. We are making increasing use of patient and family groups that meet regularly with a therapist to deal with specific issues, such as chronic illness, chronic pain, low self-esteem, and depression. Joint visits with the primary care provider and a therapist with the patient and family are common, often in the context of regular medical visits. Home visits are made when indicated by the physical and mental health care team.
>
> More recently, we have developed and are testing an innovative intervention for "distressed high utilizers" at our center. These patients are often somatizers who have unmet mental health care needs. A "health care counselor" meets with the primary care provider, patient and family to review the patient's problems and together come up with a comprehensive treatment plan that addresses all the patient's needs. The patient, famil,y and health care team meet periodically to track the progress of the treatment plan. We believe that this collaborative approach will improve health outcomes and be cost effective.

Make Families Part of the Healthcare Team

Collaborative care is not just about the relationship between the mental health and medical health professionals. Doherty and Baird described the "illusion of the dyad" in health care and offered in its place a "triangular model of medicine."[11] The illusion of the dyad refers to an exclusive preoccupation with the clinician-patient relationship as the context of care. This model neglects the tremendous impact that families have on health and illness, and how the patient relates to health care providers. Including the family as part of the treatment team creates the treatment triangle.

Family members must be respected as partners in care rather than treated as objects of care. Beliefs and practices about health and illness are woven into the fabric of family life and history in ways that influence health behaviors for a lifetime. The family is the first circle of health specialists to respond when a family member is sick or disabled, and they constitute the context of care for a majority of chronically ill patients. Patients and families are the experts on their own experience and they are the sources of many of their own solutions. Effective collaboration often rises or falls on the capacity of professionals not only to engage the family but to identify, awaken, elicit, and utilize the patient's and family's own natural resources—their own strengths, competencies, and accomplishments. This can be a challenge to the traditional one-up position of helpers and the one-down position of helpees. Therapists, physicians, nurses, nurse practitioners, and other health professionals who learn to engage families as partners will be far more successful in providing care and facilitating change. The following case example illustrates the importance of collaborating with the patient and family.

> A few years ago I consulted on a hospitalized patient who had been on a ventilator for months. The physicians were convinced she'd never get any better, while the family kept pushing for her cure. This was a vicious and expensive cycle.
>
> At one session with the family, the patient, the primary care doctor, and the in-hospital health care team, it became evident that the family was attempting to hold its grief at bay because this woman's adult children had lost their father less than a year earlier. They didn't want their mother to die on them too. The patient was caught between choosing to be off the ventilator, even if that meant death, and wanting the timing to be right for her children. Once the health care team could understand the family's perspective, they quit pressing to take her off the ventilator. And once the staff was more supportive, the family began communicating about their mother's impending death instead of complaining about the insensitive medical care.
>
> It was amazing to watch, as the parties came to understand each other, how the tone in that hospital room shifted from that of a war zone to a community of caring. It certainly is not always this easy, but in this case, within a week the patient and her family chose to discontinue the ventilator, and the patient died peacefully soon thereafter.[8]

This case illustrates what many other cultures have long understood, that patients and families are not only customers for our services who must be respected and understood, but they are also "colleagues" in their own health care.[12] They are our partners. And as such, they are a critical ingredient in any primary care model of collaboration.

Develop a Common Mission

For collaborative care to work successfully as a partnership between patients, families, primary care practitioners, and mental health professionals, a crucial task is to develop a common mission—a goal that everyone in the partnership can agree to. Many common problems in health care—from "non-compliance" (that means patients don't do what clinicians tell them to do), to conflict between the family and the physician, to problems that result in malpractice suits—arise from the failure to establish a common mission. The physician may have one mission, the patient another, and the family therapist yet another. In this situation, everyone in the system expends considerable energy working at cross-purposes. This gets to be both frustrating and expensive.

How does one facilitate the development of a common mission? A good biopsychosocial provider must be both translator and diplomat. All the parties involved must feel enough agency to be able to define and communicate their point of view about the illness, and feel sufficiently confident that their points of view will be acknowledged and respected. Thus, the simple act of listening is a major part of encouraging agency. Beckman & Frankel published a study demonstrating that as patients began to describe their concerns at the beginning of primary care office visits, their physicians waited an average of eighteen seconds before interrupting them; moreover, the patients' full agendas for the visit were never elicited.[13] The biopsychosocial provider must be patient- and family-centered, rather than physician-centered.

An important aspect of listening while negotiating a common mission involves understanding the patient's and the family's health beliefs. John Rolland[14] and Lorraine Wright[15] have both been articulate about the importance of understanding the meaning of the symptoms to the patient and the family. It is when these health beliefs conflict that problems occur.

> Several years ago a Family Medicine patient was admitted three days after suffering a heart attack. "Kathy" was a 58-year-old woman who had refused medical intervention until after her pain subsided. Her ten adult children were enraged with her. Once in the hospital, Kathy refused most treatment and basically drove the health care team nuts. Her previous doctor had moved, so one of our residents picked her up on discharge from the hospital. As a way to get to know her and her family and to get some help with the treatment plan, the resident invited them to join him at a session of his residency's seminar on family interviewing and assessment.
>
> In the first five minutes of the session, which were near bedlam, it became clear that a war of health beliefs was going on. The children were terrified and wanted their mother to accept medical intervention because their father had died a year before and they couldn't stand the thought of losing their mother as well. Out of the ten children, three of them had had coronary heart disease themselves and one had already had a heart attack, so their identification with their mother's health problems further fueled their anxiety. The physician and the health care team quickly sided with the children because the professionals' health beliefs closely matched those of the children: "go to the doctor immediately and accept intervention."
>
> The patient, on the other hand, was resistant and depressed. Her depression had not been recognized previously because of her oppositional behavior. Her husband had been chronically ill for forty years. With her husband dead and her ten children now grown, her job as caregiver was over. Moreover, her view of the future was bleak: all the men in her family had died of cardiac disease very early, in their 30s and 40s, while the women tended to live into their 80s and 90s debilitated for years by severe arthritis. Kathy had

decided that she did not want to live in a wheelchair, dependent on everyone else for basic care, for years and years.

At the onset of her symptoms she did not believe she was having a heart attack; her symptoms did not resemble those of her husband, her brothers, or her sons. Then, when it did begin to occur to her that she might be having a heart attack, she decided she much preferred cardiac illness and a shorter life to arthritis and a long life of disability. Not someone to accept the patient role with any ease, she stubbornly stayed home for three days.

My [the consulting family therapist's] role in this session was to create space for each party to tell his or her story, and to help them negotiate a common mission. This involved supporting the patient's agency—her right to make choices about her life and her treatment—while at the same time challenging some of her beliefs. After a few attempts by the resident to convince Kathy that she really should call if she had chest pain, he took the cue from me to back off. (She was tough: He had told her to call him if she needed to take three nitroglycerin tablets, so she would only take two, so as not to have to call him!)

During the forty-minute session, we eventually negotiated a common mission that included recognizing the value Kathy placed on her own independence, as well as everyone's desire for her to be as healthy as possible. We began by empathizing with everyone's tremendous sense of grief at the loss of their husband and father, and then moved toward negotiating a plan that included more home care and a greater sense of control than is typical for a patient with a recent MI. We also worked out a more functional system of communication among the children so they could inform each other about their mother's condition and her decisions, and commiserate if need be, without drawing her into an interaction that would inevitably lead to a power struggle. In this case, the joint session, with the therapist and the primary care physician in the room together, was tremendously helpful in developing a common mission and reducing the conflict.

In seeking to negotiate a common mission, practitioners must be mindful of the differential in the pace at which patients, family members, and even health care providers accept a difficult diagnosis. Family members may occupy disparate points on the continuum between denial and acceptance. This diversity may be viewed as an asset—a division of responsibility. In many families, it is one member's role to deny the illness and advocate for life to go on, and another's to help the family grapple with the hard realities. Adaptation then results from a dialog between these points of view. The process gets off track when the family members become polarized, move into major conflict, and stop the dialog. The professional's role is to avoid becoming embroiled as yet another party in the conflict and to assure that each viewpoint can be heard and respected, so the dialog can move forward once again. Respecting the others' defenses, the professional can also provide the information needed for informed decision making.

In the case above, Kathy denied much of her illness. The more she did that, the harder her children and her health care team tried to convince her that she was sick. At one point in the session, one of her children said, "Mom, there's such a thing as a 'silent heart attack.'" Kathy replied, "I know, they gave me six different papers on it in the hospital." Her problem was not one of ignorance, but rather her degree of acceptance and a meaning of the illness that differed from the others'.

For members of the health care team, negotiating a common mission is easiest when professionals clearly define and negotiate their roles and those of their patients. Power and influence must be thoughtfully balanced according to roles and expertise. Ideally, each member of the team would feel a sense of agency, a sense of respect for his or her skills, and confidence that communication across the team will be functional and clear.

Conclusions

Collaborative care has much to offer. The partnerships it engenders bridge the mind-body split and challenge old models of treatment. Collaborative care stimulates new ways of thinking and innovations in the care of patients and populations. It forges a common sense of purpose that both mobilizes and aligns the strengths and wisdom of patients, families, and an interdisciplinary team of health care professionals. Building relationships is the key to removing old barriers and working together to create new structures for primary health care. We believe that the seeds for a very bright future for health care are not out there tomorrow, but are right here, today.

References

1. Weiner JP. Forecasting the effects of health reform on US physician workforce requirements. JAMA 1994;272:222–30.

2. Narrow W, Regier D, Rae D, et al. Use of services by persons with mental and addictive disorders: Findings from the National Institute of Mental Health Epidemiologic Catchment Area Program. Arch of Gen Psych 1993;50:95–107.

3. Regier D, Goldberg I, Taube C. The de facto US mental health services system. Arch Gen Psych 1978;35:685–93.

4. Barrett J, Barrett J, Oxman T, et al. The prevalence of psychiatric disorders in primary care practice. Arch Gen Psych 1988;45:12:1100–6.

5. Miranda J, Hohmann AA, Atkinson CC, Larson DB (Eds.) *Mental Disorders in Primary Care*. San Francisco: Jossey-Bass Publishers, 1994.

6. Wells K, Stewart A, Hays R, et al. The functioning and well-being of depressed patients: Results from the Medical Outcomes Study. JAMA 1989;262:914–19.

7. American Psychiatric Association, Diagnostic and statistical manual for mental disorders: Fourth edition, Washington, D.C.: American Psychiatric Association Press, 1994.

8. McDaniel SH, Hepworth J, Doherty WJ. *Medical Family Therapy: A Biopsychosocial Approach to Families with Health Problems*. New York: Basic Books, 1992.

9. Engel G. The need for a new medical model: A challenge for biomedicine. Science 1977; 196:129–36.

10. Doherty WJ, Baird M, Becker L. Family medicine and the biopsychosocial model: The road toward integration. Marriage & Family Review 1987;10:51–70.

11. Doherty WJ, and Baird MA. *Family Therapy and Family Medicine*. New York: Guilford Press, 1983.

12. Kleinman A. *The Illness Narratives: Suffering, Healing, and the Human Condition*. New York: Basic Books, 1988.

13. Beckman H, Frankel R. The influence of provider behavior on the collection of data. Ann Intern Med 1984;101:692–6.

14. Rolland J. *Families, Illness, & Disability*. New York: Basic Books, 1995.

15. Wright L, Watson W, & Bell J. *Beliefs: The Heart of Healing in Families & Illness*. New York: Basic Books, 1996.

VII Spiritual Care: The Partnership Covenant

Mary-Therese B. Dombeck and James S. Evinger

The spectacular advances in medicine and health care in the twentieth century have been accompanied by bewildering problems for patients and clinicians. The tendency in the modern world view to separate mind from body, and spirit from matter, has been accompanied by a concurrent tendency to separate different aspects of healing from each other. The result is the creation of specialists who engage in healing the body, the mind or the spirit. It has not always been so. Historically, the vocations of nursing and medicine were not separate from religious vocations.[1] Even after these disciplines had evolved into distinct professions, each attended to the needs of a person by treating the spiritual, physical, mental, and social as interrelated. In fact, in all ages, including the present one, there have been competent and caring clinicians who have provided holistic care.

Those who would offer holistic care in the contemporary milieu, however, are immediately confronted with a conceptual and practical problem: the modern models have not generated specific guidelines for spiritual care. The dominant biomedical model, for example, omits any diagnostic category for the assessment and treatment of spiritual distress. And while selected nursing theories and diagnostic categories include the spiritual aspects of persons,[2,3,4] in practice these needs are typically either consigned to clergy, chaplains, and other spiritual "experts," or subsumed implicitly and downgraded as psychosocial aspects of persons. Thus, the separation of body, mind, and spirit continues along the lines of professional disciplines. Competent and caring physicians, nurses, and therapists who are committed to treating the whole person find few or nebulous means to address the spiritual aspects of care, while the spiritual aspects of persons are often either ignored or relegated to clergy whose contribution is seldom considered integral to health care anyway. This arbitrary and artificial division of labor diminishes the inherent potential for the patient and clinician to be fully present to each other in the partnership between patient and clinician.

It is our purpose to explore spiritual care as part of this partnership. We propose that spiritual care is not treatment, *given* to another, but rather it is a dynamic process that one person *shares* with another. The patient and the practitioner are not only in reciprocal roles with each other, but they also collaborate in a covenant that involves the full "personhood" of each one. Furthermore, we propose that spiritual partnership is a useful model for the holistic health care of persons. We present partnership in four ways. First, the distinctive features of spiritual partnership are described; second, the dimensions of spiritual partnership are traced in a case from a literary source; third, the dynamics of spiritual partnership are identified in a case of a client in a psychotherapeutic relationship with a nurse therapist; and, finally, specific practical guidelines for offering spiritual care are outlined.

Features of Spiritual Partnership

The word "spirit" derives from the word "breath" and "air/wind" (spiritus in Latin). Moreover, because of its historical connection to the word "anima" (Latin) and "nephesh" (Hebrew), it is often related etymologically to "soul" or "self." It literally means whatever gives life to (or animates) a person. Figuratively, it connotes whatever is of central and ultimate importance in all aspects of one's life.[5] Spirituality is not merely one discrete aspect of persons. Rather, it is at the core of one's physical nature, psychological awareness and functioning, social relations, and connectedness with one's world. The spiritual aspects of persons are inclusive of all aspects of human life. For example, the fact of one's mortality and vulnerability is uniquely expressed by particular individuals in particular societies; at the same time it draws on what is universally true for all humans. It is not difficult to see how, in the context of a health care encounter in which the threat of a major change or loss in life (or meaning) is at issue, the spiritual aspects of life are intrinsically involved. Health care clinicians often find themselves faced with spiritual issues, whether or not they are willing or able to deal with them.

The word "partner" was introduced into the English language through the word "parcener" (Anglo-Norman), which remains in existence as a legal term meaning "joint heir."[5] The word "partner" designates persons or parties that work, play, dance, or live together. The participation of persons in partnerships usually assumes an understanding that designated rights, responsibilities and profits are shared for the duration of the partnership. A spiritual partnership, therefore, like a business partnership, involves an agreement of close cooperation between parties having specified rights and responsibilities toward each other. What distinguishes the spiritual partnership is its convenantal nature that is rooted in the personhood, and hence spirituality, of both the clinician and the patient.

The word "covenant" has its roots in "convenire" (Latin), and literally means "coming together."[5] The word "contract" (from the Latin) "contranere" literally means to "draw together". The word covenant has always retained the element of two persons coming together *in person*, in contrast to being drawn together by something. In biblical times the word referred to God's covenant with persons. Although God and humans were each considered to have "personhood", they were not equal to each other. God gave all things freely and humans responded by giving gratitude and fidelity. The faithful were admonished to bind each other covenantally in generosity, love, and fidelity. May[6] examined the physician's relationship with patients as more like a covenant than a contract. He considered the Hellenic Hippocratic Oath to have covenantal aspects.

Spiritual partnerships involve aspects of persons that are crucially important and unique to them. In addition, they are universally focused on humans' ultimate concerns. Therefore, involvement in the spiritual aspects of another's life cannot be abstract, detached, or impersonal. The scope and depth of a patient's spiritual needs, when heard by the clinician, will draw out the clinician as a person to share in what is universally human, and, at the same time, transcends each individual human life.

Dimensions of Spiritual Partnership: A Case from Literature

The spiritual partnership is not readily entered into by either the clinician or the patient. In their practices, clinicians do not always recognize the spiritual questions of patients. If heard, spiritual concerns may be redefined into other and less sympathetic diagnostic categories. Patients, as well, may not be prepared to identify their concerns as spiritual. And if they are ready, they may be reluctant to express them as such. These difficulties constantly threaten to hinder or even prevent creating a spiritual partnership between patient and clinician. When the obstacles are overcome, the healing benefits are tangible in substance and transformative in quality. Consider, for example, the medical case of the 45-year-old Russian municipal judge in the novella *The Death of Ivan Illyich* by Leo Tolstoy.[7]

To trace the sudden course of Ivan Illyich's disease requires discerning the ways his disease is so thoroughly interwoven with him as a person, who, as a spiritual being, assigns meaning and value to his experience of being ill. Married with two children, he pursues the undistinguished career of a civil servant. His outward respectability and crafted propriety form the frail foundation of his superficial contentment. He believes himself to be in good health, despite an occasional discomfort in his left side. Then, quietly, inexorably, and quickly, the irritant emerges as a constant pressure that leaves him ill-tempered and easily provoked. Soon after, dull, gnawing pain becomes his consuming companion. So begins Ivan Illyich's ineluctable descent toward death.

Yielding to his wife's insistence, he consults a celebrated physician. The examination is abstract and detached, conducted with an affected air of importance. (Illyich recognizes in the physician's manner the same indifferent and distanced way by which he administers justice in his court.) The unidentified physician refuses to assign a conclusive diagnostic name to the symptoms. More importantly to Ivan Illyich, the question he most wants answered, "Is my condition serious?" is ignored as idle and inappropriate, not worthy of consideration. The impersonal examination of the patient has depersonalized the man. Dismissal of his identity is paralleled by, and culminates in, denial of his anxiety about his prognosis, a fate he intuitively knows and dreads.[8]

In sparse, declarative language, a linear narration moves the story on. Seeking relief for his deteriorating condition, Ivan Illyich consults other doctors and a homeopath, all to neither medical avail nor comfort for his distress. The affliction of his body progresses unabated, and soon the decorum of Ivan Illyich's world collapses. Those around him maintain their conventional deceits of empty condolences and remorseless grief in order to avoid his pain. However, their familiar diversions ironically fail, and instead, heighten his agony. Lies no longer suffice to distract him from the emptiness of his life. Because the illness exposes the inevitability of his death, the continuing denial by others of his physical condition only perpetuates and worsens his existential agony. The downward spiral irreversibly quickens: Ivan Illyich's nascent spiritual distress now intensifies his physical suffering. No physician's medication can alleviate this pain.

Finally desperate at her husband's condition, Praskovya Fyodorovna urges Ivan Illyich to receive the sacrament of communion: "It can't do you any harm, and it often helps. Really, it's such a small thing. And even healthy people often. . . ." He relents, and soon an anonymous priest arrives, hears his confession, serves him the sacrament, and leaves. No dialogue between the two is recorded. The rite inspires short-lived hope in the dying man. This, however, is abruptly shattered when his wife's hypocrisy about the sacrament painfully reminds him of the banality and futility of an existence constructed out of falsehoods. The lies co-mingle with his pain and fear, each exacerbating the other. The religious rite of the sacrament cannot touch this wound. Like the nameless physicians before him, the anonymous priest fails to honor the "personhood" of Ivan Illyich.

Illyich's Spiritual Distress: Partnership as Engagement

We have seen the lost opportunities where providers fail to enter into the spiritual distress of Ivan Illyich. Now, we retrace the story to discern where spiritual partnership is created. In the third month of his illness, amid the shame and humiliation of his wasted and naked body, Illyich discovers a healing partner in the unexpected figure of Gerasim, a vital young peasant who serves as the household pantry boy. He exudes warmth without judgment or revulsion as he daily empties Illyich's filthy chamberpot. These acts and attitudes evoke trust, prompting the vulnerable Illyich to request Gerasim's help to clothe him and position his body, and then uphold his legs in order to ease his pain. Some nights, Gerasim maintains this vigil at Illyich's legs until dawn.

The caring impact of this comfort is much deeper than the physical level of pain. Willingly, even gladly, Gerasim complies, experiencing no indignity. He acknowledges to Illyich that he can anticipate the day when he, too, will seek support from another as he lays dying. It is this overt expression of empathy, of Gerasim actively identifying with the circumstantial and existential condition as presented so poignantly by Ivan Illyich, that creates their spiritual partnership. What marks this partnership as uniquely spiritual is Gerasim's direct engagement of the ultimate, hence spiritual, concerns at the core of Illyich's being. He openly addresses the spiritual distress that is precipitated by the cluster of acute assaults on the center of this dying man: Illyich's physical existence, his identity as a person, his primary relationships, and his purpose for living. The ruddy-faced house servant becomes the touchstone of care by whom Illyich measures all others. Gerasim is the one person who is honest with Illyich and directly acknowledges that he is dying. Gerasim is also the one person who freely offers the understanding and pity that Illyich seeks. By assuming, literally, the physical stance of a partner, Gerasim relieves the physical pain in Illyich's legs. As a fully engaged spiritual partner who enters compassionately into the existential and emotional meaning of another's dreadful experience of illness, Gerasim heals the suffering isolation of Illyich's spiritual distress.

The case of Ivan Illyich reveals that spirituality entails multiple dimensions.[9] There are four primary aspects to examine as one enters into partnership at the point of another's spirituality: religious beliefs;[10] religious practices and rites;[11] significant experiences that bear important meaning; and, primary relationships central to one's life. This multidimensional understanding of spirituality honors the complex matrix of an individual's uniqueness.

For Illyich, the first two dimensions, those that are inherently religious, are not critical to connecting with his spirituality. In terms of religious beliefs, he holds to none. Formal doctrine will be inadequate to comfort this distressed man. The second dimension, religious practices and rites, functions similarly for him. After the priest hears his confession and offers him communion, the efficacy of these rituals withers swiftly and painfully in the harsh reality of the hypocrisy and lies about him. The key to Ivan Illyich's spirituality, and the opportunities for spiritual partnership, must be discovered in the other two dimensions, those that are not inherently religious.

The third dimension, that of significant and meaningful experiences, is very important to Illyich's spirituality in both negative and positive valences. In the failure of the physicians and the priest to recognize both his "personhood" and the meaning of his illness, we see that his spiritual distress increases in proportion to how impersonally he is treated. Conversely, the ability of Gerasim, regardless of his humble role, to establish a partnership with Illyich displays the healing power of the experiential dimension of spirituality. While the significant experience that bears meaning can be positive or negative, its spiritual implications are always important.[12]

Looking at primary relationships, the fourth dimension of spirituality, we find how lonely Illyich truly is as his spouse and friends approach him superficially. This relational dimension of Illyich's spirituality, is one of void and deep yearning to belong. Similarly, but again with the opposite valence, we discern in the relationship created by Gerasim's partnership its significance for Illyich's spiritual state. This relational dimension of spirituality, whether manifested as a cause to celebrate or as a source of pain, discloses another critical entry point by which to establish spiritual partnership.

The case of Ivan Illyich demonstrates the need to examine a patient's spirituality in its multiple dimensions in order to find those most appropriate for partnership. In the end, it was the non-religious dimensions of Illyich's spirituality that Gerasim was able to engage and thereby offer healing for the spiritual distress.

Dynamics of Spiritual Partnership: A Case from Psychotherapy

In exploring the dynamics of any relationship, we are concerned with the forces that produce motion and change within it. More specifically when exploring spiritual partnerships, we are concerned with the following questions: What is the impetus for or against such a relationship? How is partnership sparked or started? What are the forces that keep it going? What happens when persons are engaged in spiritual conversation?

In the case of Ivan Illyich, his vulnerability and the fact of his impending mortality became impossible to ignore, first by himself, then by Gerasim, and finally by his wife, his physician, and the priest. These latter three were aware of and observed his predicament from the perspective of their social obligations toward him. In terms of their roles, they cannot be said to have acted inappropriately. If they considered their own mortality, they did not share that process with him, nor did they allow their understanding of themselves as fellow humans to be part of their relationship with him. They remained apart. While they acted dutifully and gave him their services, they did not share themselves from the perspective of a fellow human who will also die one day. On the other hand, Gerasim entered into spiritual partnership with Ivan Illyich when he anticipated the day that he also would seek to be heard and helped by another as he lay dying. It is this capacity to enter into another's most meaningful experiences from the perspective of one's shared humanity that begins a spiritual connection and makes a partnership possible.

Jana's Spiritual Distress: An Invitation to Partnership

In this section, we describe the dynamics of spiritual partnership as presented in the relationship between Jana, a forty-two year old woman, and her nurse therapist. We describe first how it began when Jana challenged the practitioner to hear her human spirit within the context of the psychotherapeutic relationship.

The issues that prompted referral for therapy were: the loss of her father to cancer shortly following their reconciliation after being estranged for many years; Jana had also discovered that her teenage daughter was sexually active; moreover, Jana had developed troublesome somatic symptoms, specifically palpitations, a "nervous stomach," a tendency to hyperventilate, and one very frightening, full-blown panic attack. Her physicians suggested that work with a psychotherapist might help her identify the sources of her anxiety. She was referred specifically to a nurse in order that the physical nature of her symptoms would be addressed. It was also expected that the nurse would teach her how to relax.

She was an attractive, petite, neatly dressed woman. She looked keenly intelligent and sensitive as she scrutinized the therapist's face. She told her story with anxious animation, beginning with her anxiety about her daughter. She explained that although her husband was a good provider, he seldom talked to her or to her daughters about anything more than perfunctory topics. He did not know what to say about the daughter's behavior, much less what to do about it. The therapist empathized, agreeing with her that adolescents always chose the worst time to act out their independence. Jana then told about her father's illness and death, starting with the resumption of their relationship. (She had been raised by a stepfather.) Within six months of her first trip to see him as an adult, he was dead. She remained composed as she said that fate was cruel and God was cruel.

Jana agreed with the therapist to meet weekly to implement strategies for dealing with her daughter, and to talk about her father's illness and death. The therapist proceeded with the intake interview by taking her family and health histories. The comment about the cruelty of fate and God was not pursued. The session settled into a comfortable collaborative rhythm with the therapist asking the necessary questions and Jana congenially responding. When Jana told her age, the therapist volunteered that she did not look as old as she was. Jana stiffened. "Actually, I feel ten years older than I am! I feel like an old woman. Everyone assumes that looking younger is a compliment. Well! It isn't to me," she said. The therapist waited, but Jana had ended her outburst and clearly wanted to smooth over the incident. The session ended with resumption of their former cordiality. The interview concluded pleasantly, as Jana agreed to work with the therapist for a few weeks.

The therapist was left with four impressions. First, that her own tendency to be agreeable and helpful too early in the interview was still in need of careful monitoring; second, she felt that Jana had baited her; third, she sensed that in Jana's expression of anger, there was a profound revealing of suffering. The fourth impression was inaccessible; it was a discomforting memory just beyond apprehension. During the next week, the therapist was too busy to spend much time thinking about Jana's session. She wondered briefly with a measure of uneasiness whether Jana would return to therapy. She also realized that Jana vaguely reminded her of someone, but could not recall the specific person or incident.

The spiritual partnership covenant begins with an invitation into the core of another's most crucial life issues. The metaphor of core is appropriate not only because it connotes inwardness, but also (and especially) because it connotes uniqueness. No two persons are exactly alike. The pathway to each person's unique self is as distinct as each person's experience, and the process of discovering that pathway in another is usually mirrored by a similar process in oneself.

The Kindling Spark of Memory: Entering into Partnership

It was not until Jana returned the next week that the therapist remembered the specific incident and memory that Jana had triggered in her. She was reminded of a young woman in her late twenties who had come for psychotherapy more than a decade earlier. The therapist had suspected that there was spiritual distress related to her diagnosis, but she was neither sure how to recognize it nor did she feel remotely prepared to deliver "spiritual comfort". She resolved to get to that issue later. However, she never had the opportunity to do that. The client never returned. The therapist realized that although she had heard the client's words, she had not heard her spiritual need, nor had she responded to an opportunity for spiritual partnership.

All these events came back to the therapist as she let Jana in, desiring and determined to be open to any spiritual issues that might unfold. She had already allowed Jana to reach out to her by listening to the memories that were evoked in herself. Reflection on one's own processes is an important quality for effective nursing care and for psychotherapy, but it is absolutely essential when attempting to connect with the spiritual dimension of persons. Spiritual care is not given to another, rather it is shared with another. The metaphor of peeling an onion is useful in describing the process of penetrating the core issues in one's life. However, it only fits the description of this process when the health care provider realizes that the onion one peels might be one's own.

Jana's next few sessions were concerned with the critical problems with her daughter and her impatience with her husband for not being more involved in parenting. She would occasionally be close to tears when mentioning her father, but quickly went on to other topics. She attempted to follow the therapist's instructions about relaxation, and was learning to become aware of her breathing in order to slow it down. It was shortly after the therapist's affirmation of her efforts and in the context of giving a compliment that Jana delivered her next baiting statement: "You are quite understanding—for a nurse." This time the therapist did not postpone the invitation to respond. "Nurses have not listened carefully in the past," she ventured. This recognition and framing of Jana's experience opened the floodgates for a torrent of vituperative complaints about her medical treatment. She railed at the doctors who had not fully explained a surgical procedure and the consequences of a surgical experience (a hysterectomy). She thought they seemed annoyed by the persistence of her symptoms, and treated her like a "nut case." The nurses (whom she identified as any female caregiver in a hospital, clinic, or over the telephone), stupidly and carelessly planned everything for the benefit of doctors and the hospital, instead of the patients. For the rest of that session, and for many more, Jana wept and screamed about the loss of her womb and the loss of her father. She thought the hysterectomy would relieve her abdominal discomfort. Nobody explained about the persistence of other discomforts and hormones until later. Since she had consented to this procedure, she regarded herself as an accomplice in aging her own body. Now, she was no longer a daughter, nor a woman. She felt like an old woman. The therapist repeated Jana's earlier words: "Fate is cruel, and God is cruel!" Jana again expressed anger. "Don't speak to me about God!" she said. "It makes me so angry! The doctors and nurses are bad enough." At the end of the session, she wiped her swollen eyes and thanked the therapist for listening.

The memory of the former client had helped the therapist recognize that they were in the arena of spiritual discomfort. Jana's expressions of anger were an invitation to the therapist to hear her at a deeper level. The therapist accepted the invitation when she responded to the client's presenting concerns in the context of the total situation. This included the client's questions about her health care, and the profound disappointment she felt beyond that, in the forces that she called "Fate" and "God." Her spiritual distress sparked the therapist's reflexive engagement of the memory of another woman's similar struggle. Spiritual issues have a connexional quality that breaks barriers between persons. Both women's life situations were different. They were even separated in time, yet each became a reminder to the therapist of the fundamental tenacity and universality of the human spirit represented in each of them, and of the profound spiritual distress that could be experienced by humans everywhere. Memory is not only important for learning, but also for spiritual reflection.

Jana continued her tirades against the health care professionals for many sessions. The therapist found it difficult to listen to her. She seemed to make the same points repeatedly: that she had trusted them, but they betrayed her; that she thought they knew a lot, but, in truth, they knew less; that it was hard for her to trust anyone; that she no longer trusted her own body since it was always playing tricks on her and she had no control of the symptoms that plagued her. (She would say these things even though she was beginning to learn how to relax.) The therapist began to see that health care providers, especially physicians, represented science and a certainty that seemed lost to her. Jana said it explicitly: "They think they are God, but they are not!" The therapist responded, "You are very angry with them. Are you also angry with God?" "I told you I did not want to talk about God," she said. The therapist reminded her that she had introduced the topic of God in earlier sessions. She was silent for a long time. "I used to be afraid of being angry with God. Now, I'm indifferent because I don't think it matters how I feel," she said quietly.

Personal Memories: The Partnership Dance

During the next week, the therapist pondered the session. It had been cathartic and psychotherapeutically beneficial to Jana, who had been able to reach below her anger to her feelings of grief and helplessness in the presence of another person. It had also sparked the therapist's own thoughts and feelings about crucial events in her life that had redefined her own understanding of human experiences and human "personhood." The events in her own life unfolded briefly in her memory. Like Jana and others, she had been faced with the consequences of human incompleteness, suffering and mortality. The failure of trusted persons and structures had provoked anger, suffering, and helplessness and had also helped her to become aware of her own search for hope, transcendence, and connectedness. Now, she was encountering a woman who saw little hope for stable structures in her universe, who could not transcend her severe anxiety, and who was searching for meaning in her suffering. The therapist had been invited to be connected to her as a professional health care provider and as a fellow human.

The spiritual partnership covenant is like a dance[13] in which both partners are involved in movement toward their awareness of issues of ultimate concern. It is most successful when the inviting and leading partner is the client. It is not successful when the health care professional takes the client where she does not want to go, when the pace of the client is not respected or when the health care professional refuses to participate fully in the dance.

Participation in Jana's dance included listening to her concerns in the context of her human spirit. For the therapist, this meant not only being professionally challenged, but also being personally moved. For Jana, God had been the symbol of order in the universe. Now that the certainty seemed absent, she was having to redefine her experience. Was it permissible to be angry with God? Where was God? In the past, there had been certainty, and even an image of God. Now everything seemed nebulous. Was there a God? These were all questions that Jana posed in later sessions. The therapist was connected to her not only by the psychotherapeutic contract as a professional health care provider whose responsibility it was to help her deal with her daughter and her anxiety, but also as a fellow human who had heard the same questions in herself and in other humans.

Guidelines for Spiritual Partnerships

Three of the most common barriers to participating in another's spiritual struggle and quest are: first, the reluctance or difficulty to identify and engage one's own spirituality in one's work; second, the assumption that one needs to be an expert in religion and/or spirituality; third, the belief that being professional requires that one become personally detached.

By exploring the experiences of Ivan Illyich and Jana, we have demonstrated that spiritual partnerships are multidimensional. In both those cases, only those who were willing to be engaged personally in a relationship that included the spiritual dimension became a healing resource for the sufferer. Professional and social relationships were not, in themselves, sufficient, even if they were functionally appropriate, socially normative, and professionally adequate. Moreover, in order to be a spiritual resource to another in spiritual distress, one does not have to be an expert or have answers to spiritual questions. It is enough that one listen with openness and respect by offering non-judgmental acceptance. Acknowledging the profundity of the sufferer's questions in a spirit of connectedness rather than detachment is more valuable than offering one's own answers to another's spiritual quandaries. While it is important to be competent and expert in one's profession, it is also important to remember that the needs of the human spirit are universal. There are really no experts, just partners.

The third barrier to spiritual partnership is an important one. It arises for health care providers out of the desire to be professionally competent and responsible. There is a quandary about whether engaging someone personally in the context of a health care encounter might violate professional boundaries. In the dynamics of a professional partnership, the roles are reciprocal and complementary; for example, it would be inappropriate in the context of a psychotherapeutic or other health care encounter for the professional provider to take up the client's time talking about the provider's own story, or to be otherwise involved in meeting the needs of the provider. At the same time, it is not only appropriate, but also helpful to the client when the full experience of the provider is involved and used as a resource in the encounter. In the nurse therapist's earlier case, it was not helpful to the client when the therapist postponed addressing her spiritual needs. However, in the case of Jana, the client was empowered by the therapist's instrumental use of personal reflections. The therapist's memory of, and insight from, the previous case, as well as the events in her own life, were used to open up Jana's spirituality in a way that proved beneficial to her.

Conclusion

The features, dimensions, and dynamics of working with another's spiritual issues in the health care context are instructive for creating a partnership between the practitioner and the patient. In establishing such a covenant, a context is provided in which the provider is aware of her or his own spiritual issues, listens with openness, offers non-judgmental acceptance, and remains open to the possibility of spiritual regeneration.

References

1. Marty EM, Vaux KL (eds.). *Health/Medicine and the Faith Traditions*. Philadelphia: Fortress, 1982.

2. Neuman B. *The Neuman Systems Model.* Norwalk, CT: Appleton & Lange, 1989:29–33.

3. Gordon M. *Nursing Diagnosis, Process and Application.* New York: McGraw-Hill, 1982.

4. Carpenito LJ. *Nursing Diagnosis, Application to Clinical Practice* (5th edition). Philadelphia, J.B. Lippincott Company, 1993:746–763.

5. Ayto J. *Dictionary of Word Origins*. New York: Arcade Publishing. Little Brown & Co., 1990.

6. May WF. *The Physician's Covenant.* Philadelphia: Westminster Press, 1983.

7. Tolstoy L. *The Death of Ivan Illyich.* Solotaroff L. (Trans.) New York: Bantam Books, 1981.

8. Analyzing Tolstoy's story, Julia E. Connelly concludes as a physician that Ivan Illyich "... probably had pancreatic cancer. The sudden onset, the location, the persistence of the pain, and the rapidity of his demise all suggest that diagnosis." Connelly JE. The whole story. Literature and Medicine 1990;9:150–161.

9. A good example of a multidimensional approach to spirituality that is also clinically astute in Fitchett G. ed. *Assessing Spiritual Needs: A Guide for Caregivers*. Minneapolis: Augsburg Fortress, 1993.

10. Overt religious beliefs can be considered for their substance, the nature of their doctrinal meaning within a particular religious tradition or community, and the individual's appropriation of the content.

11. Like overt religious beliefs, (9) above, explicit religious practices and rites can also be considered substantively.

12. A dynamic approach to spirituality assesses how a person's belief system functions for or against wholeness and well-being. It considers the verbal, emotional, behavioral, and symbolic aspects of spirituality, for example how people assign meaning to developmental framework of Fowler JW. *Stages of Faith: The Psychology of Human Development and the Quest for Meaning*. San Francisco: Harper & Row, 1981, and in the clinical orientation of Pruyser P. *The Minister as Diagnostician: Personal Problems in Pastoral Perspective*. Philadelphia: Westminster Press, 1976.

13. The metaphor of the dance is effectively used in a poem, "The Dance of Grief," Kidd B. International Association of Human Caring Newsletter, Vol. 2, No.6., p.6.

VIII Partnerships in Do Not Resuscitate Decision-Making

Laura H. Brachman and Timothy E. Quill

The Do-Not-Resuscitate (DNR) decision is one of the most important decisions a person will ever make about his medical care. It is a decision which, regardless of what choice is made, may determine the quality and nature of the end of one's life. This decision is made between the patient, his physician, and often, his family. For such a decision to best reflect the interests, values, and beliefs of the patient, it is critical that a well-established partnership exist between patient and physician. In this chapter, we will explore these partnerships as they are reflected in DNR decision making. First, we will briefly discuss the history and nature of cardiopulmonary resuscitation (CPR) and DNR decisions. We will then look at how these decisions are made in the context of a physician-patient partnership, and consider some of the issues which arise within this partnership. Finally, using case examples, we will consider some conflicts which arise in making such decisions, and present suggestions aimed at resolving them.

Cardiopulmonary Resuscitation

Cardiopulmonary resuscitation (CPR) was first described in 1960 to be used in the management of cardiac arrest.[1] It was initially developed for use only in sudden cardiac death from cardiac standstill or respiratory failure. It has gradually become standard care for all deaths, including those resulting from chronic debilitating illnesses, unless there has been a previous and explicit directive by the patient that such an intervention not be used. Yet studies have consistently shown that patients with multiple chronic illnesses derive little benefit from CPR. Forty-three to seventy-eight percent of hospitalized patients die immediately after or within 24 hours of receiving CPR. An additional 10–41% die before hospital discharge. Of all those who receive CPR, from 0 to 24% survive to discharge, depending on selection criteria.[2–4] Only 11% of hospitalized patients receiving CPR in one series were alive at 6 months.[2] Half of those who survive to discharge live the remainder of their lives in nursing homes. Subgroup analyses show that cancer (particularly that which is metastatic), severe left ventricular dysfunction, renal failure, liver failure, and multiple chronic diseases all portend a poor prognosis.[2,3] On the other hand, those patients with relatively uncomplicated heart disease, even with advanced age, seem to have the highest likelihood of benefiting from CPR.[5]

If CPR were a completely benign procedure, then perhaps its routine implementation would be reasonable; although it may seldom succeed, patients will surely die without it. However, CPR is an invasive, harsh, and often violent treatment. It puts patients at risk for broken ribs and involves repeated blood draws; tubes placed in the mouth, nose, and bladder; and intravenous lines in the arms, neck, and groin. It is unlikely that patients can fully grasp what this experience will be like, despite the most graphic descriptions. Furthermore, those performing CPR are at risk for contracting infectious diseases (HIV, Hepatitis B) to say nothing of the demoralization from inflicting such a harsh end on a dying patient. Given the nature of CPR, coupled with its lack of efficacy in many clinical situations, it is of paramount importance that its application be based on informed choice and clear communication.

Prior to 1980, decisions regarding who would get CPR were solely the province of physicians, and often were made at the critical moment of cardiac arrest, with no prior conversation between doctor and patient regarding its use. Though such decisions were made with beneficent intent, they were sometimes arbitrary and idiosyncratic, often as dependent on the physician's values and beliefs as on the patient's medical condition. These decisions were rarely derived from a mutual, informed decision-making process characteristic of a partnership. The only beneficent, albeit paternalistic, aspect of this process was that patients and families were spared the difficult deliberations over a treatment that was likely to be both invasive and ineffective. Nevertheless, some patients may have been deprived of a potentially effective treatment that they might have chosen if fully informed of its potential, and others may have been subjected to invasive treatment they would not have wanted had they been fully informed. Beginning in 1980, physicians were first encouraged through practice standards and then mandated by law to inform patients of their options regarding CPR, and patients were allowed and then required to make their own choices. This decision-making process requires a working partnership between doctor and patient.

Partnerships and the Decision-Making Process

The goal of the partnership is to make a joint and informed decision about CPR or DNR that best reflects the patient's goals and values. Decisions of this magnitude require a doctor-patient relationship characterized by mutual honesty, trust, and respect. Ideally, such a relationship would already have formed as patient and physician navigated together the course and management of the patient's pre-existing illness. If this is not the case, a conscious effort must be made to develop such a relationship prior to discussion of CPR. Only in this way can the distinct steps be taken which are necessary in making a satisfactory decision regarding DNR.

The doctor-patient relationship can be developed in the course of the initial days of hospitalization. First, the patient and doctor must establish a shared understanding of the patient's condition and prognosis. The physician should be fully knowledgeable about the patient's disease, and its prognosis and management. This information needs to be clearly communicated to the patient, and some assessment of the patient's understanding should be made. This can be done by asking, "Tell me what you know about your illness," so that the patient can translate into his own language what may have been explained to him using medical terminology. It is important to address any misconceptions about the illness or the patient's prognosis at the outset.

Within the context of the illness, the patient's values, beliefs, wishes, and goals must be identified and clearly understood. For instance, does the patient desire any treatment option available in the hope of cure, or is relief of discomforting symptoms his goal? Is his priority to avoid hospitalization regardless of the outcome, or simply to avoid admission to the intensive care unit, with all of its attendant invasiveness? Is there an upcoming family event that he wants to attend, with no particular goals after that? The patient's goals and values must be known from explicit conversation, not assumed. A patient's beliefs cannot be deduced from the length of his medical history, his medical prognosis, the number of medications he takes, or his age.

The next step is to negotiate a treatment plan. All potentially effective treatment options are explored, and an analysis is made as to which options are likely to offer the greatest benefit with the fewest side effects and risks. Within the context of the patient's beliefs and goals, a joint decision regarding treatment is made. At this point, it is important to make sure that the patient's goals are realistic in light of the diagnosis and the patient's prognosis.

The DNR Discussion

Once a relationship between the patient and physician has been established, the option of CPR should be addressed. The timing of the DNR discussion should not be left to chance. If the topic is raised before the patient has begun to accept a new diagnosis, or before he has an adequate understanding of that illness, talking about foregoing CPR may cause him to lose hope unnecessarily. On the other hand, if discussion is delayed, he may not be aware that there are choices other than doing everything possible. In actual practice, the timing of DNR discussions is not always within the physician's control. Catastrophic events requiring emergency room evaluation or ICU admission cannot usually be predicted. All too frequently, DNR decisions are addressed only when death is imminent. Such circumstances do not lend themselves to careful deliberations. In these circumstances previous discussions regarding a patient's beliefs and goals, or reference to a living will (see below) can be useful.

A DNR discussion does not need to occur with every patient during every hospitalization. Rather, primary care physicians should explore their patients' general attitudes about advanced directives and end of life care, perhaps as part of a routine discussion about naming a health care proxy to be conducted with every patient, regardless of health status. And certainly in a clinical setting in which the odds of successful CPR are small or when the burdens of disease become high, the patient and physician should revisit the topic of end of life care and explore the option of a DNR order. Without this discussion the patient may receive CPR when he might not otherwise have chosen it, or when it offers him virtually no benefit.

A key initial step is for the physician to clearly define what DNR means. Euphemisms such as "heroics" or "keeping you alive as a vegetable through artificial means" should be avoided. A patient should clearly understand that DNR means that if his heart stops beating or his breathing becomes inadequate to sustain his life he will not be revived with either medications or machines.

It should also be made clear to the patient what DNR does *not* mean. Patients often fear that after a DNR order is entered, they will no longer receive medical care of any kind, the staff will withdraw from them and they will be abandoned. These fears are supported by disturbing reports that medical and nursing staffs sometimes do withdraw both physically and emotionally from patients once a DNR decision has been made. Yet a DNR decision in no way addresses other ongoing interventions that a patient may or may not want. Patients who have chosen DNR may receive a full spectrum of treatments, ranging from aggressive care such as intensive care unit admission, dialysis, artificial feeding, and antibiotics all the way to purely palliative measures, according to their preferences. Patients must understand that DNR is only one specific decision they make about their health care, and has no implications for other types of treatment.

Just as with any other treatment, there is uncertainty regarding the outcome of CPR, and this needs to be communicated to the patient. This uncertainty should be expressed in probabilistic terms whenever possible, to enhance mutual understanding. Since patients will make different choices depending on the way outcome rates are expressed,[6] the odds of both success (i.e., 1/10, 1/100, 1/10,000) and of failure (i.e., that 9/10, 99/100, 9999/10,000 will not survive) should be presented.

Very low survival rates raise the question of whether CPR can be regarded as futile and whether there is an obligation to offer futile treatments (or an obligation not to). A full discussion of this complex issue is beyond the scope of this chapter but can be found elsewhere.[7–10] There are very few situations in which there is unanimous agreement that the survival rate following CPR is 0%, hence literally and statistically futile. Patients' definitions of futility tend to focus not on survival per se, but rather on the likelihood of resuming a satisfactory quality of life (as defined by the patient) or of attaining a hoped for quantity of life (for example, surviving until a grandchild's birth or a child's wedding). Some would find a few additional days, even if spent in the hospital, to be acceptable. For others, a month or two more at home might not be acceptable if this time is likely to be spent dealing with new and significant limitations or imposing further burdens on themselves or their families. It is often helpful to consider previous experiences that patients have had observing terminal illness and CPR in family members or friends, and to explore how those experiences shaped their opinions and preferences about their own death.

Physicians may also have opinions about the outcome of CPR in particular situations. If they believe that CPR is overwhelmingly likely to be ineffective, they can and should present both the poor odds and their recommendation of a DNR order. However, what is seen by the physician as a one in one thousand chance may be seen by the patient as a glimmer of hope. Since futility is a concept with varying shades of gray, it may be useful to think of it as a term to be mutually explored and defined by physician, patient, and family, particularly until there is an agreed upon public policy definition.

Physician authority and patient autonomy are often regarded as mutually exclusive in the decision-making process, yet this need not be so. Inherent in the physician's role is the task of presenting all options of medical benefit to the patient and making a recommendation based upon knowledge of the medical circumstances and of the patient's goals and values. The patient may or may not accept the physician's recommendation. The informed patient's right to refuse effective treatment, even if such a refusal results in his death, is well established in both the legal and ethical communities. Therefore, patients may choose DNR even when CPR has a reasonable chance of success. Conversely, patients may request CPR even when the odds of success are very poor (particularly with futility being so difficult to establish). Serious disagreement between physicians and patients about CPR and DNR warrant ongoing mutual exploration of perspectives and values. In the interim, patients who choose near futile CPR against the advice of the medical staff should receive CPR until it is determined to be ineffective.

The DNR discussion should involve not only the patient and physician, but also those family members whom the patient wants to be involved. Families do not have authority in these decisions, provided the patient is competent, but their participation in the discussion increases the likelihood that they will support the patient's decision and reduces the opportunity for subsequent conflict, either within the family or between the family and the health care team. If the patient is not competent, a surrogate (often a family member) is asked to make the decision. Advance directives—documents such as living wills and health care proxies which are created while a patient is competent and which clearly articulate the patient's values and wishes about CPR and other treatments—are extremely useful to surrogates (a full discussion of these documents can be found elsewhere).[11] When such documents are not present, the family is asked to provide "substituted judgement," that is, to articulate what they believe the patient would have decided had he been capable of understanding his condition and prognosis.

DNR Issues for Patients, Families, Physicians, and Staffs

DNR decisions raise a number of complex and interwoven issues for patients, their families, the attending physician, and the clinical staff. Acknowledging, exploring, and resolving these issues are integral to building and maintaining doctor-patient and doctor-staff partnerships as decisions about DNR are faced together.

Fear of abandonment may be a major concern for patients who choose DNR. As described above, patients may fear that once the DNR decision is made, the physician and staff will withdraw from their medical care. Such patients are often both physically and psychologically vulnerable, and may be facing their mortality for the first time. Fear of abandonment may be particularly prominent if other potentially life-prolonging and possibly desired treatments are not simultaneously addressed, or if palliative care options are not explored. Patients may also worry that death will have to be faced alone, and will be painful, prolonged, and filled with uncertainty and fear. The mistaken view of some doctors that a DNR decision is a choice between "doing everything" and "doing nothing" further reinforces patients' fears of abandonment. Physicians need to reflect on their own willingness and ability to stand by patients throughout their illness and then communicate this commitment clearly to their patients.

DNR decisions compel patients to face their mortality. They must come to grips with incurable illness and with death, neither of which they can control. Such considerations can be initially frightening, humbling, and depressing, yet they may eventually be very liberating. While DNR is often initially equated with "giving up," the manner in which patients will continue to fight, find meaning, and find some peace in acceptance of their mortality may emerge from the ensuing discussion. Avoiding difficult DNR discussions permits patients to cling to false expectations of recovery, and deprives them of the opportunity to fully address, understand, and accept their situation. Patients' thoughts about their mortality are usually not far below the surface; gentle probing by the physician can encourage the patient to talk about his hopes and fears.

DNR discussions may initially elicit anger at the medical profession for not having cured the disease and for not being as powerful as had been expected or hoped for. There may be uncertainty about whether the patient's impending death is the result of physician inadequacies or inherent limits of medical treatment. On the other hand, there may be relief associated with the DNR discussion because the patient is given a forum to discuss what has likely been on his mind all along. Patients often have opinions and concerns about how and when they will die, yet they may fear that it is wrong to talk about death. Some patients may be eager to explore alternatives to the most aggressive technological care, particularly when they are already severely burdened by their illness. Discussing a DNR order creates the opportunity to explore more meaningful, less technologically-oriented options regarding end-of-life care. This may include a hospice approach in which all interventions are geared toward the patient's comfort, personal contact, and symptom relief, as opposed to a focus on disease control or cure.

DNR discussions can also raise difficult issues among the patient's family members. Previously unrecognized misunderstandings about the patient's disease and prognosis, or differences of opinion and attitude about end-of-life treatment may be uncovered. Disagreements about who has a say in making decisions about CPR may be significant, even when patients are competent and able to make their own decisions. Families often have complex patterns of group decision-making and sometimes may give preferential power to family members other than the patient. The DNR discussion may trigger reemergence of previously unresolved family issues that lay dormant until the threat of death forces them back to the surface. If the patient is competent, the physician may need to remind the family that it is the patient's prerogative to make whatever decision he feels appropriate for himself. If the patient is not competent and there is no designated health care proxy, it is useful for the physician to request that a spokesperson be chosen for the family, preferably the person who knows the patient the best. In this way the physician is less likely to become enmeshed in family battles. If problems remain and patient's best interests are not being served, expert help may be needed—a mental health consultant, patient advocate, or member of the clergy jointly selected by the family.

Just as the discussion of end-of-life treatment and DNR orders can be emotionally charged for patients, so can it be for physicians. As physicians help their patients work through the DNR decision, they must also reflect on their own issues with these decisions. Some physicians find that working through these decisions with patients to be the most significant thing they can offer to their patients, given the import of such a life and death decision. Others may prefer to avoid DNR discussions to protect patients and their families (or possibly themselves) from anticipated hopelessness and despair. Gaining a clear understanding of who is being protected in such decisions can be challenging. Working through the DNR choice with patients may force physicians to face their own mortality, and may lead them to think about what their own choices would be in similar circumstances. Some physicians feel that the DNR decision should be their exclusive domain. They may feel that patients are not capable of making such decisions, particularly in complex medical circumstances. Conversely, the DNR discussion may cause physicians to question their own abilities, especially if treatment during the course of the patient's illness has been unsuccessful. Physicians may also have strong biases for or against the technological fight for life. Such biases may influence how they discuss DNR with the patient, yet may have little to do with the particulars of the patient's medical condition or what the patient wants.[12]

Conflicts Within the DNR Decision

Conflicts that arise as the DNR decision is being made may have little to do with the DNR discussion itself. Careful reflection upon the quality of the doctor-patient relationship and conversations that preceded the discussion of CPR may reveal the true source of conflict. Table 1 shows levels of interactions that are useful to review when seeking to identify the source of conflict. Within each level are shown several questions to be asked which may help to understand the nature of the conflict. Once conflicts have been identified they must each be addressed in sequence. In general, agreement must be achieved about the nature of the patient's problem and its prognosis before complex questions about goals and treatment options can be

Table 1: Questions for identifying the nature of conflicts in DNR decision making.

Doctor-patient relationship

- Did the patient and physician have a previously established relationship?
- If so, is this relationship based on mutual respect and commitment to honest communication?
- If the patient-physician relationship is a new one, what are the circumstances of the initial meeting (a routine outpatient visit? In the emergency room or ICU at the time of a medical crisis?)
- If patient and physician are new to each other, was adequate time taken to establish and build the relationship before DNR decisions were addressed?
- Is there an unrecognized struggle for power and control between physician and patient (and/or family members)?

Understanding of the illness necessitating the discussion of CPR

- Do both patient and physician have a mutually-agreed-upon understanding of the illness?
- Do the patient and physician have a mutually-agreed-upon understanding of the patient's prognosis?
- Is the patient fully aware of what treatment options exist, as well as the limitations of those treatments?
- Was a treatment plan put into place which was acceptable to the patient?
- Has some assessment been taken with the patient of what this treatment did and did not accomplish?
- Have the doctor and patient talked about death?

Exploration of the patient's goals

- Was the patient given ample time and encouragement to explore and express his goals and wishes. Were these heard?
- Were the physician's goals considered?
- What does the patient hope to achieve with CPR?

Definition and methods of DNR

- Does the patient understand what interventions a DNR order does and does not refer to?
- Once DNR status was chosen, was attention also paid to additional treatments the patient would want?
- Was the family in agreement with the DNR decision?
- Was the family supportive of the patient?
- Was the physician in agreement with the DNR decision?
- What is the meaning of DNR to the patient and family?

Other

- Were the patient's fears adequately explored (including fear of abandonment, fear of dying, fear of dying in pain, anger with the physician, anger with the family)?
- Should a member of the clergy be included in these discussions?
- Has the patient reflected on where he looks for hope? Is hope defined as living until a certain date (i.e., a family event)? Is hope found by ensuring a peaceful death without aggressive medical interventions?
- How do the patient and family see the future?

adequately addressed. If disagreements that arise early in the decision-making process are not fully explored and resolved, they are likely to resurface as the process continues. Taking the time to backtrack and to resolve areas of potential disagreement will ultimately allow the patient and physician to move forward.

Case Examples

The implementation of the principles described above can best be presented in the context of actual cases. The first case, involving a competent patient, illustrates the processes of establishing an adequate patient-physician relationship before pursuing a discussion about a DNR order, eliciting the preexisting knowledge and attitudes of the patient and his family, and communicating clearly information about the outcomes of CPR.

Mr. F, a 75–year-old man, was hospitalized for treatment of complications resulting from metastatic prostate cancer. He chose a city hospital seventy miles from his home for treatment. He was admitted to the service of a urologist with whom he had a previous relationship, but was seen in consultation by an internist whom he had never met. Because of the advanced nature of his disease, the consulting internist felt it imperative to talk with the patient about both DNR and advance directives, subjects that had never been broached with him. Since his condition was not critical at the time, the physician chose to allow a few days for the patient, his wife, and herself to get acquainted and to lay the foundation for having these difficult discussions. After beginning to establish a trusting relationship, the physician asked Mr. and Mrs. F what their understanding was of the disease and its prognosis. They seemed to understand that his condition was serious, that the radiation therapy he was receiving was only palliative, and that his life expectancy would likely be measured only in terms of months. The conversation then turned to their understanding of advance directives and the DNR decision. When it became clear that they had never considered any of these issues, the physician gave them copies of a living will and health care proxy form to read. They agreed to meet the next day, at which time their daughter, son, and daughter-in-law would join the conversation. That evening, the patient and his family had a chance to discuss the kinds of end-of-life treatment he would and would not want. The next day, after reviewing these issues again, the patient chose his wife as his health care proxy, and his philosophy about the end of life was explored.

The physician then raised the issue of CPR. She presented it as a set of interventions used in the setting of sudden cardiac or respiratory arrest, and made it clear that in patients with widely metastatic cancer it was virtually always ineffective. She described it as a harsh treatment that, if successful in restoring cardiac or pulmonary function, would likely do so for only a matter of hours or days; there would be little hope that the patient might ever leave the hospital. At this point, the patient's wife recalled that both of her brothers-in-law had died after receiving CPR, and that after watching them and her sisters being subjected to these experiences she would never want that for her husband. Mr. F took these comments to heart and, in keeping with his understanding of the terminal nature of his disease, chose DNR status. They then discussed what his goals were in light of his disease and prognosis, and discussed other potentially effective treatment options available to him.

Both the patient and his wife said they were humbled by the process of having this discussion. Yet they also seemed freed to have subsequent discussions about treatment options such as pain control and radiation therapy. Interestingly, the patient, with the support of his family, subsequently chose to undergo aggressive chemotherapy in another city. Although he ultimately died of his cancer, and it is unlikely that the treatment he chose prolonged his life, it was important to Mr. F to try anything that might give him a chance, no matter how small that chance was. Yet in the end his feelings about CPR did not change. He died at home free of pain, and with his family around him.

The second case illustrates the decision making around DNR in a patient whose illness rendered her unable to make a DNR decision for herself. It illustrates the process of helping a family to provide substituted judgement, and to overcome internal conflict.

Mrs. A was a 72-year-old woman admitted with a very large stroke involving most of the left side of her brain. She could not speak and her entire right side was paralyzed. She became progressively less alert over the first twenty-four hours of her hospitalization. A CT scan showed substantial bleeding in the left side of her brain surrounded by considerable swelling. Surgical treatment of the internal bleeding was not possible. As the patient deteriorated, the potential need for mechanical ventilation and CPR needed to be addressed, but the patient had no advance directives. A family meeting was scheduled so that the attending and the family could come to grips with the situation and make decisions on the patient's behalf.

The patient's husband was dead, but four of her five children attended the meeting. Although they were very upset by her deterioration, they all felt very strongly that she would not want mechanical ventilation or CPR given the severity of her stroke. They cited her past distress about her husband's death in the hospital ICU, saying she would never want that type of treatment. They also reported her disdain for potentially having to live in a severely compromised state. Although they all felt strongly about what their mother would want, they wanted to wait for the oldest sibling to come in from out of state to make a final decision. They were informed that if she had a cardiac arrest in the interim she would receive CPR. They understood this but remained fearful of making a decision without consulting their older brother.

Fortunately, the patient stabilized temporarily. When her eldest son arrived, he was very angry, insisting that more consultants be involved and demanding that his mother be sent to a tertiary care center. The remaining family members were intimidated by him, and at first acceded to his demand that all aggressive treatments including artificial ventilation and CPR be given. Rather than confronting his decision, we asked this son to spend as much time as possible in his mother's room and to reflect on what she would want with regard to CPR if she could speak for herself. Though he remained angry with all the doctors he encountered, he was able to establish rapport with a member of our hospital clergy. It became apparent that he had a lot of guilt about his father's death, feeling that not enough had been done, including his own inability to visit his father before he died. He eventually was able to agree with the rest of the family that his mother would not want CPR or mechanical ventilation if she were able to make the decision herself, and the appropriate DNR directives were completed.

Three days later, the patient died peacefully, attended by her family. The eldest son left town immediately, and did not stay for the memorial service.

Conclusion

A working partnership between the doctor and patient is critically important in DNR decision making. The DNR decision governs whether or not the patient, in the event of a cardiovascular collapse, will receive the most aggressive technological interventions available or whether he will be allowed to die. Since this treatment is highly invasive and its effectiveness is determined to a large extent by the patient's underlying conditions, it should be the subject of truly informed consent. Yet because it is applied in an emergency, when momentary delay can be fatal, consent is assumed if the question has been avoided by doctor or patient. Many issues arise as a result of the DNR decision-making process that affect the patient, family, physician, and the staff members who perform CPR. These issues must be acknowledged and addressed in advance whenever possible in order to maintain healthy partnerships.

When making a choice about CPR, a number of conflicts can arise that may hinder making a decision that is in the best interest of the patient. These conflicts may be related to diagnosis and prognosis, to the values and goals of patient or physician, to the efficacy of the treatment (CPR), or to the doctor-patient relationship itself. The process of identifying and working through these conflicts about DNR can often lead to better, more informed decisions that are in the patient's best interests, and to richer, more multilayered partnerships between doctor and patient.

References

1. Kouwenhoven WB, Ing, Jude JR, Knickerbocker GG. Closed-chest cardiac massage, JAMA 1960;173:1064–7.

2. Bedell SE, Delbanco TL, Cook EF, Epstein FH. Survival after CPR in the hospital. N Engl J Med 1983;309:569–76.

3. Murphy DJ, Murray AM, Robinson BE, Campion EW. Outcomes of CPR in the elderly. Ann Intern Med 1989;111:199–205.

4. Von Gunten CF. CPR in hospitalized patients: When is it futile?. Am Family Phys 1991;44:2130–34.

5. Quill TQ, Bennett NM. The Effects of a hospital policy and state legislation on resuscitation orders for geriatric patients. Arch Intern Med 1992;152:569–72.

6. Tversky A, Kahneman DK. The framing of decisions and the psychology of choice. Science 1981;211:453–58.

7. Schneiderman U, Jecker NS, Jonse AR. Medical futility: Its meaning and ethical implications. Ann Intern Med 1990;112:949–54.

8. Tomlinson T, Brody H. Futility and the ethics of resuscitation. JAMA 1990;264:1276–80.

9. Younger SJ. Who defines futility? JAMA 1988;260:2094–95.

10. Lantos JD, Singer PA, Walker RM, et al. The illusion of futility in clinical practice. Am J Med 1989;87:81–84.

11. Quill TQ. *Death and Dignity: Making Choices and Taking Charge*. New York: W. W. Norton & Company, 1993.

12. Schneiderman U, Kaplan RM, Pearlman RA, Teetzel H. Do physicians' own preferences for life-sustaining treatment influence their perceptions of patients' preferences? J Clin Ethics 1993;4:28–33.

IX Friends as Patients and Patients as Friends: A Challenge to Professional and Personal Boundaries

Rebekah M. Wang-Cheng and David L. Schiedermayer

Although most patients do not come to a physician seeking friendship per se, a survey in the United Kingdom found that about 30 percent of patients do consider their doctor to be something of a personal friend.[1] Physicians have also indicated that friendship with patients makes their work enjoyable.[2,3] Suchman and Matthews have coined the term "connexions" to describe the relational bond which results in an enriching experience shared between clinicians and patients.[4,5]

Although the value of a "friendship model" for the physician-patient relationship has been debated by bioethicists[6,7], the practicalities of friendships with patients outside of the office have not been addressed. Can professional care and friendship co-exist comfortably? Where should appropriate boundary lines be drawn? The psychiatric literature abounds with references to boundaries, particularly sexual, but not much has been written for nonpsychiatrists until the recent overview of professional boundaries by Gabbard and Nadelson.[8]

Through the use of personal vignettes, we will share our perspectives on the challenges and privileges inherent in physician-patient friendships and suggest guidelines for maintaining proper boundaries. Boundaries are important for preserving the integrity of the physician-patient relationship, preventing physician burn-out, and enhancing patient care.

Potential Pitfalls in Friendships with Patients

Every relationship has certain risks; the physician-patient relationship is no exception. Prior to initiating friendships with patients or providing medical care for friends, physicians should be aware of challenges in several areas (Table 1).

Loss of Objectivity

An inability to be objective was cited as the primary reason for refusing to treat family members by 465 physicians at a suburban hospital.[9] Respondents described

Table 1: Potential Pitfalls in Physician-Patient Friendships

- Loss of objectivity
- Inability to discuss personal details or perform exam
- Physician countertransference
- Excessive guilt or anger at bad outcomes

other physicians who had lost their objectivity as "not performing procedures needed to investigate serious illness," or "pursuing care considered futile."

Close friends can be compared to family; in fact emotional closeness may be even greater with friends. Does caring too much for a patient impair clinical judgement? Talcott Parsons, a sociologist, insisted that affective neutrality, which implies no emotional involvement with patients, should be one of the principles governing the physician's role. We would counter that being "objective" at all times with patients is neither achievable nor even desireable, but the closer the relationship the more difficult it is to set limits without feeling that you might be jeopardizing your friendship. I (RW) will share two anecdotes in which I did not feel "objective."

> Rose Cantelli (all patient names have been changed) is a 64–year-old woman whom I've known for several years. She worked as our housekeeper for several years until she and her husband retired and moved out of state. During a recent visit she bemoaned the fact that she had gained considerable weight, and mentioned that she had just read a report on how safe fenfluramine is for weight loss.
>
> Knowing what was coming, I began to struggle inwardly with what to do. She begged for "just a month's supply so I can get started." Since I hesitated, she played her trump card. "Sherry [her daughter and my medical school classmate] suggested that I talk to you about this."
>
> I wanted to just say "no"; instead I agreed to think about it and call her. The next day I told her "no", but felt uncomfortable.

I think it was easier for me to refuse because she was going to leave after the weekend. Although there are reasons not to prescribe a medication when you are not providing on-going care, the "friend" part of me wanted to make her happy, save her a physician visit, and please her daughter as well. Maybe she would have benefited from using the diet suppressant. My inability to be objective was linked to my anger that she would put me in an awkward position. I do know that if she had not been a friend I could have made the decision more easily because of the absence of conflicting emotions. Maybe that's what objectivity is about—not the absence of emotion but the absence of ambivalent emotions. In the next incident involving a friend I see regularly, I did not refuse.

> Phyllis Baker called me (RW) early one morning at home. "I really hate to ask you this, but . . ." Instinctively I prepared myself—be firm; you are not face-to-face. "I'm flying

today to my parents' 50th anniversary party, and I'm feeling very uptight. Since I can't get in to see my doctor, could you give me some Xanax?"

Because of her time crunch, I caved in. "Sure, what's your pharmacy number?" I left for work feeling not particularly good or bad, but wishing that I had some guidelines to follow. As it turned out, she didn't take any of the Xanax, after all, but felt reassured just knowing that she had it available.

Was the Xanax an example of an innocuous favor? Would it be analogous to asking her husband to give us a car discount? In both cases was my reluctance based on fear of setting a precedent for more informal care or does it reflect a genuine concern for well-being or both?

These examples raise questions for which there are no easy answers.

Inability to Discuss Personal Details or Perform Complete Exam

Frankel[10] writes of an internist who interviewed a patient in front of students. She was a woman in her 50's with rheumatoid arthritis whom he clearly liked and admired. After the patient left the classroom, a student asked the internist about her sexual functioning and possible marital difficulties. He replied that he suspected stress in her marriage, but he couldn't really talk with her about it because he knew her too well. Her husband (who was also his patient) was a frequent golf and fishing partner. The internist viewed them both as close friends. To pursue her problems might create discomfort in his friendship with her husband.

The reluctance to discuss personal and intimate details can be shown by the physician, the patient, or both, and often for good reason. Questions about sexual functioning might make both physician and patient uneasy at the time and later. If a female patient tells her male physician-friend of marital difficulties, this can lead to emotional closeness that persists outside of the professional context.

Discomfort may also arise with certain parts of the physical exam, such as the breast, genital, and rectal exam. This is particularly true if the physician and patient are of opposite gender. An internist related an incident from his residency days when the wife of a fellow resident came to the ER with acute abdominal pain. While he ordinarily felt quite competent evaluating abdominal pain, the fact that she was a friend, and the wife of a friend, intensified his sense of responsibility. While trying to be reassuring to her, he was also nervously trying to decide if a pelvic exam was indicated. Ultimately, he did not perform one (which may not have been good medicine), but fortunately it turned out that she did not need it after all.

Physician Countertransference

Physicians are just as human as patients and find that they like some patients and dislike others. Some of these feelings are overtly conscious. Other feelings while more unconscious, influence our behavior nonetheless; maybe even more so because we are not able to consciously control them.[11] These feelings of the physician, known as countertransference, can be negative or positive. When the patient is also a close friend, the physician's inherent feelings may be intensified.

Strong negative feelings towards patients are probably rare, but they do occur. In a survey of 59 physicians, Smith and Zimny[12] found that the greatest negative reactions resulted from perceived threats to physician integrity or self-esteem. These physicians' reactions did not appear to diminish with experience or age. What if the young woman with somatization disorder is also a friend? As a friend, one may patiently listen to complaints, but as a physician, it is easier to become frustrated because of the difference in expectations. As the friend, I am not expected to "cure," but as the physician I am being paid to "fix" the patient. These distinctions blur in a dual relationship and can threaten the friendship.

With some patients, physicians may experience a good "chemistry" because of shared interests, common backgrounds, or mutual acquaintances, and a greater potential for friendship may exist. If the patient is physically attractive, physiological responses may evolve with strong emotional arousal.

> Luci Jenkins, a 25–year-old , thin, shapely woman, came to see me (DS) with bronchitis. She also spent time outlining difficulties at work. As I was finishing the exam, she said in a low voice, "so, how are youuu?"
>
> I was dismayed at my reaction. It was the most heartfelt inquiry into my well-being that anyone had made of me that entire day. Actually, I was not doing too well. I had been at several difficult meetings, and I was dead tired after being in clinic all afternoon. I had been gaining weight lately and felt particularly unattractive. When she asked the question, I felt my heart warm. "All right," I said. "A bit tired."
>
> "Are you planning anything this weekend?" she asked. Getting a grip on myself, checking my wedding ring, thinking of my children's faces, I said, "Well, maybe some skiing."
>
> "With your grandchildren?" she asked, in a matter-of-fact tone, her voice now not even remotely husky.
>
> Since I am only 39, this was a bitter blow. I chuckled and said, "Well, with all three kids."
>
> I have seen her since and now realize that when she asks "How are youuu?" which she does on every visit, she means only and exactly that. The seduction was only in the doctor's mind.

It's easy to look back and laugh at how quickly I jumped to conclusions based on my own needs at the time. The feelings of attraction are in themselves not wrong or even uncommon among professionals who work closely with people. Being aware of one's internal emotional milieu can provide early warning signals before lines are crossed. Having a trusted colleague with whom one can confide such feelings and encounters helps place them in perspective and may prevent any actions that could lead to embarrassment or harm.

In the physician-patient relationship where intimate feelings are often shared, the potential for romantic or sexual attraction is great. However, any romantic involvement is unwise, and crosses into potentially damaging and unethical territory. The American College of Physicians Ethics Manual states that a relationship with a former patient is still unwise, but that the delay necessary to protect a patient varies with the people involved and the intensity of the prior professional relationship.[13] Others have suggested a wait of at least two years after the professional relationship ends.[14] Whatever the length of the waiting period, the possibility of undue influence exists.

Sexual relationships with patients are unethical, but they occur in a variety of medical fields. In a nationwide survey of 10,000 U. S. physicians,[14] including family physicians, internists, obstetricians, and surgeons, a surprising 9 percent of the respondents admitted to sex with at least one patient, and 4 percent had been involved with more than one patient. Most states have legal prohibitions against sexual contact between mental health professionals and current patients, but similar legislation for nonpsychiatric physicians is not widespread. The American Medical Association has highlighted the importance of reporting offending colleagues to licensing boards.[15]

Excessive Guilt or Anger at Bad Outcomes

Patients and their families place an enormous amount of trust in their physicians. In turn physicians have a fiduciary responsibility to provide the best care for their patients. When the patient is also a close friend, the physician may have a heightened sense of responsibility, and a bad outcome may result in excessive guilt or anger on either side.

> Sally was the daughter of our children's "grandmother figure." She was in her mid-50s when I (DS) became friends with her. We were at their house for Thanksgiving dinner when Sally told me about her problem. She had been treated for breast cancer ten years earlier and now had shortness of breath, chest heaviness, and swollen legs. Since it was a holiday, it made sense for me to initiate the work-up. Although Sally was first my friend, that day she became my patient. A chest x-ray revealed a massive pleural effusion. A pleural biopsy and bone scan were both positive for recurrence, and chemotherapy was instituted.
>
> Although the cancer's progress was checked, she developed congestive heart failure. As a physician (and ethicist), I knew I should discuss terminal care with her. As her friend, I found this extremely difficult. I was invested in keeping her alive. Our friendship made the day-to-day decisions in her care very complex.
>
> Sally had a respiratory arrest several months before she died. During the intubation (which she had requested), the resident couldn't find her cords. As her physician, I was at her bedside in the ICU while layers of anesthesia hierarchy were summoned. As her friend, I was angry with these physicians. Why couldn't they intubate her?
>
> Amazingly, she survived the event and appeared in clinic several weeks later. Although I felt terrible about the whole episode, she felt reasonably well and insisted on receiving aggressive care should it recur.

Six months later I was present at her final code. I asked the residents to stop after a half hour without a heartbeat. After the room emptied, I whispered, "Sally, it's OK to go."

Although I knew we had done what we could, I felt guilty that she had died. If she had not been my friend, would I feel this guilty about a death which could no longer be forestalled? Would I have kept pushing with treatment? Her mother brought us a crate of oranges every Christmas for a year or two, but we drifted out of touch. Sally's death was a painful memory between us.

Guidelines for Physician-Patient Friendships

There are certain situations in which it might be difficult for a physician to avoid friendships with patients, such as a university, small town, or rural setting. For physicians who are in a dual role with patients as friends, we suggest some guidelines (Table 2).

Determine a personal comfort zone for the depth of friendship

Some physicians and patients develop a close friendship which remains confined to the office context. In addition to provision of medical care, there will be mutual sharing of information about children, values, goals, thoughts, and feelings. Other physicians may extend the interaction to shared activities, such as engaging in athletics or hobbies together, and some may feel comfortable visiting in each other's homes.

Rourke[16] has described six different possible combinations of physician-patient relationships:

1) physician and patient are friends only;
2) physician and patient are friends, but patient inappropriately tries to get medical advice;
3) physician and patient are friends and have an overlapping professional relationship;
4) physician and patient have dual relationships but keep them totally separate;
5) physician and patient have professional relationship with some degree of friendship; and
6) physician and patient have professional relationship only.

Whatever the degree of friendship, the physician should be alert to any signs that medical care is being compromised or that decisions are being affected. Fear of losing the friendship should not influence medical decision-making. The physician must be willing to make and stand by difficult decisions with which the patient may disagree. The patient should agree to defer to the physician's professional opinion

Table 2: Guidelines for Physician-Patient Friendships

- Determine a personal comfort zone for the depth of friendship.
- Set clear expectations of where and when medical problems will be addressed.
- Be sure that both parties recognize the risk of unintended bad outcomes that could fracture the friendship.
- Discuss whether or not the friendship will be continued if the professional relationship is terminated or vice-versa.
- Establish a clear agreement about the meaning and extent of gifts.

without using the friendship as leverage. At times it may be necessary to refer to another physician for a more "neutral" opinion. If the physician frequently feels uneasy, the relationship needs to be reevaluated.

Set clear expectations of where and when medical problems can be addressed

This helps avoid a sense of intrusion into the physician's private time. Are discussions limited to office hours or are social occasions all right? Although physicians have many other roles in life, the physician role is not easily dropped. Most physicians have been approached at a social event for free, convenient medical advice. While such an episode can produce momentary discomfort, often a simple answer or humor will suffice. Legally, it is unwise to offer "off-the-cuff" medical advice which may be considered the beginning of a doctor-patient relationship. As concerned physicians, we sometimes have difficulty maintaining boundaries of time and place.

> Randy Adams is a 24–year-old man known by me (DS) on several levels. I saw him as a patient before I realized we would be in a study group together. We had also spent a day together hiking in the snow. He is an extremely anxious person, requiring reassurance that he is in perfect physical condition (which he is).
>
> His call about a neck lump came at 4:30 p.m, and I told Randy that I 'd see him the next morning. That evening, I received an anxious call asking if there wasn't some way for me to see him right away; otherwise, he said he wouldn't get any sleep. I didn't feel like making a house call, so I suggested that he come over to my house—a "reverse" house call.
>
> His mass was not a node, which I had presumed on the phone, but a small anterior nodule. I discussed the possibilities with him allowing that although it might be a tumor mass, it was probably a cyst or some other benign thyroid abnormality. We went through a medical text at the kitchen table.
>
> Mr. Adams went home less anxious than when he came, but very concerned that he see the thyroid specialist as soon as possible. He left some vegetables as a gift. The thyroid specialist saw him the next day: the diagnosis was "not cancer."

In this instance I deliberately decided to allow this patient to come to my house because of his overwhelming anxiety. Such a gesture can set a precedent and become an expectation in the future unless the prior agreement is respected.

Be sure that both parties recognize that bad outcomes could disrupt the friendship

The nature of medicine (and life) being what it is, bad things sometimes happen to good people. Most professionals (and most people for that matter) do not like to face failure or loss of control, but at times in the course of illness this is unavoidable. When the patient is also a friend, the sense that we have "let them down" can cause distress. The patient will also need to be careful to avoid the natural tendency to place blame somewhere, particularly on the physician who may already have a heightened sense of guilt.

Breaking bad news to, then caring for, a terminally ill friend can be especially difficult. The physician may struggle with ambivalent feelings in decisions regarding palliative care. In addition, the physician may also have to deal with pressure from the patient's family. Long after the patient's demise, he may be avoided because his presence is a painful reminder. On the other hand, a physician who is a close friend can also be especially comforting.

Discuss whether or not the friendship would continue if the professional relationship were terminated or vice-versa

A medical editor wrote about how the retirement of his family internist came as a relief.[17] He had grown progressively unhappy with his physician's lack of responsiveness to questions and concerns, but since he considered him a friend he did not want to hurt his feelings by leaving him. The editor was willing to tolerate suboptimal medical care for the sake of friendship.

Both the friendship and medical care suffer under such circumstances. Often the friendship and professional relationship may be more closely linked than realized, and both will gradually taper off together.

> Grace Nachreiner was in her early 50s when she established care with me (RW). After her second visit she invited my family to spend a weekend with them at the coast. My husband had misgivings, but we went and had a lovely time. Since we were a generation apart, I wasn't sure why she continued to invite us for dinner. One time, a good portion of the conversation centered around her diarrhea. I reassured her that it was her irritable bowel and tried to move to more pleasant topics. After all, I was there as a friend.
>
> Soon after that, I took a hiatus from seeing private patients, and our friendship also dissolved. I'm not sure how "deep" our friendship was since it ended so abruptly.

Establish a clear agreement about the meaning and extent of gifts

Gifts, in the context of friendship, are often fraught with misunderstanding and, within the physician-patient relationship, may even be cause for ethical concerns. I (RW) recall early on in my practice naively accepting $100 from a male patient who had just lost his wife to cancer. I really tried to return the money, but he insisted that I keep it. He said he was so grateful that I had helped him to quit smoking. Unfortunately, more gifts followed, and he finally admitted that he was attracted to me even though he was decades older.

In general, occasional, small, nonmonetary gifts serve as genuine and heartfelt expressions of gratitude for excellent professional care above the call of duty.

> Cora Johnson had suffered anoxic encephalopathy from recurrent seizures. Her husband had been taking care of her at home for about a year when I (DS) began making house calls. As Mrs. Johnson's condition slowly deteriorated, she required constant attention. I had to make house calls weekly to manage her fluid status.
>
> Mr. Johnson required much support, and I came to know his voice immediately on the phone. As we visited in his living room, talking about the things he had seen and his views on issues, we became friends. He asked me about my work, and we talked about the meaning of work and duty.
>
> Eventually, Mrs. Johnson died peacefully at home. I pronounced her and attended her funeral. Three weeks later when I started my car in the garage, I heard a strange beep—a car phone had been installed! Next to it was a note in Mr. Johnson's familiar script. "Doctor," it read, "thank you so much for your care for Cora. I couldn't have done it without you. You will need one of these if you keep making house calls."

Perhaps I should feel guilty about the lavish gift, but I don't. I gave her lavish care, and I use the phone happily.

The Privilege of Close Physician-Patient Relationships

Friendships with patients pose challenges to personal and professional boundaries. While physicians need to be conscious of the obstacles and pitfalls, they should also take care not to be excessively rigid in defining their limits. Going beyond the usual patterns of interaction to be fully present to a patient can lead to experiences that are mutually rewarding.

> One morning at the office, I (RW) received a phone call from a patient, Mildred Van Gelder, a 75–year-old woman, who in spite of many medical problems, remains active as a music teacher. That morning she sounded quite distraught and said she really needed to talk to me. Since it did not concern medical matters, she wanted to take me to lunch. Although this was not something I usually did, I sensed her great distress and agreed to meet her.
>
> During lunch she tearfully told me that she had just learned of her daughter's decision to leave her husband and two children for a female lover. Mildred shook her head with a mixture of disbelief and horror. "I can't believe that my dear Jane is a . . . I can't even bring myself to say the word! Her dad and I are heartbroken and we can't tell a soul. You are the one friend that I knew I could confide in!"
>
> I sat there and remembered the physician visits over the years in which I had adjusted her medications for her angina, hypertension, or arthritis. Somehow during those visits,

I had shown enough warmth and concern to earn her trust, and to have her regard me as her friend.

Keeping patients at arm's length and holding our emotions in check may help avoid painful situations, but it also deprives us of experiencing the real joy of medicine and of growing as persons. Certainly, there are situations and times when detachment is appropriate, when too much emotional involvement is harmful for our patients or leads to burnout for us. Finding a safe middle ground with boundaries that protect the patient and the professional is important. We have touched on just a few issues and hope that readers will examine their own relationships with patients to find their own comfort zone in the face of ever-changing economic and societal forces.

References

1. Cartwright A, Henderson R. *General Practice Revisited.* London:Tavestock, 1981, 45.

2. Richards JG, McPherson MJ. General practice—the state of the art: part 1. New Zealand Med 1981; 93:78–81.

3. Linzer, M. Blueberry muffins and mystery novels. Ann Intern Med 1994; 121:56–57.

4. Suchman AL, Matthews, DA. What makes the patient-doctor relationship therapeutic? Exploring the connexional dimension of medical care. Ann Intern Med 1988;108:125–130.

5. Matthews DA, Suchman AL, Branch WT. Making "connexions": enhancing the therapeutic potential of patient-clinician relationships. Ann Intern Med 1993;118:973–977.

6. Illingworth PML. The friendship model of physician/patient relationship and patient autonomy. Bioethics 1988;2:22–36.

7. James DN. The friendship model: a reply to Illingworth. Bioethics 1989;3:143–46.

8. Gabbard GO, Nadelson C. Professional boundaries in the physician-patient relationship. JAMA 1995; 273:1445–49.

9. La Puma J , Stocking CB, La Voie D, Darling CA. When physicians treat members of their own families. N Engl J Med 1991; 325:1290–4.

10. Frankel BL. Affective neutrality. JAMA 1986; 256: 515.

11. Zinn, WF. Doctors have feelings too. JAMA 1988;259:3296–3298

12. Smith RC, Zimny GH. Physicians' emotional reactions to patients. Psychosom 1988; 29:392–397.

13. American College of Physicians Ethics Manual. Ann Intern Med 1992; 117:947–960.

14. Gartrell NK, Milliken N, Goodson WH, Thiemann S, Lo B. Physician-patient sexual contact: prevalence and problems. West J Med 1992; 157:139–143.

15. Council on Ethical and Judicial Affairs, American Medical Association. Sexual misconduct in the practice of medicine. JAMA 1991;266:2741–2745.

16. Rourke JTB, Smith LFP, Brown JB. Patients, friends, and relationship boundaries. Can Fam Physician 1993;39:2557–2565.

17. Rubin W. A patient's musings on switching internists. The Internist 1993 (September): 29–31.

X Building Teaching Partnerships to Address Bias in Healthcare

Alexis D. Abernethy, Elizabeth H. Naumburg, and David M. Siegel

Introduction

The role of culture in the provider-patient relationship is receiving increasing attention.[1,2] Another emerging focus is the examination of attributes of the provider and the patient that may influence this partnership.[3,4] This perspective acknowledges that the patient and the provider hold culturally-bound world views that color their interactions.[5,6] In the process of developing a one-session workshop intended to increase providers' ability to examine their own biases in their work with patients, we found that we were engaged in a parallel process in which we examined the influence of our own world views and cultural backgrounds as we approached this task.

In response to the interest of the pediatrician member of the team in providing cultural competency training for pediatric residents, a partnership was formed between Dr. Siegel, a pediatrician and internist, Dr. Naumburg, a family physician, and Dr. Abernethy, a clinical psychologist. Dr. Siegel had previously offered cultural sensitivity training for pediatric house officers for several years with mixed results, and was interested in improving both the process and outcome by working with a multidisciplinary, multiracial team. Dr. Naumburg had experience leading workshops on racial disparities in health care and influence of health beliefs for family medicine residents and teachers. Dr. Naumburg was interested in both collaboration and scholarship. Dr. Abernethy had offered experiential and didactic sessions to psychology trainees and medical students and was interested in research. Collaboration with physicians provided Dr. Abernethy an opportunity to adapt her approach to cultural competency training to the culture of medicine.

In the process of this evolving partnership with common and differing perspectives, a 1½ hour workshop, "Rethinking Partnerships: Bias in Health Care," was developed. This workshop provided an opportunity for participants to explore their own biases, to hear personal stories related to bias, and to obtain examples of effective intercultural communication. The structure of the session was: 1) an overview of terms; 2) Black-White disparities in health care delivery; 3) a self-awareness exercise; 4) group discussion; 5) live role play; and 6) debriefing of the role play, discussion of bias, and alternative communication techniques.

Philosophy

During initial team meetings, despite differing professional and personal backgrounds, a common commitment to the importance of cultural competency training was established. Cultural competence was defined as the possession of key attitudes, behaviors, and policies that enable mutually effective interactions between people from varying cultural backgrounds.[7] Carter and Qureshi's[8] five philosophical assumptions that underlie approaches to multicultural training were considered: 1) universal (race should be transcended); 2) ubiquitous (increase awareness of others' cultures and expose stereotypes); 3) traditional (experience new culture through exposure); 4) race-based (learn about racism and racial identity development); and 5) pan-national (learn about psychology of oppression and domination). Dr. Naumburg's perspective emphasized the role that race and socioeconomic status play in health status, as well as a commitment to working at the individual and institutional level. Dr. Siegel was attuned to the oppressive power of stereotypes on health care delivery, particularly the lack of insight on the part of physician trainees regarding the influence of lifelong held stereotypes. Dr. Abernethy emphasized the countertransferential reactions of providers and was committed to the value of experiential exploration of one's personal background as a prerequisite to cultural competence. After discussing the advantages and disadvantages of each perspective, the team agreed to examine race and prejudice using Carter and Qureshi's typology.

Background: Developing The Partnership

In exploring potential biases of health care professionals, the first and most important step was to address the issue in a way that is growth-enhancing rather than threatening. Discussions of race and bias in diverse groups can be powerful and affectively charged.[6] Consequently it was important to prepare workshop facilitators to manage their affect in addition to handling emotional responses of the participants.[9]

During the partnership development of this team, it became clear that members needed to explore and discuss their own cultural and professional biases. Although the authors were clear regarding their focus on racial prejudice, they were not in agreement regarding how direct this focus should be. Dr. Abernethy's background as a psychologist and an African-American woman influenced her preference of dealing directly and frankly with the feelings associated with race and racism. However, her upbringing in predominantly white academic settings equipped her with skills to address these issues in non-threatening, effective ways. Dr. Abernethy's experience sensitized her to the value of discussing race in an environment that was considered safe and non-judgmental. Dr. Naumburg's background as a Jewish female physician with personal work on the issue of bias enhanced her willingness to deal directly with the power of race, sexism, and socioeconomic status in interactions. Her experience combining information from relevant medical literature and gentle personal exploration had persuaded her that direct discussion of race was workable and productive. Dr. Siegel's experience as a white Jewish male physician with a safe and comfortable upbringing left him with strong feelings regarding the importance of the position of privilege held by white males in our society. He was aware of his own biases and the ways inequities in health care delivery are perpetuated in medicine. However, based on his experience with pediatric residents of African and European descent, his perspective was that an approach perceived as too direct or even accusatory could interfere with learning. On the other hand, indirectness might perpetuate denial regarding the power of race and prejudice. After reviewing these differing experiences and perspectives, the team agreed to use the more neutral term *bias* rather than *prejudice* as a less provocative means of introducing the focus of the workshop.

As the partnership developed, members became more comfortable sharing honest reactions. This increasing sense of trust and willingness to take risks with one another enhanced our understanding of this content on an affective level. Members of this new partnership agreed to use some of their reflections as a part of the workshop presentation. The process involved in these discussions had sensitized the team and enhanced their ability to prepare workshop participants to engage in similar discussions in fruitful and constructively challenging ways. In an effort to present the *bias* in a manner that would motivate participants toward self-examination, the partners decided to introduce this topic from a neutral cognitive perspective.

Method

Bias associated with racial prejudice was the focus of our training module. A training approach proposed by Sue, Arredondo, and McDavis[13] outlines three major dimensions central to the development of cultural competence: 1) self-understanding of assumptions, values, and biases; 2) understanding the world view of others; and 3) acquiring the necessary skills. Based on this approach, our training module for providers was designed to increase awareness of bias in the delivery of health care, to demonstrate the dynamics of difference and power, and to discuss strategies for overcoming bias. The components of the module address each of these educational goals.

The introduction of the influence of bias on the provider-patient relationship was based on the premise that bias often interferes with the development of partnership between provider and patient. Bias may affect the provider's perception of the patient, the presenting problem, the content of care, and the quality of the relationship. Bias may also interfere with patients' willingness to trust and to share sensitive information. The concept of prejudice was also introduced with the explanation that prejudices can be held toward different groups based on any characteristic. Negative ethnic prejudice was defined as "an antipathy based upon a faulty and inflexible generalization. This prejudice may be directed toward a group as a whole, or toward an individual because (s)he is a member of that group."[10] Jones[11] defines racism as the "transformation of race prejudice and/or ethnocentrism through the exercise of power against a racial group defined as inferior, by individuals and institutions with the intentional and unintentional support of the entire culture" (p. 28). Therefore, the combination of power and race prejudice result in racist behavior. However, less conscious and unthinking pathways to racism are more prevalent in the patient-provider relationship. Providers' reactions to patients may include their imprinting of racial stereotypes. A stereotype is defined as a cognitively-based response to another person that describes a person's beliefs about an individual based on group membership.[12] These unconscious stereotypes, combined with the inherent power differential of the provider-patient relationship, may lead to racist behavior. The unconscious nature of this process frequently makes it less amenable to change.

As new partners, our approach would be to present information from the medical literature that revealed evidence of racial bias in health care delivery. The dynamics of difference and power would be explored in dyads, then this experience would be processed in the larger group. Finally, a role play would be used as a trigger for a discussion that identified specific behaviors that might unwittingly communicate bias.

Preparation of the Role Play

As partners, we discussed potential scenarios for the role play which led to further examination of each individual's assumptions, values and biases. This process further enhanced the cultural competence of our partnership as understanding of self and each another deepened. Dr. Abernethy's frustration with frequent stereotyping inclined her to want to depict a physician who was extreme in his racism, but she tended to ignore how the patient's biases might influence the interaction. Dr. Siegel offered practical examples of provider insensitivity where racial prejudice colored interactions. Dr. Naumburg played a moderating role as she encouraged our portrayal to be real and believable, but not too potent or dramatic.

After discussing our varying perspectives and teaching points, this new culturally sensitive team enacted the role play. It was decided to use roles that were most familiar. Dr. Siegel played the White male pediatrician, Dr. Abernethy played the female African-American adolescent, and Dr. Naumburg was the female moderator. As these role assignments were decided, our new team of partners became aware that each played stereotypic roles, and that from participants' viewpoint, this might endorse assumptions that might not be supported by these newly evolving partners. We decided that our ability to address bias, prejudice and stereotyping with one another and in the workshop would model an equal partnership and consequently serve as a valuable teaching opportunity.

The Role Play

A seventeen year old African-American female attended her first visit with a 40-year-old white pediatrician at a community clinic. Below are excerpts from the role play that include a description of our teaching points and commentary regarding our collaboration.

Doctor: Hello, my name is Dr. Siegel. How are you?
Patient: Okay.
D: So what brings you here today?
P: The bus.
D: I'm sorry, what are you here for?
P: I came to see, you know, you.
D: Any specific thing that you want to be sure we cover today.
P: No, you know I just used to go to my baby doctor and so I don't want to go to the baby doctor any more so I came here.
D: You're not a baby anymore.
P: Right.

Clarifying the purpose. The pediatrician used colloquial medical language to determine the presenting problem, but he did not expect his communication to be taken literally. An assumption was made that the patient would understand what was meant by this open-ended question. The rephrasing of the question in terms of anything that she wanted to cover was a clearer, less ambiguous statement. Although this question yielded a more productive answer from the patient, the provider did not take advantage of the opportunity to build on the patient's communication, for instance, to inquire regarding the reasons for her visits to her baby doctor. Instead (in a condescending tone) he stated the obvious regarding her not being a baby anymore and in so doing lost credibility.

Partnership commentary. In the process of beginning this role play, we stumbled on an unexpected example of miscommunication. When Dr. Siegel asked the patient about what brought her here, he expected Dr. Abernethy to give a vague response. He did not expect to be misunderstood. This patient's "concrete" response highlights the ambiguity of his question and his assumption that he would be understood. Dr. Siegel was genuinely surprised by the patient's response to his initial question.

D: Okay, I understand. How did you hear about coming here.
P: Well my friends come here.

D: And they like it here?

P: Um-hum.

D: Well that's good. What do they come here for?

P: Lots of different things.

D: Nothing specific comes to mind?

P: When they're sick, whatever, check-ups.

D: Right. So for you there's not any special reason you're here today.

P: No.

D: Just because of my experience in seeing other people who come in, I feel like I should ask you whether you might be interested in contraception. Is that a reason you're here, you know, birth control?

P: No.

D: You're 17 and you don't think birth control is something that we need to talk about then, right?

P: No.

The search for a connection. In recognition of this awkward beginning, the pediatrician tried another approach. This flexibility was helpful, however, the opportunity to establish rapport was lost as his bias became evident. He assumed that this patient's avoidance was due to her ambivalent feelings regarding her desire for contraception. Gathering information regarding sexual activity would be a standard component of any history. However, the assumption of her interest in contraception prior to basic inquiry regarding sexual activity and intimacy revealed this pediatrician's bias. Furthermore, the insistence by the physician after the patient's expressed lack of interest revealed the strength of this bias. For many physician trainees, the readiness to "get quickly to the point" and to ask about contraception is perceived as a strength and a sign of progress in developing skills with adolescents. They may see their reluctance to deal with sexuality earlier in their training as the only important obstacle to overcome. Therefore, a pediatric resident might conclude that the important content of an interaction with a 17 year-old African-American woman is for the provider to push hard on the subject of sexuality and contraception. The importance of exploring where that individual is in her own sexual development is often lost in the face of epidemiologic data concerning adolescents, sexuality, and pregnancy.

An alternative approach would be for this pediatrician to share some of the types of problems that pediatricians commonly handle. This list could include headaches, feeling down or depressed, infections, sexually transmitted diseases, contraception, and physicals. Then, the patient could be asked whether any of these issues were of concern. The provider might also need to step backward in the process and try again to establish rapport by gathering a general social history.

Partnership commentary. As Dr. Siegel made assumptions about this patient's reason for seeing him, Dr. Abernethy found herself feeling angry at him despite the realization that this was a role play. Dr. Abernethy felt disconnected from Dr. Siegel and did not want to reveal anything about herself. She knew what he wanted to know, but she did not want to make it easy. She was amazed that despite understanding the context of the role play, it generated real and powerful feelings. Dr. Abernethy shared these feelings and Dr. Siegel was affected as well. Despite many years of providing care in an adolescent clinic setting, he found Dr. Abernethy's character difficult to connect with and was challenged to salvage the relationship. We were all struck by the compelling experience of participating wholeheartedly in a role play and working to be "in character."

D: Well, actually let me find out about your home. Who lives at your house? Your mom's at home, do you have any brothers and sisters?

P: Yes, I have a brother and a sister.

D: How old's your brother?

P: He's 13.

D: And he manages to stay out of trouble?

P: Um-hum.

D: Is he passing in school?

P: Yeah.

D: That's great! What an unusual family. What about your sister?

P: She's fine.

D: Is she doing okay at school?

P: Um-hum.

D: So what's it like not having your father around.

P: Well you know I talk to him, he's not there at home but I talk to him.

D: Sort of check in with him once a year or so?

P: No, I talk to him definitely once a week if not more.

D: Really. Well that's great. So he's really involved in your life.

P: Yeah.

Relationship building. In an effort to move away from an unsuccessful attempt at defining the chief complaint, the pediatrician pursued less threatening topics and attempted to get to know the patient. His explicit statement regarding his intention to establish a co-equal relationship ("get to know each other") was hollow in its sincerity and failed to enhance trust. His assumptions regarding her lack of academic achievement, family support, misconduct by her brother, and father's absence reflected stereotypes and generalizations. His condescending and patronizing statements further weakened this tenuous relationship and engendered anger in the patient. This interaction provides the opportunity to discuss with workshop participants again the appropriate use and incorporation of population-derived epidemiologic information into the provider-patient partnership. Sensitivity to the patient and self-awareness of personal biases allows this data to inform but not to direct an interaction with a patient.

Partnership commentary. Just as the patient began to feel that perhaps Dr. Siegel cared about her, he made assumptions about her family. She felt totally misunderstood and stereotyped. Dr. Siegel's intent in portraying this pediatrician was to present a provider who possessed an awareness of the importance of family in working with adolescents. However, he repeatedly undermined his own attempts to explore this and form a connection with the patient. His assumptions about her teen brother and father demonstrated an inability to integrate constructively sociocultural information into an individual interaction. The young woman is left feeling that she and her family have no personal identity, but are categorized based on their presumed sociocultural membership.

D: Good. That's really good. So what do you think about doing after you graduate from high school and get a job? What kind of job are you thinking about?

P: Well I'm thinking about being a nurse.

D: A nurse! Well I think that's a good career plan. I think being committed to helping people like that as a young woman and becoming a nurse is commendable. I think that's something you would really like. Are you thinking about getting an Associates Degree at one of the community colleges? Being a Practical Nurse?

P: Well I don't know, I know there's nursing school and then sometimes too I think about being a doctor maybe, so I'm not sure which one.

D: That's pretty hard.

A lost opportunity for role modeling. The pediatrician's biases not only influenced his current assessment, but his future expectations for this patient. He assumed that she would not go to college. The opportunity to inspire her in her career goals and to build on the more positive dimensions of the relationship was lost as inadequacy was assumed. Both race and gender bias were operating here.

Partnership commentary. This was an interesting exchange as we had different reactions. The patient was quite relieved that there was finally a positive response from the physician and overlooked the negative stereotyping. Dr. Abernethy was impressed by the subtle cumulative effects of bias that reduced her expectations for positive feedback. Dr. Siegel, again, found his character caught in the bind of wanting to do the right thing (support a positive career direction), but was compromised by his generalizations and stereotyping (i.e., this African-American adolescent could never undertake the challenge of pursuing medical school).

Successes

This workshop offered several dimensions that contributed to its success. A team teaching approach that included culturally diverse (ethnically, professionally, and mixed gender) members allowed the preparation and delivery to be informed by a broader perspective. Presenting specific medical literature regarding racial bias in health care at the outset added credibility to the presentation and communicated with providers in their own language. Our partnership paralleled the process of self-exploration and learning about others' experience that we requested of workshop participants. This process increased our sensitivity to the challenges that participants might experience as they addressed these issues.

The small group exercise decreased resistance and increased openness so that participants were able to observe the role play critically and to join productively in the discussion. The use of a specific case, acted out in role play, that was discussed objectively, increased receptivity to concepts and allowed for a less threatening discussion of biased communication.

After observing the role play, participants in the workshop were able to analyze a common clinical situation without feeling vulnerable or exposed. It was important for demonstration purposes to select a case that was realistic by portraying the provider as having some strengths as well as some problems; his behavior was neither too subtle nor too blatant. Interactions used in workshops such as this one must be believable and contain unambiguous manifestations of bias in order for participants to take it seriously.

Obstacles

The greatest obstacle to exploring bias in the provider-patient relationship is guilt. This may be due to specific educational or personal experiences in which participants felt unfairly labeled and/or blamed, or the general collective guilt that may be associated with white privilege in our society. In either situation, attention must be paid to making the group experience a safe experience, and powerful emotions must be carefully managed. As new partners in this process, we purposely used structured exercises and designed them to minimize the likelihood of negative experiences. We observed group and individual interactions to assure that personal boundaries and defenses were not violated as the workshop moved into the active discussion phase. One participant out of fifteen raised persistent questions which impeded our presentation. However, we were most successful in directing her when we responded to the *process* rather than the *content* of her questions.

When proposing this type of workshop for providers, initial support may be obtained; however, resistance may arise as the workshop is anticipated. Concern is often expressed that participants will feel uncomfortable or threatened. In response to these objections it can be helpful to state clearly that if the topic did not raise some anxiety and uneasiness, it would not need to be addressed. In addition, the pervasiveness of bias in our culture and its effect on health care delivery warrants special attention.

Appropriate goals and expectations must be made based on assessment of group needs and readiness, but without some strong expression of feeling and insight during the workshop, it becomes questionable as to what has been accomplished and to what extent participants have begun to personalize the content. A single session is, of course, inadequate to address this complex issue, but an initial experience should motivate members to continue the process of self-exploration.

Conclusion

The focus of this workshop was racial bias, but the principles outlined here can be applied to other types of bias in the health care partnership. Achieving cultural competence has become a stated goal in many settings, including business, industry, and human services organizations. The importance of providing cultural competency training for providers is clear. However, determining the most effective approach for particular groups is an area for continued development. The endeavors of this new partnership of providers and the discussion of the development of this workshop suggests that a combination of factual background material, controlled personal disclosure, and critical analysis of typical clinical encounters can serve as an effective approach. Successful teaching of cultural competence requires that partnerships risk offended feelings and expose ignorance for the purpose of increased self-understanding and sensitivity to others. We have been challenged, transformed, and stimulated by our journey together.

References

1. Burkett GL. Culture, illness and the biopsychosocial model. Fam Med 1991;23:287–291.

2. Kleinman A. *Rethinking Psychiatry: From Cultural Category to Personal Experience*. New York: Free Press, 1988.

3. Kavanagh KH, Kennedy PH. *Promoting Cultural Diversity: Strategies for Health Care Professionals*. Newbury Park, CA: Sage Publications, 1992.

4. Patcher LM. Culture and clinical care. J Amer Med Assoc 1994;271:690–694.

5. Like RC, Steiner RP, Rubel AJ. Recommended core curriculum guidelines on culturally sensitive and competent health care. Fam Med 1996;28:291–297.

6. Pinderhughes E. *Understanding Race, Ethnicity and Power: The Key to Efficacy in Clinical Practice*. New York: Free Press, 1989.

7. Cross TL, Bazron BJ, Dennis KW, Isaacs MR. *Towards a Culturally Competent System of Care*, vol. 1. Washington, DC: Child and Adolescent Service System Program Technical Assistance Center, 1989.

8. Carter RT, Qureshi A. A typology of philosophical assumptions in multicultural counseling and training. In JG Ponterotto, JM Casas, LA Suzuki, CM Alexander, Eds., *Handbook of Multicultural Counseling* (pp. 239–262). Thousand Oaks, CA: Sage Publications, 1995.

9. Abernethy AD. Managing racial anger: A critical skill in cultural competence. J of Multicult Couns 1995;23:96–102.

10. Allport GW. *The Nature of Prejudice* (25th anniversary ed.) Reading, MA: Addison-Wesley, 1979.

11. Jones JM. The concept of racism and its changing reality. In BJ Bowser, RG Hunt Eds., *Impact of Racism on White Americans* (pp 27–49). Beverly Hills, CA: Sage, 1981.

12. Fiske ST. Controlling other people: The impact of power on stereotyping. American Psychologist 1993;48:621–628.

13. Sue DW, Arrendondo P, McDavis RJ. Multicultural counseling competencies and standards: A call to the profession. J of Couns & Develop 1992;70:477–486.

XI Promoting Decision-Making Partnerships Using the Analytic Hierarchy Process

James G. Dolan

Partnership: A relationship between individuals . . . that is characterized by mutual cooperation and responsibility . . . .[1]

The importance of making medical decisions in the context of a therapeutic partnership between patient and physician has long been recognized as ethically and legally correct.[2] Recently, the more tangible benefits of therapeutic partnerships have been demonstrated in studies showing that active patient participation in medical encounters results in improved clinical outcomes.[3] Achieving a partnership-based model of decision-making, however, can be a difficult task. The discussion between doctor and patient must include not only the biomedical facts such as risks of side effects and expected rates of cure, but also the values that will be used to guide the decision making process.[4] After the important issues have been identified and the data reviewed, a logically consistent decision must be made and the rationale for the decision documented for future reference. Because most clinical decisions are made in busy clinical settings, the decision-making process needs to be both efficient and effective. Since current models of medical decision-making contain no formal provisions for accomplishing any of these steps, new techniques for enabling patients and physicians to achieve and maintain decision-making partnerships in clinical encounters are clearly needed.

One promising technique is a decision-making method called the Analytic Hierarchy Process (AHP). The AHP has been described as a way " . . . to model problems incorporating knowledge and judgments in such a way that the issues involved are clearly articulated, debated, and prioritized."[5] Although it is an extremely powerful tool for decision-making, the AHP is remarkably easy to learn and use. This combination of attributes suggests that the AHP could serve as a readily accessible tool that doctors and patients could use to help them achieve a decision-making partnership.

The results of previous work with the AHP support this hypothesis. When AHP was used to investigate the role of routine diagnostic tests in the management of patients hospitalized for acute upper gastrointestinal bleeding, investigators found that all 22 primary care physicians and 25 of 26 patients who participated were able to successfully complete a realistically complex analysis in a reasonable amount of time.[6] In a subsequent study of patient preferences for colon cancer screening strategies, we found similar results in a group of 20 volunteers from an ambulatory medicine clinic. Moreover, in the latter study, 95% of the participants indicated that they would be willing to use the results of the analysis to help them make a screening decision and 90% indicated that, if they could, they would use this method to help them make other medical decisions.[7]

To illustrate how the AHP could be used in a clinical context to foster the development of a decision-making partnership between doctor and patient, the author will describe how it could be used to help make a decision regarding post-menopausal hormone replacement therapy. Detailed explanations and practical applications of the AHP have been published previously.[8–11]

The decision regarding post-menopausal hormone therapy has many potential effects on a woman's health. The benefits of hormone replacement include a decreased chance of developing the complications of osteoporosis, a decreased chance of developing coronary artery disease, and an increase in life expectancy. The risks of hormone replacement include higher rates of endometrial and, possibly, breast cancer, vaginal bleeding, and medication-related side effects. A variety of other effects, both good and bad, are also possible. Adding a cyclic progestin to the hormonal regimen decreases the risk of endometrial cancer associated with estrogen replacement. Unfortunately, this change may also reduce the degree of cardiovascular protection afforded by hormonal replacement therapy. Because of the complicated mix of benefits and risks, a recent guideline published by the American College of Physicians concluded that, while all women should consider post-menopausal hormone replacement therapy, the decision needs to be individualized for each patient.[12]

Example of Patient-Provider Partnership Using AHP

Suppose Mrs. Johnson, a 52–year old white woman at average risk for osteoporosis, coronary heart disease, and breast cancer—is seeing her physician, Dr. Samuel, for a routine health maintenance visit. She mentions that her periods are beginning to be irregular and that she is experiencing some minor hot flashes. Dr. Samuel decides to use this opportunity to use the AHP to help establish a decision-making partnership with Mrs. Johnson.

The first step in the AHP is to make a diagram or model of the decision. The goal is placed at the top, the alternatives under consideration at the bottom, and the criteria that will be used to determine how well the alternatives meet the goal in the middle. Making such a model forces the decision-makers to define the considerations that will influence the decision making process. This step gives the doctor and patient a logical starting point to begin their discussion of the decision to be made and serves as a way to record the decision-making process both during deliberations and for archival purposes.

Previously, after reading the ACP guidelines, Dr. Samuel created the decision model shown in Figure 1. At the top is the goal of the decision: maintain health and well being for as long as possible. He chose this rather broad goal for two reasons: a) to emphasize the fact that although the decision is quite complex the motivation behind it is simple, viz., to promote health and prevent disability and premature death, and b) to give patients a broad initial focus that can then be narrowed as needed. Under the goal are several pertinent criteria, some of which are further broken down into subcriteria. The purpose of this subgrouping is to facilitate the subsequent analysis (described below) by making the criteria on each level of the model easily comparable and reducing the number of required steps. At the bottom of the model are the three treatment alternatives: no replacement therapy, replacement therapy with estrogen alone, and replacement therapy with estrogen and a progestin.[13]

Dr. Samuel begins the discussion of post-menopausal hormone replacement therapy by briefly discussing the menopause. He then turns specifically to the decision regarding hormone replacement therapy:

Dr. Samuel: The reduced amount of hormones your body makes affects your health in a number of ways. We know that estrogen is very important for maintaining the calcium in your bones. After menopause women's bones lose calcium and become weaker. This process is called osteoporosis. If the bones become too weak, they break. Osteoporosis is a common cause of broken hips and fractures of the vertebral bodies in your backbone. Estrogen is also important for protecting your blood vessels from atherosclerosis, or hardening of the arteries, the disease responsible for many heart attacks and other heart diseases in older women. Taking estrogen supplements after menopause can substantially reduce your risk of developing osteoporosis and heart disease. It will also prevent the hot flashes associated with menopause.

Mrs. Johnson: That sounds wonderful, but I thought there were significant risks associated with hormone replacement therapy.

Dr. Samuel: There are. Women who just take estrogen have a higher chance of developing cancer of the lining of the uterus, which is called endometrial cancer. Taking both estrogen and progestin avoids this, but adding a progestin may decrease some of the beneficial effects of hormone replacement therapy on heart disease. Unfortuately, no one is really sure if it does or not. Hormone replacement therapy may also increase your risk of breast cancer, will result in regular bleeding—like a menstrual period—at least for a while, and in some women causes side effects such as headache, breast tenderness, and bloating.

Mrs. Johnson: What do you think I should do?

Dr. Samuel: That's a difficult question because the best choice depends on the tradeoffs between the advantages and disadvantages of the three treatment alternatives. I would have to know how you would make these tradeoffs before I could make any sound recommendation. Does that make sense?

Mrs. Johnson: Yes it does. I think I would feel better if I really understood the choice that needs to be made and had a chance to be sure that the decision we make is consistent with the things I think are important.

Dr. Samuel. OK. To help us talk about this, let's use this diagram. (Here he refers to

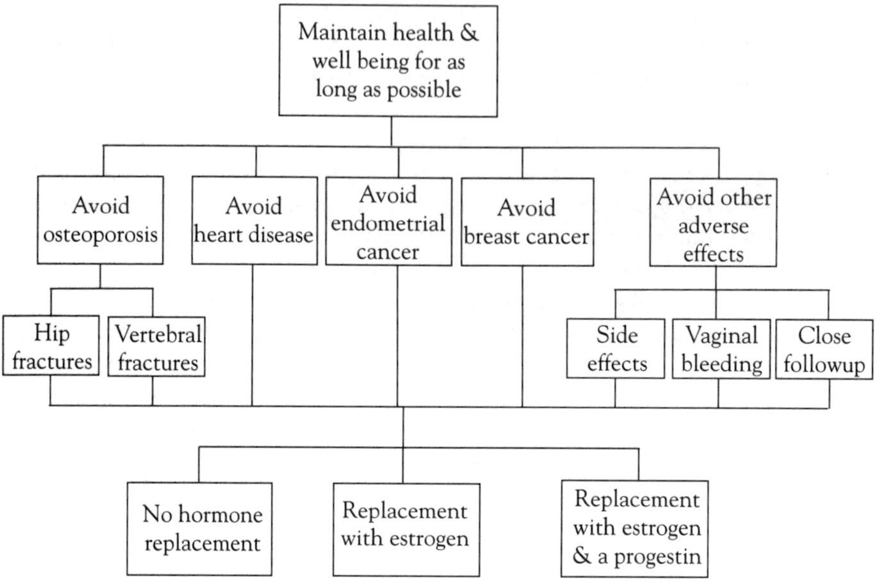

Figure 1.

the AHP model in Figure 1. At the top is our goal. Right now it says that the goal of our decision about hormone replacement is to keep you as healty and happy for as long as possible. Does that sound OK to you?

Mrs. Johnson: I think so. This is such a complicated issue, I find it really helpful to have a simple goal to keep in mind.

Dr. Samuel: OK, then we will leave it the way it is. At the bottom of the diagram are the three treatment options available: to take no replacement therapy, to take replacement therapy with estrogen alone, and to take replacement therapy with estrogen and a progestin. In the middle are a number of criteria that can be used to judge how well the alternatives meet the goal. As you can see, these are the issues we discussed earlier. On the left is decrease your risk of developing osteoporosis. Listed below osteoporosis are the two most serious consequences of osteoporosis: hip fractures and fractures of the vertebral bodies. Next is decrease your risk of heart disease followed by avoid endometrial cancer, avoid breast cancer, and avoid other side effects. Included in the last category are vaginal bleeding, the need for close monitoring for the possible development of endometrial cancer, and the physical side effects that are associated with taking the hormones such as headaches and bloating.

Do you think there are any considerations listed here that are not important to you in making this decision?

Mrs. Johnson: No, they are all important to me.

Dr. Samuel: Are there any considerations that are missing?

Mrs. Johnson: The cost of the hormones is one, but looking at these other issues I think the cost is a minor consideration. Don't hormones have other effects as well?

Dr. Samuel: Yes, I'm glad you asked. There are lots of other effects but we don't know as much about them. For example, we know that women who take hormone replacment therapy may have a higher risk of gallbladder disease and probably have a lower risk of developing urinary incontinence. Because so much less is known about these other effects, it has been recommended that, at least for now, decisions about hormone replacement be based on the considerations that I have listed. If you like we could review the other considerations later.

Mrs. Johnson: OK, that sounds fine with me.

The second step in the AHP consists of breaking the decision down into a series of smaller pieces that are simpler and easier to work with. In the AHP this usually involves making comparisons among the alternatives relative to each of the criteria and comparisons among the criteria relative to the goal. The comparisons are performed between two elements at a time and continue until all possible combinations have been evaluated. The comparisons can be made using a numerical scale running from 1 to 9, a corresponding verbal scale (1 = both elements are equal, 3 = one element is moderately preferred, 5 = strongly preferred, 7 = very strongly preferred and 9 = absolutely preferred), or one of several graphic methods.[14] Once the comparisons are made they are put into a "comparison matrix." A numerical estimate or "score" indicating the relative desirability of the elements being compared is then calculated using a standard matrix algebra procedure (calculating the left principal eigenvector); this process is conceptually similar to averaging the individual pairwise judgments. If a set of alternatives was compared relative to a criterion, the scores indicate how well each of the alternatives meet the criterion; if the comparisons were among a set of criteria relative to the goal, the scores indicate the relative importance of the criteria in achieving the goal. By convention these scores are adjusted mathematically so that they add to one.[15]

Dr. Samuel: What we will do now is to break the decision down into smaller parts that will be easier to work with. First we will see how well each alternative fulfills each of the criteria. Then we will look at the criteria to see how important you think they are in meeting the goal. The final step will be to put everything back together again to see how well each of the alternatives can be expected to meet the goal. We will use the computer to help us keep track of things as we take the decision apart and then put it back together again.

Let's start by looking at how the three alternatives will affect your risk of hip and spine fractures. As you can see on the chart (Table 1), if you don't take hormone replacement therapy, you have about a 15% chance of breaking your hip. This means that your chance of not breaking a hip is about 85%. Put another way, of seven women like you, one would experience a hip fracture, and the other six would not. If you take estrogen your chance of a hip fracture would fall to about 13%. This means that instead of needing seven women like you to find one who would have a hip fracture, you would need to find eight. Adding progestin to the estrogen is not believed to affect these figures so that the risk of hip fracture is the same if you took both estrogen and progestin.

The question that we need to answer is how much better you think the hormone replacement regimens are in terms of preventing hip fractures. Do you think they are equal to no replacement therapy, moderately better, strongly better, very strongly better, or absolutely better?

Table 1. Sample data used for the analysis. Adapted from Grady et al.[12]

Criterion	No treatment	Estrogen	Estrogen & Progestin
Hip fracture	about 1 in 7 (15%)	about 1 in 8 (13%)	about 1 in 8 (13%)
Vertebral fractures	1 in 10 (10%)	1 in 20 (5%)	1 in 20 (5%)
Coronary heart disease	about 1 in 2 (46%)	1 in 3 (34%)	1 in 3 (34%)
Endometrial cancer	1 in 33 (3%)	1 in 5 (20%)	1 in 33 (3%)
Breast cancer	1 in 10 (10%)	about 1 in 8 (13%)	between 1 in 5 and 1 in 8 (13–20%)
Irregular vaginal bleeding	low	about 1 chance in 3 (35%)	about 1 chance in 3 (35%)
Side effects: bloating, breast tenderness, headache	no chance	between 1 in 10 and 1 in 20 (5%–10%)	between 1 in 10 and 1 in 20 (5%–10%)

Mrs. Johnson: Well my chance of hip fracture is less if I take hormone replacement so I don't think they are equal. On the other hand, the difference is only two percent which isn't all that much. So I would say that hormone replacement is somewhere between equal and moderately better.

Dr. Samuel: OK, lets call that slightly better. We enter the judgment into the computer and translate it into a score that indicates how well you thought the alternatives meet this criterion. Both estrogen and estrogen plus progestin have a score of 0.4, and no replacement therapy has a score of 0.2. You can see that the scores accurately reflect the initial judgment: the two hormone replacement therapy alternatives are equally good and both somewhat better than no hormone replacement. The purpose of transforming the judgments into scores is to put into common units all of the information we will use in the course of analyzing this decision. This enables us to break the decision down into small pieces and still be able to reassemble them again to come up with an overall solution.

We will now repeat this process for each of the criteria. So let's go on and compare the alternatives according to how they will affect your chances of developing vertebral fractures. . .

Dr. Samuel and Mrs. Johnson then proceed to make comparisons in this fashion among the alternatives regarding the rest of the criteria: avoid heart disease, avoid endometrial cancer, avoid breast cancer, avoid resumption of vaginal bleeding, avoid hormonal side effects, and avoid the need for close monitoring for endometrial cancer. The results are summarized in Table 2.

Dr. Samuel: We have now determined how well the alternatives meet the criteria. Now we will use the same technique to determine how important you think the criteria are in meeting the goal of maintaining your health and well being for as long as possible. The first question is: relative to the goal of maintaining your health and well being, is it more important to decrease your risk of broken bones due to osteoporosis or to decrease your risk of heart disease?

Mrs. Johnson: That is a hard question because both are important, but I think that heart disease is slightly more important.

Dr. Samuel: OK. The next question is: relative to the goal of maintaining your health and well being, is it more important to decrease your risk of broken bones due to osteoporosis or to avoid endometrial cancer?

They then proceed to compare the rest of the criteria in a similar manner. The results are summarized in Table 2.

Dr Samuel: OK, now we are all done. We have looked at how well the alternatives meet the criteria and how important the criteria are in meeting the goal. We will now put the pieces back together to see how well the alternatives meet the goal based on everything you have told me so far.

The third step in the AHP is to reassemble the decision problem by combining the results of the subsidiary analyses. This is done using a method similar to calculating a weighted average by multiplying each alternative's score relative to a criterion by the score achieved by the criterion relative to the goal and then adding the results across all of the criteria.[16] The results indicate how well each of the alternatives meet the goal based on the series of pairwise comparisons made during the analysis.

Dr. Samuel: These results are scores for the alternatives that are similar to all of the other scores that we have been looking at. The best option is hormone replacement with estrogen and progestin with a score of 0.371, followed by estrogen alone with a score of 0.317 and no replacement therapy with a score of 0.312.

Mrs. Johnson: That was very helpful. The comparison process really helped me to think about what I really think is important, and apply it directly to the choice among these treatment options. I also appreciate having had the opportunity to explore my values and incorporate them into our decision making process. I feel that I have a much better understanding of the issues involved in this decision now, and I can understand why you were reluctant to just tell me what to do when I asked earlier—there is no way you could have known how I viewed some of these tradeoffs!

Table 2. Results of Mrs. Johnson's comparisons.

A. Avoid osteoporosis/hip fractures	No HRT	E	E+P	Score
No hormone replacement (No HRT)		1/2	1/2	0.20
Estrogen (E)			1	0.40
Estrogen + progestin (E&P)				0.40

B. Avoid osteoporosis/vertebral fractures	No HRT	E	E+P	Score
No hormone replacement (No HRT)		1/3	1/3	0.143
Estrogen (E)			1	0.429
Estrogen + progestin (E&P)				0.429

C. Avoid heart disease	No HRT	E	E+P	Score
No hormone replacement (No HRT)		1/3	1/3	0.143
Estrogen (E)			1	0.429
Estrogen + progestin (E&P)				0.429

D. Avoid endometrial cancer	No HRT	E	E+P	Score
No hormone replacement (No HRT)		5	1	0.455
Estrogen (E)			1/5	0.091
Estrogen + progestin (E&P)				0.455

E. Avoid breast cancer	No HRT	E	E+P	Score
No hormone replacement (No HRT)		2	2	0.50
Estrogen (E)			1	0.25
Estrogen + progestin (E&P)				0.25

F. Avoid other adverse effects/ Vaginal bleeding	No HRT	E	E+P	Score
No hormone replacement (No HRT)		3	2	0.55
Estrogen (E)			1	0.21
Estrogen + progestin (E&P)				0.24

G. Avoid other adverse effects/ side effects	No HRT	E	E+P	Score
No hormone replacement (No HRT)		3	3	0.60
Estrogen (E)			1	0.20
Estrogen + progestin (E&P)				0.20

H. Avoid other adverse effects/ need for close monitoring for endometrial cancer	No HRT	E	E+P	Score
No hormone replacement (No HRT)		7	1	0.467
Estrogen (E)			1/7	0.067
Estrogen + progestin (E&P)				0.467

I. Avoid osteoporosis	HIP	VERT	Score
Hip fractures (HIP)		1	0.50
Vertebral fractures (VERT)			0.50

J. Avoid other adverse effects	VB	SE	ECM	Score
Vaginal bleeding (VB)		1	4	0.444
Side effects (SE)			4	0.444
Endometrial cancer monitoring (ECM)		0.111		

K. Goal	OSTEO	HD	EC	BC	OAE	Score
Avoid osteoporosis (OSTEO)		1/2	2	1/2	4	0.198
Avoid heart disease (HD)			2	1	4	0.297
Avoid endometrial cancer (EC)				1/2	3	0.140
Avoid breast cancer (BC)					5	0.308
Avoid other adverse effects (OAE)						0.057

At this point the initial AHP analysis is complete. The decision-makers can then decide to choose the alternative rated most highly in the analysis, disregard the results and choose another alternative, or think it over for a while. In addition, most AHP software programs make it possible to explore easily how changes in the judgments made during the analysis affect the results–a process called sensitivity analysis. This type of analysis increases the complexity of the overall process but is likely to be worthwhile in situations where some of the pairwise judgments were particularly difficult to make, especially among the criteria.

Dr. Samuel: I'm glad you found this process helpful. Our time is almost up; are you ready to make a decision now, or would you prefer to think about it for a while.

Mrs. Johnson: I don't know. I think I'd like to give hormone replacement therapy a try, but I'm a little hesitant because I really had a hard time making some of the comparisons among the criteria; I'm wondering if I made the right choices.

Dr. Samuel: We can use what we have done so far to get some additional perspective on the decision that may help you resolve some of your doubts. With the help of the computer we can quickly determine how different judgments affect the results using a series of graphs. The first (Figure 2a) shows what happens as the importance assigned to preventing osteoporosis is changed from 0, meaning that it has no importance, to 1, which means that it is the only consideration affecting the decision. You can see that as you vary the importance you assign to preventing osteoporosis—and don't change your mind about the relative importance of the other four criteria—the results of the decision don't change until you get down to a priority of 0.05. At higher priorities hormone replacement with estrogen and progestin always has the highest line of all the alternatives, indicating the highest score, and therefore is the best choice. Below 0.05, no hormone replacement therapy has the highest score and is the best choice. The point where the lines cross, and the results of the analysis change, is called the decision threshold.

Based on your original comparisons, you gave avoid osteoporosis a priority of 0.2—

Table 3. Calculation of the overall scores indicating how well the alternatives meet the goal.

	Avoid osteoporosis (0.198)					Avoid other adverse effects (0.057)			Overall Score*
	hip fractures (0.099)	vertebral fractures (0.099)	Avoid heart disease (0.297)	Avoid endometrial cancer (0.140)	Avoid breast cancer (0.308)	vaginal bleeding (0.025)	side effects (0.025)	close monitorin g (0.006)	
No replacement therapy	0.2	0.143	0.143	0.455	0.5	0.55	0.6	0.467	0.312
Estrogen	0.4	0.429	0.429	0.091	0.25	0.21	0.2	0.067	0.317
Estrogen & progestin	0.4	0.429	0.429	0.455	0.25	0.24	0.2	0.467	0.371

*–The overall score for the No replacement therapy option is equal to:
(0.099)(0.2)+(0.099)(0.143)+(0.297)(0.143)+(0.14)(0.455)+(0.308)(0.5)+(0.025)(0.55)+(0.025)(0.6)+(0.006)(0.467) = 0.312.
The other scores are calculated similarly.

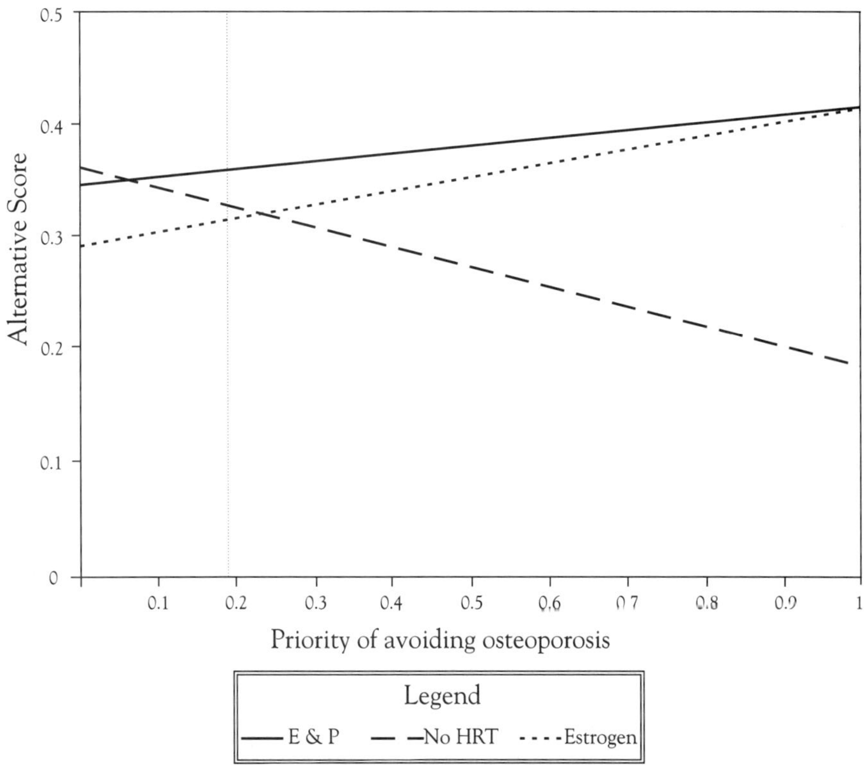

Figure 2-a.

quite a bit more than the threshold of 0.05. Do you think that you overvalued the importance of avoiding osteoporosis relative to the other criteria by a large amount?

Mrs. Johnson: Maybe a little, but not a lot.

Dr. Samuel: Since we now know that little changes won't change the results, I think we are OK so far. Let's go through the other criteria.

This next graph (Figure 2b), shows what happens when we do the same thing with the criterion avoid heart disease. Here the threshold is equal to 0.19. Below this point no hormone replacement is best; above it replacement therapy with estrogen and progestin is best. Based on your comparisons, you gave avoiding heart disease a priority score of 0.30. Do you think that you overvalued the importance of avoiding heart disease relative to the other criteria?

Mrs. Johnson: No, if anything I was wondering if I undervalued it.

Dr. Samuel: OK, then I think we are still OK. Let's go to the next criterion: avoid endometrial cancer (Figure 2c). Here, unlike the others, there is no threshold point: hormone replacement with estrogen and progestin is the best alternative all the time—so small changes here won't matter.

They then review the other two sensitivity analysis graphs (Figs. 2d and 2e). In

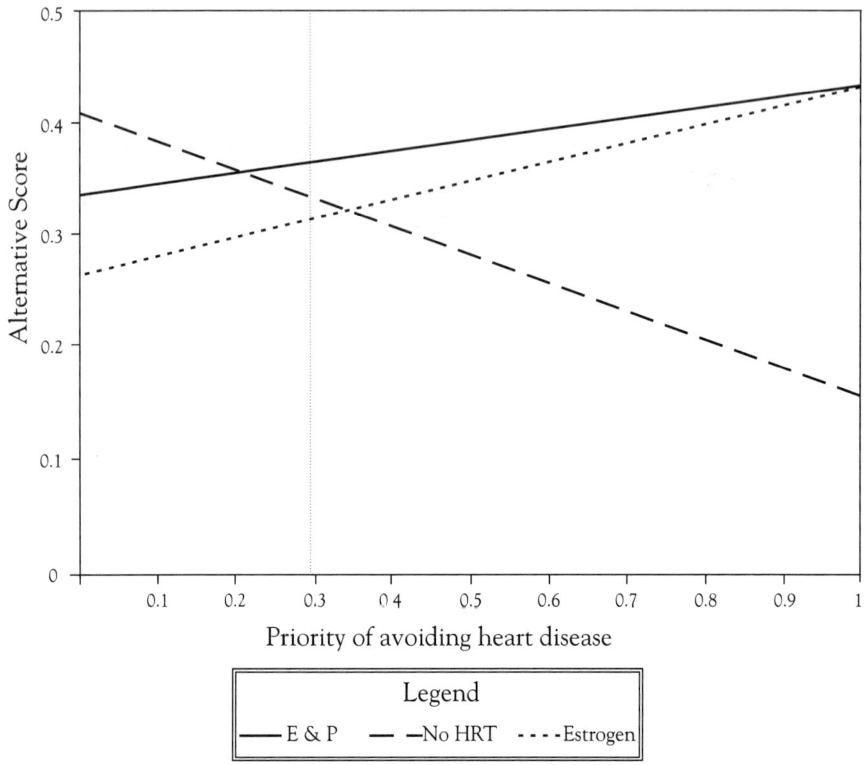

Figure 2-b.

each case Mrs. Johnson decides that her doubts about her initial judgments were not strong enough to change the results of the analysis.

> *Mrs. Johnson:* That was helpful—I feel reassured little changes in my initial judgments won't drastically change what we decided to do. But before I make a final decision I'd like to hear what you think I should do.

The Working Relationship of the Patient/Provider Partnership

An important aspect to a decision-making partnership is the working relationship between the parties involved. The doctor and patient should be free to negotiate what role each plays in the ultimate decision whenever a decision needs to be made. An AHP based analysis can also serve to facilitate this negotiation.

> *Dr. Samuel:* After looking at the judgments you made during the analysis, I am quite comfortable recommending that you follow your inclination to take estrogen and proges-

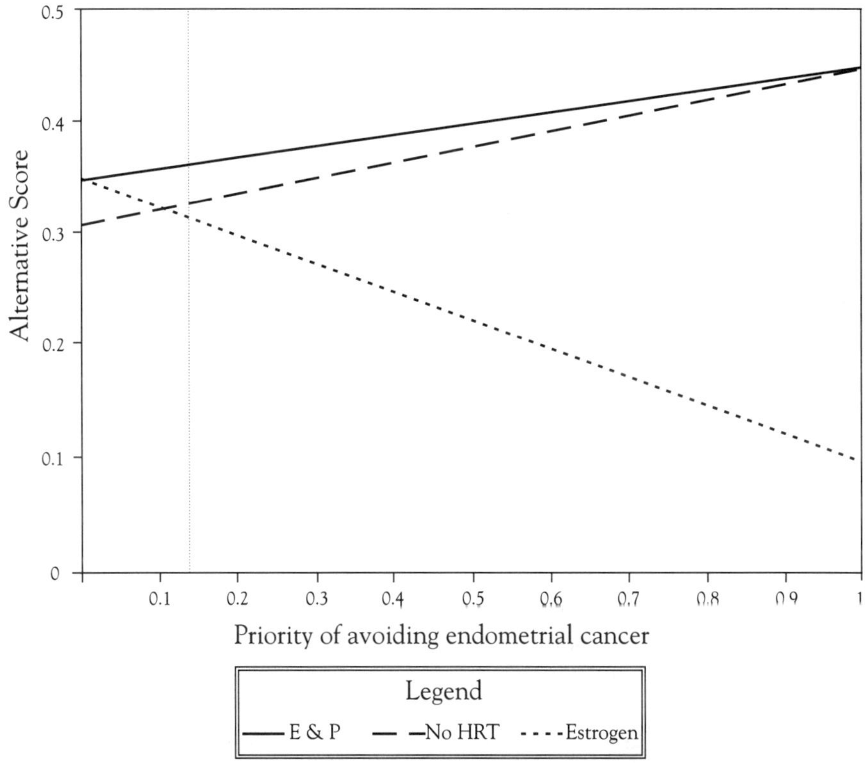

Figure 2-c.

tin. Your decision making was consistent and seems to be based on what you really want. I think you should try it—we can always reconsider the decision later.

I will give you a copy of the information we reviewed and the results of your analysis to take home with you. I will also put a copy in your medical record so that we can refer to it in the future if we like. Here is a prescription for the hormones. Wait right here, and I'll ask the nurse to come in and go over with you how to take the hormones and the side effects you should watch for.

In the introductory chapter to this book, the editors have summarized both the potential advantages of adopting a partnership-based approach to health care decision-making and the difficulties to be overcome, including the need to develop a new way of communicating that facilitates the identification of shared goals, the definition of individual values, the discussion of differences, and the adoption of a shared decision-making process. The large and ever-increasing amount of information affecting even common medical decisions is an additional problem that needs to be overcome. As illustrated above, the AHP provides a readily available tool for surmounting many of these problems. In addition to providing a framework for

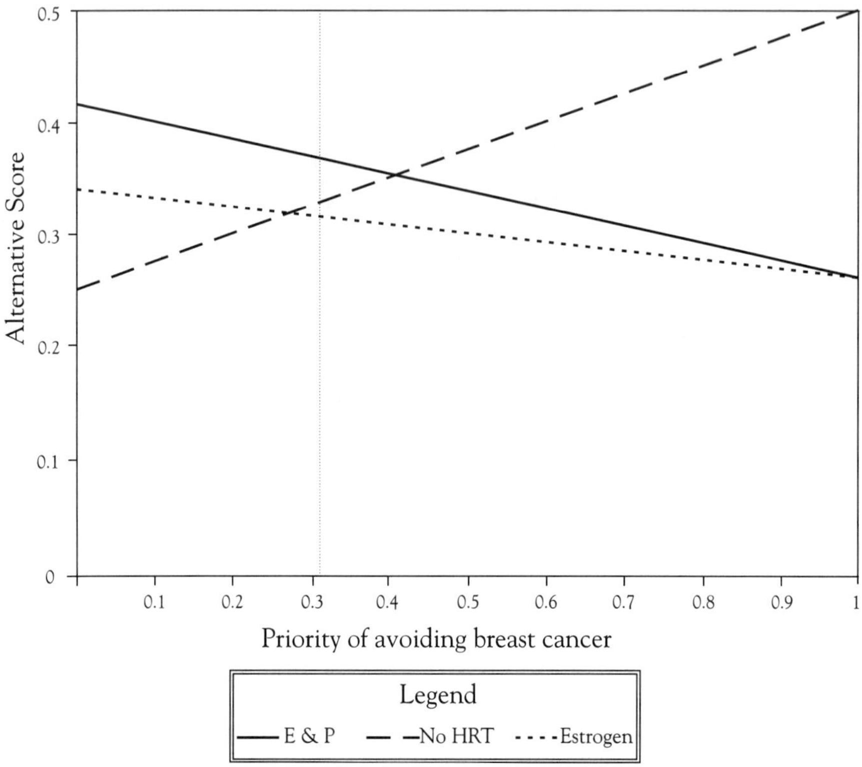

Figure 2-d.

eliciting and discussing values and summarizing large amounts of information, it also promotes effective communication by helping to identify disagreements, assess their importance, and create compromise solutions. Based on these characteristics, and the preliminary work cited earlier, the author believes that a clinical decision-making process based on the AHP has an excellent chance of proving to be an effective method for fostering and achieving health care partnerships.

References

1. *The American Heritage Dictionary of the English Language*, 3rd edition. Boston: Houghton Mifflin, 1992:1321.

2. President's Commission for the Study of Ethical Problems in Medicine and Biomedical Research. *Making Health Care Decisions: The Ethical and Legal Implicaitons of Informed Consent in the Patient-Practitioner Relationship*. U.S. Government Printing Office, 1982;Volume 1.

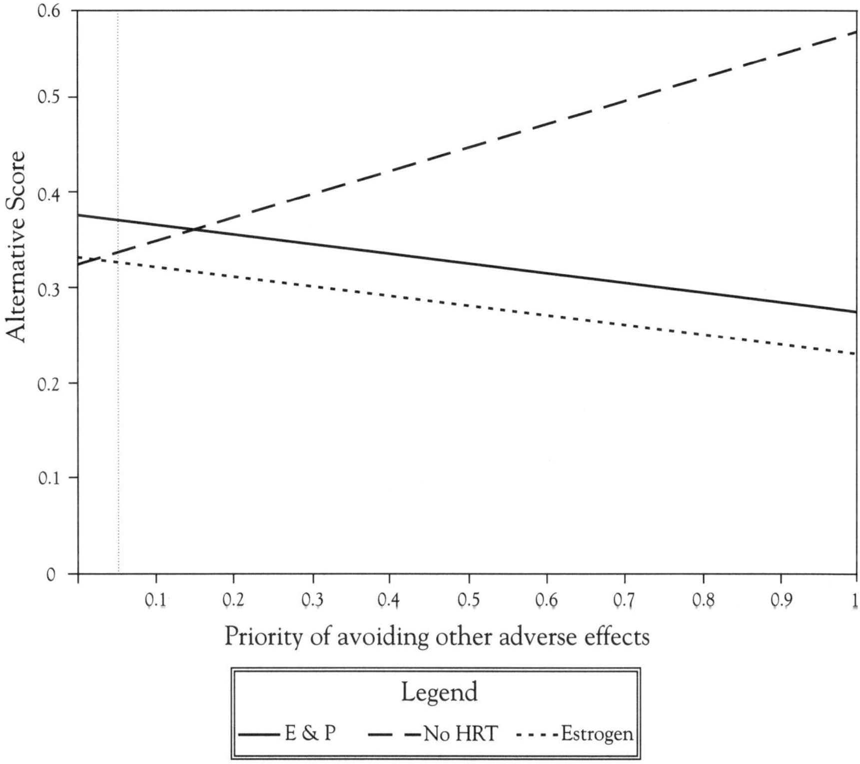

Figure 2-e.

3. Kaplan SH, Greenfield S, Ware JE. Assessing the effects on physician-patient interactions on the outcomes of chronic diseases. Medical Care 1989;27 (3,Supplement):S110–S127.

4. Forrow L, Wartman SA, Brock DW. Science, ethics, and the making of clinical decisions. JAMA 1988;259:3161–3167.

5. Saaty TL. *Multicriteria Decision Making: The Analytic Hierarchy Process*. Pittsburgh, PA: RSW Publications, 1988:xi.

6. Dolan JG, Bordley DR, Miller H. Diagnostic strategies in the management of acute upper gastrointestinal bleeding: patient and physician perspectives. J Gen Intern Med 1993;8:525–529.

7. Dolan JG. Are patients capable and willing to use the Analytic Hierarchy Process to help make clinical decisions? Med Decis Making 1995;15:76–80.

8. Dolan JG, Bordley DR. Involving patients in complex decisions about their care: an approach using the Analytic Hierarchy Process. J Gen Intern Med1993;8:204–9

9. Dolan JG, Isselhardt BJ, Cappuccio JD. The analytic hierarchy process in medical decision making: a tutorial. Med Decis Making 1989;9:40–50.

10. Harker PT, Vargas LG. The theory of ratio scale estimation: Saaty's analytic hierarchy process. Management Science 1987;33:1383–1403.

11. Saaty TL. *Fundamentals of Decision Making and Priority Theory with the Analytic Hierarchy Process–Vol VI*. Pittsburgh: RWS Publications, 1994.

12. American College of Physicians. Guidelines for counseling postmenopausal women about preventive hormone therapy. Ann Intern Med 1992;117:1038–1041.

13. Other alternatives are available—for example progrestin can be added to the regimen cyclically or used continuously. Since the main purpose of this example is to illustrate the process, I will simplify matters and use only the three basic alternatives.

14. Although there is evidence that the graphic scale is the most accurate for use in patient care decisions, it is difficult to adequately describe a graphic analysis in a paper such as this. Therefore, for this illustration, I will describe all pairwise comparisons using the verbal scale.

15. Although it is possible to do an AHP analysis by hand, a computer is a practical necessity for performing the necessary calculations, documenting the analysis, and for some of the graphic comparison techniques. Several AHP computer programs are available commercially.

16. There are several ways to reassemble or synthesize the results of an AHP analysis. All are conceptually similar in that they rate the alternatives according to how well they fulfill the criterial taking into account the relative importance of the criteria. A full discussion of this issue is beyond the scope of this paper—further information can be found in the AHP references cited in the text. In the illustration, for clarity, I have used the simplest method: a weighted average.

PART 3

Partnership in Healthcare Teams

When we reflect on the current state of partnership on health care teams, the core principles which we advocate in this book—respectful relationships, recognition of the other's perspective and shared, flexible decision making—seem more honored in their breach than in their fulfillment. Stereotypes and projections create distance between professionals and prevent dialogue. Hierarchy inhibits free flowing communication and creative problem solving. Decision rights are tightly defined and jealously protected.

As we saw in the Botelho and Walker chapters in the Perspectives on Partnership Process section, the current health care environment includes imperatives to improve partnership and factors that undermine it. On the positive side, the demand for accountability regarding clinical outcomes, quality of service, and cost requires integrated networks of professionals and institutions. In the current fragmented configuration of our health care system, no single component has the information, resources, or authority needed to monitor and respond to performance of the system as a whole. The Darwinian, competitive forces of the marketplace will reward systems which help physicians to relinquish their expectations of absolute autonomy and adopt instead a more collective team-oriented mindset. Successful systems of the future will work smarter by empowering and supporting work teams. They will harness the full potential of individual and interdisciplinary diversity by valuing differences and promoting dialogue. The sense of community engendered within learning organizations will enhance the motivation, service orientation, personal commitment, and satisfaction of the individuals within.

On the other hand, divisive elements are legion. The implementation of the Resource Based Relative Value Scale as the basis for physician compensation creates significant shifts in income and pits specialties against one another. Efforts to decrease referrals further alarm specialists. Implicit in the "gatekeeper" role of primary care providers is the expectation that sometimes the gate will be closed—access will be denied. Controversy also rages over the creation of interdisciplinary work teams which transfer certain responsibilities from one set of professionals to another set that is less highly trained and less well paid. Thus physicians resist the privileging of nurse practitioners while nurses decry the interdisciplinary teams of "Patient Focused Care" as the "deskilling" of inpatient units. The threatened loss of income or of one's job would not seem to be a conducive context for practicing partnership. Yet it is just this capacity for partnership, along with the ability to let go of the status quo and seek new solutions, that maximizes individuals' value to their organizations and minimizes their risk of loss.

The four papers in this section address various aspects of partnership within health care teams. Using both a literature review and clinical vignettes, *Ron Epstein*'s contribution documents the nature and consequences of problematic communication between primary care practitioners and specialists. Problems range from process issues (for example, the failure to engage in any communication at all or using an inappropriate medium) to content issues (for example, lack of clear goals for the consultation or failure to communicate what has been told to the patient). The cultural differences between generalists and specialists that arise from their differences in focus (the *person* versus the *disease*), tolerance for uncertainty, and spectrum of their clinical experience (early undifferentiated symptoms versus advanced pathology) may give rise to mutual misunderstanding and prejudice. Dr. Epstein argues that this gulf can be bridged only by direct personal interactions and relationships. He concludes his chapter with suggestions about how generalists and specialists can improve their communication and partnership.

Despite demonstrated beneficial effects of collaboration on patient outcomes and staff satisfaction, true collaboration between nurses and doctors continues to be rare. *Judith Baggs* reviews this literature in detail with particular emphasis on the intensive care unit setting, and documents the divergence of opinions between these two professions about their decision rights, communication styles and predisposition to engage patients as partners. She describes several innovative programs that improved collaboration by enhancing mutual understanding and respect for each others' unique contributions. When nurses and doctors fail to realize the full promise of interdisciplinary partnerships, it is their patients who suffer most.

On the other end of the spectrum, *Colleen Dwyer*, *Pat Hinton-Walker*, *Peter Coggiola*, and *Tony Suchman* describe the process by which nurse practitioners and internists collaborated successfully to establish a new free standing nurse practitioner practice at a senior center. Their model identified nurse practitioners as professional colleagues of physicians rather than "physician extenders." This was underscored by including nurse practitioners on the faculty of a primary care residency program. They describe the theoretical foundations for their collaboration and the obstacles that had to be overcome. They also present two case histories that illustrate the complementarity of the professional roles of nurse practitioners and physicians.

Mary Talen, *Margaret Clark Graham*, and *Gordon Walbroehl* address the question of how to teach future health care professionals the value of and skills needed for interdisciplinary collaboration. Paradoxically, they found that the participating departments were not themselves observing the principles of collaboration that they were hoping to teach (similar "parallel process" stories can be found in Abernethy's chapter in the Patient-Clinician Partnerships section and Blandin-Clark's paper in the Educational Partnerships section). Once that situation was rectified, they were able to create an exemplary model of a community based, multiprofessional practice team that provided effective experiential learning for students from several professions. Students learned about the nature of each others' training and credentialing requirements. They compared the unique perspectives of their disciplines as they provided team-based care to patients in the community. These experiences and formal education in group process issues such as leadership styles, power, and conflict resolution positions prepared these students to be skilled collaborators.

Readers interested in health care teams and the process of promoting interdisciplinary collaboration will also find valuable insights in the chapter by *Tom Campbell* and his colleagues in the Patient-Clinician Partnerships section of this volume. This chapter presents the Collaborative Care model which brings mental health professionals into primary care settings and makes them part of the primary care team. Many of the principles of team building articulated in that chapter apply not only to mental health professionals but to all types of interdisciplinary partnership. The same issues emerge in the chapters by *Geoffrey Swain* (which also describes collaboration between primary care and mental health professionals, but in an educational context) and *Gary Eddey* (in the Community Partnerships section) which discusses partnership building between medical and educational professionals.

XII Communication between Primary Care Physicians and Consultants

Ronald Epstein

Effective partnerships between health care practitioners are essential in providing high quality medical care. Patients want to know that their health care practitioners work together effectively, involve the patient in important decisions, are in meaningful dialogue about the nature of the problem, and communicate a consistent therapeutic approach. Health care practitioners want to know that their input into patient care is used and valued; they want to be treated with respect and to foster collaborative relationships.[1]

Optimal communication between physicians is characterized by accurate gathering of information about the patient, taking the time to communicate information expeditiously that will be useful to other health care practitioners, indicating what the patient has been told, planning who will take ongoing responsibility for the patient, and keeping the door open to further communication.[2–5] Yet, despite these ideals, communication between health care practitioners is sometimes problematic. Consider the following example:

> Mr. Norman, a 70-year-old man with coronary artery disease, had experienced worsening dyspnea on exertion and orthopnea during the previous 4 weeks. He had undergone coronary artery bypass surgery 5 years previously, after a myocardial infarction.
>
> His wife had died of an arrythmia at home 13 months previously, a year after a large anterior myocardial infarction. Coronary angiography had demonstrated inoperable severe coronary artery disease. She had been under my care and had also consulted the hospital cardiology group several times after her initial myocardial infarction.
>
> I suggested to Mr. Norman that he might need further evaluation of his heart disease and discussed the possibility of consulting a cardiologist. His initial response was that he preferred that I take care of him, because he trusted me. After further discussion, he expressed reluctance to consult the hospital cardiology group because it brought up bad memories of his wife. I arranged for him to see a cardiologist in a different office, to which he agreed. The appointment was scheduled for 3 days later. I called Dr. Ruben (the cardiologist), sent a referral letter and faxed relevant records to him.

I would like to thank Tony Suchman MD for suggesting the concept of this article, Zsolt dePapp MD, Peter Franks MD, Susan McDaniel PhD, Robert Poe MD, Charles Sherman MD, and Thomas Talley MD for their helpful comments, and Deborah Fox MA for her editorial suggestions. Originally published in the Archives of Family Medicine 1995 May; 4(5):403-9.

Dr. Ruben performed echocardiography, prescribed a regimen of digoxin, and suggested a thallium test be performed with "delayed images". Mr. Norman was contacted by the cardiology technician, and the test was explained to him, including the need to remain at the hospital for ten hours. At that point, he indicated that he did not wish to have the test. Before I received a consultation letter, the technician called me to notify me that the patient refused testing, without specifying the reason.

I called the patient, and he expressed willingness to adhere to the recommended medication regimen but adamantly refused to have the thallium stress test performed. He believed that the inconvenience was not worth the therapeutic benefit that might result. I called Dr. Ruben, who believed that the test was optional and most probably would not result in a change in management.

This story reports some challenges for the primary care physician in the referral process and some successful communication between the patient's primary care physician and a consultant. However, the patient's values and wishes were lost in the communication between health care professionals in regard to performing the thallium stress test.

The Context of Consultation

Communication between primary care physicians and consultants does not occur in a vacuum.[6] Often, there are more than three parties to consider, including a primary care physician, one or more consultants, psychotherapists, non-physician health care professionals, and patients' family members.[7] All have the main goal of improving the patient's health. Although this article focuses on relationships between primary care physicians and medical consultants, many of the same principles apply equally to other parts of this complex web of relationships.[8,9]

Although some authors use the terms *referral* and *consultation* interchangeably, I consider a *referral* to imply a transfer of care for the ongoing management of a specific problem, whereas a *consultation* involves another physician for a specific, time-limited therapeutic or diagnostic task.[10] Referral and consultation comprise a broad spectrum from "curbside consultations" to a formal transfer of care; when the system works well, the patient's needs are served, all parties are informed, and there are no redundancies or omissions in care.

A variety of intrinsic and extrinsic factors influence communications between primary care physicians and consultants. Among primary care physicians, there is considerable diversity in referral patterns.[10–14] Type of insurance (or lack thereof) may restrict patient access to some physicians and demand certain kinds of communication between physicians. Financial incentives may promote or discourage diagnostic testing or referral. In addition, psychological characteristics of physicians, such as risk aversion, tolerance of ambiguity, perceived confidence, fear of exposure of their own lack of knowledge, and fear of malpractice, may influence physicians' referral patterns, style of communication, and need for information.[15,16]

Aside from consultation and referral, primary care physicians and consultants interact in other forums. In the political arena, primary care physicians and consultants promote agendas, some of which are complementary (such as vertical integration of services ensuring a referral base), and some of which come into direct conflict (such as fee discussions). In the academic setting, with a push towards generating more primary care physicians, there has been increasing contact between community-based physicians and tertiary care-based consultants in discussions about curricular time, funding, and space allocation.

The relationships between primary care physicians, consultants, patients, and their families will evolve along with changes in the US health care system.[17] Patient self-referral to consultants will become less common to reduce medical costs. In this context, effective collaboration between consultants and primary care physicians will be necessary and must be supported by the larger health care system to provide high-quality care.

Communication Barriers

Communication between primary care physicians and consultants takes place in person, over the telephone, and in writing. Virtually all of the research that exists on communication between primary care physicians and consultants has studied the content of letters exchanged, leaving the nature of other forms of communication poorly understood.

The majority of this research has been conducted in Great Britain, Australia, and the Netherlands.[18–22] Letters from general practitioners were often found to be lacking critical information, explicit statements of the reason for consultations, or plans for follow-up.[23,24] Letters from consultants were sometimes absent, delayed, "discourteous," or overly detailed. A Dutch study indicated that there was often disagreement between primary care physicians and consultants about the necessity of referrals in problematic cases.[25]

In a large US city hospital, McPhee et al.[26] analyzed request forms from primary care physicians in the hospital-based general internal medicine group practice to consultants within the same hospital. Of 464 consultations from 27 practitioners, the reasons for consultation were provided for only 77%, and only 76% of the referring physicians indicated that they wished to learn of the consultant's findings. Appropriate clinical information about patients was provided in 98% of the cases. The referring physician contacted the consultant directly by telephone in nine percent of the cases and telephoned when the request was urgent. Consultants did not provide information on the results of their consultation in 45% of the cases when the referring physician requested them. Most often the results were communicated via a shared medical record.

In 1982, Hansen et al.[27] reported that of 141 consecutive referrals from family physicians to consultants, consultation reports were returned from 88% of commu-

Table 1. Communication Problems between Primary Care Physicians and Consultants

- Patients not following through with recommendation to see consultant
- Delays in having patient seen by consultant
- Patient self-referral with or without knowledge of primary care physician
- Multiple providers, especially in clinic settings
- Refusal to care for patient
- Consultant views referral as inappropriate
- Unclear or inappropriate reason for referral
- Lack of information on referred patients
  - no referral letter or phone call
  - missing relevant laboratory results
- Unreturned phone calls
- Repetition of diagnostic tests by consultants
- Absent or delayed response from consultant
- Cross-referral (from one consultant to another)
- Consultants "stealing" patients from primary care physicians
- Response that does not answer question
- Lack of discussion as to follow up
- Family physicians not carrying out consultant recommendations

nity-based consultants, from 75% of university faculty, and from 43% of university outpatient clinic physicians. The quality of the reports, as judged by the referring physician, varied directly with the amount of referral information provided to the consultant.

In 1975, a two-physician general practice generated 233 referrals within a 6–month period, all accompanied by referral material and a request for follow up information. The overall rate of receiving follow up information from consultants was 62%.[28] Communication from the consultants to the primary care physician was more common when the consultant was in private practice than when he or she was from an academic medical center. In an older study[29] of 100 consecutive private patients who were referred to one consultant, 47 were accompanied by no or minimal clinical information. In half of those cases, the clinical data that were available to the primary care physician would have been of significant value to the consultant.

A recent study by Williams and Peet surveyed 200 family physicians and 200 consultants in Ohio.[4] Both family physicians and consultants placed a high value on information, most preferred that both written and telephone contact be used for chronically ill patients, and there was variation in preference for written or oral communication. The family physicians often commented that they wished that consultants would (a) ask the patient to return to the referring physician, (b) indicate to the patient that communication would occur between physicians, (c) avoid cross-referral (referral from one consultant to another without the knowledge of the primary care physician) and (d) avoid negative comments about the referring physician. The following narrative from a primary care physician illustrates some of the difficulties when this does not happen:

> A 55 year old truck driver was referred by his primary care physician, Dr. Wall, to a neurologist, Dr. Klein, for evaluation of dizziness and paresthesias. Dr. Wall sent a letter summarizing the patient's medical history and the consultation question. After the initial visit with Dr. Klein, Dr. Wall received a consultation letter that summarized the patient's history, along with a diagnostic work-up plan. After several months, the patient returned to Dr. Wall's office. The patient indicated that he had undergone magnetic imaging scanning, electroencephalography, nerve conduction studies, electromyelography, and a lumbar puncture. None of these results had been sent to Dr. Wall. He (uncharacteristically) wrote a letter to Dr. Klein in a sealed envelope and gave it to the patient so that he could deliver it personally to Dr. Klein in 3 days (his next appointment). It was an angry letter, noting that the lack of communication had occurred not only with this patient, but also with other patients in the past. Dr. Klein responded immediately by letter, and gave results of the diagnostic work-up. However, Dr. Wall described further communication with Dr. Klein as "obsequious." Of note, Dr. Klein was the only neurologist in the closed panel health maintenance organization to which the majority of Dr. Wall's patients subscribed.

In the study by Williams and Peet, consultants were uncomfortable when their roles were not specified by the referring physician, when referring physicians made changes in management without informing the consultant and when patients were referred too late in the course of illness.

In London, Ontario, a questionnaire survey and in-depth interviews were conducted with primary care physicians and consultants (M. Bass, unpublished data, 1991). Problems identified by the family physicians and consultants are listed in Table 1. A follow-up questionnaire indicated that both family physicians and consultants considered it important that the patient understand the reason for referral, the patient be informed of the consultant's opinion by the consultant, the family physician's continuing role be supported, the consultation be psychologically therapeutic, and relationships between primary care physicians and consultants be cooperative.

The only study relating communication to clinical actions was a 1989 prospective study of patients who received screening mammograms.[30] Written reports were sent and primary physicians' offices were called when the findings of the mammogram were abnormal and biopsy was recommended. After 2.5 months, only 37% of the patients had a biopsy performed. Only after additional phone calls and another month, did 84% of the patients undergo the indicated procedure. However, it was unclear from their study whether those recommendations were actually made to the patients who initially refused, whether the primary care physician disagreed with the need for biopsy, or whether the primary care physician simply did not make the recommendations at all.

Of note, there have been no studies on the impact of the development of vertically-integrated medical care systems and managed care on communication between physicians.

Why are There Communication Difficulties between Physicians?

There are many factors that contribute to poor communication around consultations (Table 2). From an anthropologic perspective, primary care physicians and consultants belong to different subcultures. The intellectual bases, site of practice, type of patient concerns, and prevalence of disease are different.[31] The core values of primary care physicians may differ substantially from consultants. Primary care physicians are likely to be more "person-oriented,"[32] that is, they value long-term relationships with patients and experience more distress when dealing with complex medical problems and technology. By contrast, consultants are often more "disease-oriented," that is, they enjoy complex medical problems, feel more comfortable with technologic interventions and place relatively less importance on getting to know their patients as people. As such, misunderstanding, stereotyping, and prejudice would be expected, resulting in uncommunicated expectations, unmet needs, and frustration. Problems might include the primary care physician's unwillingness to refer unless for a specific diagnostic test, or consultants ordering tests that they do not think are necessary in the belief that that is what their consultees want. Because primary care physicians commonly encounter undifferentiated illness, they may be more accustomed to dealing with uncertainty, and may pursue diagnostic certainty less vigorously than consultants. Primary care physicians may place a higher value on patient preferences, especially in longstanding patient-physician relationships.

The cultural gap is exacerbated by the fact that primary care physicians and consultants often have little contact. This is especially true between community-based primary care physicians and university-based specialty settings. Because the practice settings and clinical management styles between primary care physicians and consultants differ substantially, each may be judgmental of the other's style unless they are more familiar on a personal basis. Some consultants may view primary care physicians as unnecessary and may believe that they can provide "primary care" for a problem within their domain (such as a dermatologist treating a patient with acne). These relationships are likely to change with the advent of managed care; most likely there will be more contact, with more opportunities for healthy collaboration.[17]

In fee-for-service health insurance plans, it may be more convenient for patients to self-refer to consultants, especially if the primary care physician is not immediately available. This problem, accompanied by the problem of cross-referral, excludes the primary care physician, as well as other consultants, from the communication loop. In addition, some patients seek multiple medical opinions without the knowledge of the other physicians involved. Consider the following situation, and the difficulty in trying to communicate with multiple consultants:

> Ms. Brown has been a patient in our practice for 15 years, and I took over her care 8 years ago. She has multiple medical problems including peptic ulcer disease, osteoarthritis, major depression with a prior suicide attempt, chronic headaches, facial pain, back pain,

Table 2. Reasons for Poor Communication between Family Physicians and Consultants

- Physicians are busy
- Good communication takes time
- Communication is an undervalued activity
- "Cultural" differences between primary care physicians and consultants
- Lack of familiarity between primary care physician and consultant
- Some referrals may be for "non medical" reasons
- Physicians may feel they have "ownership" of patients
- Patients and physicians may view referral as abandonment
- Physician anxiety
- Patient self-referral
- Patient lack of adequate insurance
- Concern that referral will reveal inadequacies in care

and bunions. When I first met her, in her chart there were letters from two gastroenterologists, three psychiatrists, three orthopedists, a chronic pain clinic, a podiatrist, and two neurologists with whom she had consulted in the previous 2 years. It became evident that there were other physicians, chiropractors, and therapists whom she had seen as well either via self-referral or cross-referral from other health professionals. She had, in her home, at least six different nonsteroidal anti-inflammatory drugs, three histamine-2 blockers, a variety of migraine preparations, and narcotics, which were prescribed by different physicians who were often unaware of all of the medications she was receiving. Ten years ago, she had an episode of severe upper gastrointestinal tract bleeding due to the simultaneous use of four nonsteroidal anti-inflammatory drugs. After a few months, she requested referral to a surgeon for partial gastrectomy because of an episode of recurrent ulcer pain.

Although the delay when having the patient see a consultant was the most commonly cited problem in family physician-consultant communication in a Canadian study (M. Bass, written communication, 1991), this problem has a different character in the United States. Among well-insured patients, prompt access to care is the rule rather than the exception. The delays encountered by some patients in managed care settings may improve with increased competition, but there is not yet a plan that will reduce the often intolerable delays in finding appropriate care for uninsured patients or patients with Medicaid.

Inadequate communication between primary care physicians and consultants appears to be more common in tertiary-care clinic settings than in private practice,[28] and it is compounded by the involvement of multiple practitioners in the patient's care.[8] A patient may be seen by a physician assistant, a resident, and/or an attending physician, raising the question of whom to contact, and who is ultimately responsible for the patient's care. Inadequate secretarial and clerical staff can make the process of sending letters, copying records, and answering return phone calls difficult or impossible.[8]

A 33-year-old woman infected with the Human Immunodeficiency Virus (HIV) was hospitalized at the primary care physician's community hospital for pneumocystis carinii pneumonia. She had an initial evaluation at the Acquired Immunodeficiency Syndrome (AIDS) center after her initial diagnosis 1 year earlier, and had an annual follow up visit scheduled soon.

The patient improved enough to go home, and was seen the next week for a visit at the AIDS clinic. Finding the patient to be severely dyspneic, she was admitted directly to the university hospital. The AIDS clinic physicians attempted to contact the primary care physician, but he was on vacation, and the covering physician was unfamiliar with the case. The primary care physician returned to find his patient intubated in the intensive care unit at the university hospital (at which he did not have admitting privileges on the "staff" service) and under the care of a physician who was not her usual caretaker in the AIDS clinic. Chart notes were written by a series of residents. The primary care physician had some therapeutic suggestions and had discussed end-of-life issues with the patient, including her wish not to receive "heroic" care but was unclear with whom to share that information. The patient's family was confused as to who was in charge.

Good communication between physicians takes time and involves skills they may not have developed during training. Physicians are busy, and they may undervalue the importance of a clearly written letter or phone call. Also, there are economic considerations; some of the time spent communicating between physicians and patients, and all communication between health care practitioners, is not directly compensated.

Some referrals and consultations may be for "non-medical" reasons, which may not be communicated clearly to the consultant.[33] Patients and physicians may need reassurance from a consultant that the diagnostic and therapeutic measures under consideration are in the patient's best interest. Difficult problems may generate significant physician anxiety, raise medico-legal concerns, and may appear to challenge the physicians' authority. In these cases, the reason for the consultation may not be made explicit to the consultant; referral may be an attempt for physicians to get psychological support in dealing with patients with whom they find it difficult to communicate for a variety of personal reasons.[9]

Physicians' feelings of connection to patients form the sustenance of work that is otherwise stressful and demanding.[34,35] In the course of seeking consultation, physicians may fear losing some of that connection, just as patients are afraid of being abandoned by their physician in the process of referral. Primary care physicians may also face embarrassment or shame if a consultant discovers inadequacies in the care that he or she has given the patient.

Consequences of Poor Communication

The adverse consequences of poor communication between primary care physicians and specialists are shown in Table 3. With poor communication, data may be

Table 3. Results of Poor Communication between Primary Care Physicians and Consultants

- Disruption of continuity with primary care physician and/or consultant
- Missed or delayed diagnosis
- Repeated or unnecessary testing
- Increased iatrogenic morbidity
- Polypharmacy
- Increased risk of litigation
- Unwillingness to care for patient

lost in the referral process, patient expectations may be unclear or unrealistic, and there may be erosion of the physician-patient relationship. Inadequate communication between multiple practitioners increases the likelihood of repeated and unnecessary diagnostic testing, and, con- sequently, increases the number of false-positive test results. These, in turn, lead to more diagnostic workups, some of which may contibute to iatrogenic morbidity. At the opposite extreme, if poor communication results in patients not attending recommended consultations, diagnoses may be missed or delayed. The prescribing of medications by multiple practitioners can leave patients confused about medication regimens,[36] and can increase the risk of drug interactions and adverse effects.

One consequence of poor communication, disruption of continuity, is associated with increased medical morbidity and poor outcomes, independent of patient age, disease status, and the number of medical visits.[37] Furthermore, there are many intangibles about a patient that cannot be adequately communicated through a medical chart, including the patient's response to illness, the patient's communication style, and family and social supports,[38] making effective oral and written communication more crucial. Lack of explicit discussions about whether the consultation implies a transfer of care, who will follow up with the patient, and with what frequency may result in inconsistent messages to the patient, physician concerns that patients may be "stolen" from them with little or no follow-up, and patient concerns about abandonment.

Finally, primary care physicians may be unwilling to care for some patients unless adequate specialty backup is available. This has been dramatic in the case of HIV infection; physicians express more willingness to care for HIV infected patients in locations where subspecialty support is more available and of higher quality.[39]

Solutions and Directions for Further Research

With communication problems in consultations more clearly delineated, solutions become apparent (Table 4). The increased emphasis on understanding organizational design and the process of care, coupled with movement towards creation of vertically-integrated health care systems offers the potential to move from a *consultation* model to a *collaborative* relationship between generalists and subspecialists. This has important implications.

Discussions of uncertainty, continuity, values, patient perspectives, and patient preferences that occur between physicians and patients should also be part of the communication between physicians.[6,40] Physician, patient, and family expectations of the consultation should be addressed explicitly. Mutual understanding of core values will align all parties providing care around the patient's needs and create a respectful collaborative working relationship. Specialty consultations that are primarily to address the needs of the primary care physician are not necessarily inappropriate, and may facilitate patient care. Especially in these situations, the reason for referral needs to be communicated clearly to guide consultant input into the patient's care.

Strong patient-physician relationships will facilitate communication between physicians. It is crucial to elicit the patient's understanding of his or her illness, to respond to his or her concerns about the management of the illness, and to explore together the reason for referral and the potential benefits (and risks) of seeking a consultant's opinion. A patient who is psychologically invested in the outcome of the referral process is far more likely to follow through with the recommendation of a consultant, and to become an active participant in his or her care.

Personal knowledge and personal contact between the primary care physician and the consultant can transform an otherwise anonymous professional relationship into a working partnership. This now happens more readily in small, contained health care systems such as community hospitals, multispecialty groups, and small health maintenance organizations. Some of the features of these systems can be applied to larger, more complex health care systems. For example, creating smaller, vertically integrated administrative units within a large health care system could provide more opportunities for consultants and primary care physicians to interact. By placing offices in close proximity, casual communication that forms the basis for relationships and discussion of values and practice styles may occur more readily. Written communication could be facilitated by using standardized referral forms, and electronic networks could transfer data more efficiently. There are already innovations in some parts of the United States that serve to improve communication between physicians; these need to be developed further and disseminated.[17]

Referral letters and phone calls from primary care physicians need to include clearly stated reasons for consultations, the patient's expectations, what the patient has been told,[41] and who will take primary responsibility for further management of the problem. The telephone should be used when an urgent response is required,[26] and when there are sensitive topics, such as mental illness, HIV risk behaviors, and family information that practitioners would not want to include in a letter.[2] Letters from consultants might be written in a "friendlier," more respectful format. The letter would begin with a restatement of the consultation question followed by an answer and specific recommendations. This then would be followed by a more tra-

Table 4. Possible Solutions

- Facilitate personal knowledge and contact between primary care physician and consultant.
- Explicit discussion of uncertainty, continuity, patient understanding of illness, values and preferences.
- Have referral letter include clearly stated reason for consultation.
- Explicit discussion of roles and follow-up plans
- Use of the telephone for urgent or sensitive matters
- Improved transfer of clinical data
- "Friendlier" consultant letter format
- Outcomes research on the consultation/referral process
- Seminars on consultation/referral for trainees

ditional recording of the history, physical examination, and diagnostic test results and discussion to substantiate the recommendations. Incorporation of seminars on how to consult and refer, along with critique of actual referral letters by experienced preceptors, should be incorporated into medical training.

Finally, there is a need for well conducted outcome research on the consultation and referral process.[10] Actual face-to-face communications, letters, telephone calls, and electronic media[42] need to be examined in greater detail to determine other ways in which communication between primary care physicians and consultants can be improved. Intervention studies should examine whether improved primary care physician-consultant communication results in better patient outcomes.

References

1. Reames HR, Jr., Dunstone DC. Professional satisfaction of physicians. Archives of Internal Medicine 1989;149:1951–1956.

2. King M, Pullen I. Communication between general practitioners and psychiatrists. In Pullen I, Wilkinson G, Wright A, Gray DP (eds). *Psychiatry and General Practice Today*. London: American Psychiatric Press (for Royal Colleges of Psychiatrists and General Practitioners), 1994:251–264.

3. Kessel NA. Hospital consultant's view. In Walton J, McLachlan G (eds). *Doctor to Doctor*. London: U.K. Nuffield Provincial Hospitals Trust, 1984.

4. Williams PT, Peet G. Differences in the value of clinical information: Referring physicians versus consulting specialists. Journal of the American Board of Family Practice 1994;7:292–302.

5. Byrd JC, Moskowitz MA. Outpatient consultation: interaction between the general internist and the specialist. Journal of General Internal Medicine 1987;2:93–98.

6. Emanuel LL, Richter J. The consultant and the patient-physician relationship. A trilateral deliberative model. Archives of Internal Medicine 1994;154:1785–1790.

7. McDaniel SH, Campbell TL, Seaburn DB. *Family-Oriented Primary Care: A Manual for Medical Providers*, New York: Springer-Verlag, 1990.

8. McCue JD, Beach KJ. Communication barriers between attending physicians and residents. Journal of General Internal Medicine 1994;9:158–161.

9. McDaniel SH, Campbell TL, Seaburn DB. Working together: Collaboration and referral to mental health professionals. *Family Oriented Primary Care*. New York: Springer-Verlag, 1990:343–360.

10. Nutting PA, Franks P, Clancy CM. Referral and consultation in primary care: Do we understand what we're doing? [editorial; comment]. Journal of Family Practice 1992;35:21–3.

11. Geyman JP, Brown TC, Rivers K. Referrals in family practice: A comparitive study by geographic region and practice setting. Journal of Family Practice 1976;3:163–167.

12. Brock C. Consultation and referral patterns of family physicians. J Fam Pract 1977;4:1129–1137.

13. Mayor TR. Family practice referrals in a health mainenance organization. Journal of Family Practice 1982;14:315–319.

14. Riley GJ, Wille CR, Haggerty RJ. A study of family medicine in upstate New York. JAMA 1969;208:2307–2314.

15. Reynolds GA, Chitnis JG, Roland MO. General practitioner outpatient referrals: Do good doctors refer more patients to hospital? BMJ 1991;302:1250–2.

16. Ludke RL. An examination of the factors that influence patient referral decisions. Medical Care 1982;20:782–796.

17. Epstein RM, Beckman HB. Health care reform and patient-physician communication. American Family Physician 1994;49(8):1718–1720.

18. Roland M, Morris R. Are referrals by general practitioners influenced by the availability of consultants? BMJ 1988;297:599–600.

19. Dowie R. *General Practitioners and Consultants: A Study of Outpatient Referrals*. London: Oxford University Press, 1983.

20. Graham H, Senior R, Lazarus M, Mayer R, Asen K. Family therapy in general practice: Views of referrers and clients. British Journal of General Practice 1992;42:25–8.

21. Barrand J. The consultation in general practice. 2. Problem, solution and management. Medical Journal of Australia 1991;154:741, 744–6.

22. Barrand J. The consultation in general practice. 1. Patients, problems and resources. Medical Journal of Australia 1991;154:671, 674–6.

23. Jenkins RM. Quality of general practitioner referrals to outpatient departments: assessment by specialists and a general practitioner. British Journal of General Practice 1993;43:111–3.

24. Westerman RF, Hull FM, Bezemer PD, Gort G. A study of communication between general practitioners and specialists. British Journal of General Practice 1990;40:445–9

25. Doeleman F. Improving communication between general practitioners and specialists. Family Practice 1987;4:176–182.

26. McPhee SJ, Lo B, Saika GY, Meltzer R. How good is communication between primary care physicians and subspecialty consultants? Arch Intern Med 1984;144:1265–1268.

27. Hansen JP, Brown SE, Sullivan RJ, Jr., Muhlbaier LH. Factors related to an effective referral and consultation process. Journal of Family Practice 1982;15:651–656.

28. Cummins RO, Smith RW, Inui TS. Communication failure in primary care. Failure of consultants to provide follow-up information. JAMA 1980;243:1650–1652.

29. Kunkle EC. Communication breakdown in the referral of the patient. JAMA 1964;187:663

30. Robertson CL, Kopans DB. Communication problems after mammographic screening. Radiology 1989;172:443–4.

31. Hensel WM. A relationship built on mutual distrust. Academic Medicine 1994; 69:468–469.

32. Linn LS. Career orientations and the quality of working life among medical interns and residents. Social Science and Medicine 1981;15A:259–263.

33. Hennen BK. Family doctors consult with specialists for nonmedical reasons [editorial]. Canadian Medical Association Journal 1992;147:609–10

34. Suchman AL, Matthews DA. What makes the patient-doctor relationship therapeutic? Exploring the connexional dimension of medical care. Annals of Internal Medicine 1988;108(1):125–130.

35. Matthews DA, Suchman AL, Branch WT, Jr. Making "connexions": Enhancing the therapeutic potential of patient-clinician relationships. Annals of Internal Medicine 1993;118:973–7.

36. Lassen LC. Connections between the quality of consultations and patient compliance in general practice. Family Practice 1991;8:154–60.

37. Franks P, Nutting PA, Clancy CM. Health care reform, primary care, and the need for research. JAMA 1993;270:1449–53.

38. Emanuel EJ, Brett AS. Managed competition and the patient-physician relationship. New England Journal of Medicine 1993;329:879–82.

39. Woolever DR, Epstein RM, Frankel RM, Suchman AL. Strategies for improving primary care physicians' involvement in the care of patients infected with HIV. J Univ Rochester Med Ctr 1993;5:5–11.

40. Emanuel EJ, Emanuel LL. Four models of the physician-patient relationship. JAMA 1992;267:2221–6.

41. Newton J, Eccles M, Hutchinson A. Communication between general practitioners and consultants: what should their letters contain? BMJ 1992;304:821–4.

42. Branger PJ, van der Wouden JC, Schudel BR, et al. Electronic communication between providers of primary and secondary care. BMJ 1992;305:1068–70.

XIII Nurse-Physician Collaboration: The Challenge of Collaboration

Judith Gedney Baggs

Many physicians and nurse clinicians believe that patient care is better delivered and patients have better outcomes if all the healthcare professionals collaborate to plan and deliver care. Advocating for collaboration is one thing; implementing it is quite another. This chapter, after first addressing the question, "What is collaboration," will review selected literature on nurse-physician collaboration.

Definition and Models

The dictionary definition of collaborate is "to work jointly with others" from the Latin collaborare, *co* for together and *laborare* for work.[1] But working together is not enough of a definition. It could mean merely working in parallel, a term Webster[2] used to describe the activities of medical students and nurses working at the same bedside, each doing her or his own tasks and never acknowledging each other's presence, a form of interaction that made patients uneasy because they believed neither of the providers knew what the other was doing. Collaboration implies an interaction that is complementary, with input and responses from each participant, allowing for a synergistic building to better care delivery.

Two authors have provided ways to classify collaboration. Lewis[3] developed five models, but she concluded that in two of them, "declarative" and "pontification," the providers merely say they are collaborating when they are not. Two others, "constant contributor" and "unification," apply only to groups or agencies. Her interpersonal, collaborative model was called skill-based and built on the unique knowledge and skills of different participants. Armitage[4] formulated a description with five stages: isolation, encounter, communication, collaboration between two people, and collaboration in an organization. Lower levels of collaboration may be adequate in some situations, while more complex situations may benefit from a higher level.

One model that has been used as a basis for much of the research on collaboration was originated by Thomas.[5] Thomas's model is two-dimensional. He defined collaboration as a form of handling interpersonal conflicts that combines a high level of cooperativeness, or concern, for the other's interests, with a high level of assertiveness, or concern, for one's own professional interests. In his model, collaboration occurs when the levels of both cooperativeness and assertiveness are high. Weiss and Davis[6] proposed that collaboration is important in situations even where there is no conflict, and they adapted Thomas' model for assessing interpersonal problem solving, rather than just for conflictual interactions.

The critical attributes of collaboration are: sharing in planning, decision making, problem solving, goal setting, and responsibility; working together cooperatively; coordinating; and communicating openly.[7] Based on the critical attributes and the Thomas model, a definition of nurse-physician collaboration is "nurses and physicians cooperatively working together, sharing responsibility for solving problems and making decisions to formulate and carry out plans for patient care."[7] Collaborative interactions may be carried out in person, on the telephone, or in writing, although most research has focused on face-to-face interactions.[8]

Very little has been written about the negative aspects of collaboration. In a recent study, care providers were unable to identify risks or costs of collaboration in practice.[9] However, negative aspects that have been discussed include deference by low-status collaborators to those of higher status[3] and the need for time for communication and coordination.[10]

Diverse Perspectives on Collaboration

Does collaboration mean the same thing to all healthcare providers? The concept of collaboration is addressed more often in the literature by nurses than by physicians.[11] The *Cumulative Index to Nursing and Allied Health Literature* (CINAHL) has had a subject heading called "collaboration" since 1989. In the 1994 volume, that heading has two pages of citations. *Index Medicus* does not use collaboration as a subject heading; rather one must look under "interprofessional relations" or "patient care team," headings also found in CINAHL.

Most of the literature about collaboration between physicians and nurses is written by nurses,[6,12–18] or by nurses and physicians together.[19–21] In one of the few articles by physicians, Stein, Watts, and Howell[22] updated Stein's classic article on the "doctor-nurse game,"[23] noting that the earlier relationship between the professions has changed. Stein said in the earlier work that nurses could only have input by convincing physicians that physicians had initiated the decision making.[23] In the view of Stein and his colleagues, nurses have become more highly educated in recent years and have a more defined area of expertise, functioning more as autonomous health professionals. These authors characterize the new relationship between physicians and nurses as one of "mutual interdependency."[22] They cite movement toward collaboration and collegiality, with a less hierarchical and more open relationship than in the past. The movement toward interdisciplinary team care, a construct related to collaboration, has similarly been described by an historian as a move by nurses to make interprofessional relations more egalitarian and less hierarchical.[24]

In contrast, Styles[18] noted that physicians often feel threatened when nurses discuss collaboration. They see the process as one of invasion of their position of authority and power. Nurses, too, may feel threatened by the increased responsibility and accountability involved in collaborative practice.[25,26] For nurses to fulfill the assertive aspect of collaboration they need to assume accountability and increased authority in practice areas. It is much easier to sit back and critique someone else's decisions than to have responsibility for making those decisions yourself. For example, an ICU nurse may ask himself, "Do I want to have to decide whether to give potassium in the middle of the night for a borderline level, or would I rather call someone else?" The latter choice would allow him to "Monday morning quarterback" later, critiquing the physician's decision when the patient gets hyperkalemic. This attitude is frustrating for physicians and nurse managers who want to encourage independent clinical decision making by nurses.

There is research evidence to suggest that doctors and nurses make different observations and interpretations of collaboration. In a medical ICU (MICU), staff nurses and medical residents were asked to report how much collaboration was involved in making decisions about transferring patients out of the MICU. A standard definition of collaboration was provided at the top of each questionnaire. Although the means and distributions of response scores were similar between the professions (nurses: 4.4±2.0, residents: 4.1±2.2; $t=.09$, $p>.05$; Baggs, unpublished data), the correlation between their reports of amount of collaboration in making the same decisions was only $r=.10$ ($p=.10$).[27] In a more recent study, nurses and residents rated collaboration in decisions about level of MICU care to be provided (from comfort care to full therapeutic endeavor). Again, the correlation between ratings of collaboration was small ($r=.19$, $p=.04$) though significant.[28]

In a study involving interviews of administrators in medicine and nursing about interdisciplinary team care, the goals identified for each profession differed. Physicians expected nurses to act as physician extenders, while nurses expected to use their knowledge to direct patient care.[29] Although there was agreement about the definition of a team (people with differing expertise working together to provide patient care), there was conflict about leadership and authority in decision-making.

There is evidence of common ground. Based on interviews with 10 medical residents and 10 ICU nurses, members of both professions defined collaboration similarly and valued it highly.[9] Both were very clear that collaboration benefited patient care and made their work more satisfying. One physician said, "I think it takes less time to work as a team. You are smoother. If everybody is clear on what is going on . . . where we're going, and how to get there, then you get there faster and better." A nurse said, "Things that the MD is going to miss along the way, the nurse is going to pick up on, and then the patient will be totally cared for."

This same study revealed differences in power that might explain why providers from different professions interpret the collaborativeness of the same interaction in different ways. Several residents noted that they had ultimate authority and responsibility for patients. The residents could write orders for patient care without any input from nursing. In addition, the residents wanted nurses to be sensitive to the strain of their ICU rotation, their physical exhaustion, and multiple patient responsibilities. Thus, what nurses believe to be collaborative interaction might be perceived by residents as intrusive.[9]

In summary, the definitions of collaboration appear similar for the professions of nursing and medicine, but there are differences in the implications of the term for each. Not too surprisingly, a change in the dynamics of power would be perceived differently by the two professions. Nurses and physicians also may interpret interactions differently in assessing how much collaboration has taken place.

Collaboration in Actual Practice

One of the first groups to promote the notion of collaboration in practice, particularly between physicians and nurses, was the National Joint Practice Commission (NJPC), which was established in 1971 "to make recommendations concerning the roles of the physician and the nurse in providing high quality healthcare."[30] The Commission, founded by the American Medical Association and the American Nurses' Association, was composed of equal numbers of physicians and nurses. They proposed five guidelines for the development of what they termed joint or collaborative practice: (1) establishment of a joint practice committee, (2) primary nursing, (3) encouragement of nurses' individual clinical decision making, (4) integrated patient records, and (5) joint patient care record reviews. In studying patient care units that implemented collaborative practice, they found that providers believed care had improved, and there were increases in patient satisfaction, improved nursing job satisfaction, and reduced need for physician supervision of nursing.[30]

Despite the widespread adoption of a number of these guidelines, collaborative practice is not yet a reality for many providers. Prescott and colleagues used the Thomas model to classify the modes of handling disagreements reported by nurses and physicians in acute care settings. Only 14% of physicians and seven percent of nurses reported using collaborative interdisciplinary problem solving. The most common mode used by both providers was competition; no outcomes of the decision-making process were measured in this study.[31] Further analyses of data from this study suggested that the staff nurse role in clinical decision making about patient care was more supportive rather than collaborative, with physicians accepting nurses' input but handling final decision making themselves in most situations.[26] The nurses in the study were generally satisfied with this arrangement, provided they believed their input was listened to and valued by physicians. Many of the nurses did not want more responsibility for decision making. Physicians were only willing to cede to nursing decisions considered unimportant. In fact, they did not view them as decisions at all.[26]

Obstacles to collaborative practice come from both the nursing and medical professions. In some of the areas where we would most expect to find collaborative practice, it has been found to be missing. For example, in a study of primary care teams of nurse practitioners and physicians, where the nurses were highly educated in their area of practice, Lamb and Napodano audiotaped interactions between team members on two teams and found that only 23% of the interactions qualified as collaborative.[8] They defined collaborative as requiring face-to-face interactions, contributions from each provider, and integration of ideas to formulate a new solution. McLain, who studied similar providers, said, "The research literature is strikingly devoid of studies that have demonstrated the actual existence of collaborative practice as a predominant pattern between nurses and physicians."[15] She found both physicians and nurse practitioners engaged in distorted communication and nonmeaningful interactions that blocked collaboration in practice.[15,16]

Benefits of Collaboration

Collaboration has been linked with job satisfaction for nurses and physicians and with retention of nurses. Residents and nurses interviewed were very clear that their work satisfaction was increased by collaboration.[9] In studies of ICU nurses, residents, and attending physicians, collaboration is significantly associated with satisfaction with the decision-making process.[27,20,32] In addition, in the MICU study of transfer decisions, collaboration was very closely associated with satisfaction with the decision-making process, which predicted nurse retention a year later.[33]

More important, empirical studies suggest that collaboration improves patient outcomes. In a major research project by an interdisciplinary team, Knaus, Draper, Wagner, and Zimmerman studied more than 5,000 patients in 13 intensive care units (ICUs) and concluded that the quality of the working relationship between physicians and nurses was the most important determinant of patient mortality rates.[32] Studies of interdisciplinary teams in long-term care have demonstrated that team care leads to a slower rate of deterioration in chronic diabetics[34,35] and lower mortality.[36] Outcomes of collaboration in acute care hospital general units include lower costs[37] and increased patient satisfaction.[38] In ICUs, investigators have found collaboration associated with better than predicted mortality.[32,39] In the MICU transfer study, when nurses reported full collaboration in making the decision to transfer patients out of the ICU, patient risk of death or readmission to the ICU was 5%; when they reported no collaboration, the risk of a bad outcome was 16%.[27]

Based on both empirical and conceptual work, a number of authors have summarized benefits of collaboration.[40,41] Christensen and Larson[42] see collaboration as a way of maximizing both the sharing of information unique to different professional participants and of creating the opportunity for each to learn more about the other's knowledge and talents, which will improve future collaboration. McMahan, Hoffman, and McGee[11] noted that collaboration and teamwork have become part of the quality improvement process.

Promoting Collaborative Practice

Some providers have developed and studied new care models to promote collaboration.[43] One example is a special care unit in Cleveland developed for the treatment of the chronically critically ill.[44,45] In this unit, a nursing case management model using protocols for common patient problems, such as ventilator weaning, has decreased costs using a less hierarchical, more collaborative interdisciplinary approach to care. Care is organized around the patient, not around a physician team leader.

In another example, a group of nurses and physicians created a general medical-surgical unit at Beth Israel Hospital in Boston with the mission of developing a model of physician-nurse collaboration, defining and describing the concept and practice.[46] A group of physicians, nurses, social workers, physical therapists, occupational therapists, and hospital administrators met to develop the model and plan for the unit. They focused on patients best able to benefit from collaborative care, those needing intensive nursing care but not the technology of an ICU admission. In delivering care on this unit, it was expected that providers would take the time to develop good collaborative relationships and that interdisciplinary rounds would be held daily to plan care.

But there were barriers to overcome before this unit could be developed. A physician member of the group noted his apprehension as the project began: "At the hospital where I trained . . . nurses were not only asserting themselves, but my impression is that they also identified physicians as the enemy who had oppressed them and were fighting back."[46] He added, "Popular phrases about nurses caring and doctors curing and of nurses being patient advocates made the hairs on the back of my neck bristle. I found myself being portrayed as uncaring." His initial response to discussions about collaboration was that "what was often being said was that physicians needed to change their behaviors. Little, if anything, was said of changes that nurses needed to make to facilitate collaboration. The continued portrayal of physicians as the enemy gnawed at me and made me distrust the concept of collaboration". This same physician, committed to a true collaborative endeavor, helped develop the structure for the new unit and found that practice there produced "great personal satisfaction."

There have been many strategies proposed for promotion of collaboration in practice. At the first level are plans for educating nursing and medical students together.[47,48] In one academic medical center, new interns spend a day shadowing and assisting a staff nurse.[49] Other proposals stress use of quality improvement activities to model interdisciplinary collaboration,[50–52] and clarification for nurses of their role in team conferences.[53] Strategies proposed for the ICU, many of which may be applicable in other settings, include co-directors from nursing and medicine with equal power in the hospital organization, shared governance with nurses and physicians represented on all patient care committees, standing protocols (developed jointly) with practice guidelines for both nurses and physicians, daily interdisciplinary rounds, formal and informal conferences about patient care and ethical issues, integrated patient records, joint participation in educational and case review programs, and promotion of communication styles that reflect respect, validation of each other's feelings, and recognition of the unique contributions of each profession.[54,55] Other activities to promote collaboration could involve orientation of those new to an area (e.g., new nursing staff, new interns, new attendings) by members of both professions and role modeling of collaborative behavior by unit directors and other senior staff.

Collaboration in Bioethical Decision Making

Collaboration between nurses and physicians has been proposed to support improved bioethical decision making in healthcare.[56] Many decisions about patient care have bioethical dimensions. For example, the question about what level of care to provide to a patient is a bioethical one, as are decisions about withholding and withdrawing treatment, and about whether to intubate or when to extubate patients. When these decisions are made, input is needed not just about clinical course and probability of clinical benefit, but about the value of treatment—its good and its harm to the patient, family, and community. These values may be viewed differently by different professions, by providers and patients, or by patients and families.

In the past, the major healthcare provider involved in making bioethical decisions had been the patient's attending physician. The attending physician is responsible for knowing the patient, the illness, and potential treatments and outcomes. The attending physician also bears legal responsibility for much decision making, particularly about limitations of treatment, so he or she must be involved in bioethical decision making.

Other providers, such as nurses, have an important contributory role for influencing bioethical decisions about patient care. Nurses know patients in a different way than do physicians. Patients and families often confide in nurses. Nurses spend longer periods of time with patients. The nursing perspective may be vital in helping patients and families participate in ethical decision making.

Luce[57] emphasized the importance of inclusion of nurses in ethical decision making because of their closeness to patients and frequent communications with patients about these issues. There is evidence that nurses support patient autonomy. Research has shown that nurses are interested in disclosing more information to patients to support patient participation in decision-making.[58] As communication facilitators, nurses support patient values and beliefs[59] and, in some cases, their knowledge of patient preferences exceeds that of physicians.[60] According to nurse ethicist Amy Haddad,[61] nurses are more inclined to use a style of decision making that identifies the importance of personal relationships and to look to the specifics of the situation, while physicians are more concerned with rights and rules. The combination of both perspectives may enrich ethical decision making.

Stein and his colleagues[22] believe practice is more collaborative now than in the past. Providers value collaboration and are more open to listening to each other and to patients and families, as well. In a study conducted in 1975 in an MICU,[62] many nurses (74%) but fewer residents (33%) favored involvement of conscious patients in end-of-life decision making. The investigators found agreement between nurses and residents about the factors of importance in influencing decisions about withholding or withdrawing treatment, and that physicians generally made such treatment decisions. In a similar study conducted more recently, support for involvement of patients had increased dramatically. All of the residents in this more recent work believed conscious patients should be involved in treatment decisions as did all of the nurses.[7] Both residents and nurses reported that patients were involved in over 50% of decisions, which is a striking increase from the 14% reported by residents in the earlier study.

In the 1995 multicenter SUPPORT study of end-of-life decision making by healthcare providers and patients, researchers found poor communication about prognosis, resuscitation, and patient preferences, and lack of implementation of patient refusals of care. That study also included an intervention, having a nurse communicate about patient choices with patients, families, and attending physicians. Unfortunately, that intervention was unsuccessful in changing patient outcomes, suggesting again a lack of collaboration between care providers.[63]

Pike[64] proposed that collaboration is a way to overcome the moral distress that arises when providers encounter barriers to carrying out what they believe is the morally appropriate action. She presented two paradigmatic cases of patients admitted to the collaborative care unit at Beth Israel Hospital . In the first case, soon after the unit opened, an elderly patient who had asked not to have "heroic treatments" had become incapacitated for decision making and was designated do-not-resuscitate (DNR). However, when peripheral intravenous (IV) access became difficult, a central line was placed leading to a pneumothorax. The family and nurses were upset about continuation of invasive and painful procedures. Only after the patient's death did the nurses and physicians communicate well about the incident. Together they recognized a failure to set clear goals for care. The nurses learned that the physicians also had been deeply troubled by the care, but both groups had agonized in isolation. In the second case, several weeks later, with a similar patient, when IV access became an issue and no family or guardians wanted to be involved in decision making, the nurses and physicians discussed the case together and chose not to continue to seek IV access and not to place a feeding tube. Comfort measures were made the primary treatments. The second patient died in peace and dignity, the first in pain and distress. Both physicians and nurses believed that the collaborative management of the second patient contributed to better care.

Nurses and social workers and physicians may proactively support patient involvement in making choices, while the patient is capable of inclusion, so that there will be fewer missed opportunities if the patient later becomes comatose.[65] Despite recognition from many in healthcare that there are roles for nurses with physicians in ethical decision making, much of the information available about how clinical ethical decisions are made and who care providers believe should be involved suggests problems in interdisciplinary decision making. One study examined the decision of choice of level of care for all patient admissions in a small community hospital.[66] In that study, the researchers found a significant difference in the level of care that nurses and physicians would choose for patients and a lack of communication between the professions. Physicians indicated they had communicated with nursing staff in 564 cases about their reasons for choosing levels of care for patients, but nurses only indicated awareness of this communication in only 56 of those cases.

Studies have demonstrated that nurses perceive ethical decision making as a problematic area in their relationship with physician caregivers. Gramelspacher, Howell, and Young[67] found that nurses and physicians from an MICU and oncology floor reported that they frequently encountered interdisciplinary conflicts about ethical problems, but 21 of 26 nurses reported conflicts with physicians about ethical issues, while only 1 of 24 physicians perceived conflicts with nurses. Collaboration cannot be an approach to conflict resolution if half the team does not know there is a conflict.

Risks of Failing to Collaborate

There are risks to making care decisions in a noncollaborative fashion. When care providers do not collaborate and come to mutual decisions, there is potential for a decrease in the quality of care given and estrangement between the professions. Providers who believe care should be more (or less) aggressive may go beyond or (deliver less than) the level that has been ordered or may withdraw emotionally from the patient. There are several examples of such a breakdown in care in the literature. Parker[68] tells "Mike's" story" about her interaction as a staff nurse with a diabetic carpenter. This man, disabled by a stroke and chronic emphysema, had undergone amputations of both legs and now lived in a nursing home. He was admitted to her ICU after removal of his right femur and had a large wound extending from his rib cage to his pelvis. Despite his aphasia, she came to understand what Mike knew and wanted, using cues such as his fearful expressions when it was time for dressing changes, his moaning, and his efforts to move away. She discussed the possibility of withdrawing care with him and he seemed to be at peace for the first time. When she tried to explain Mike's wishes to the attending physician, she could not explain using the moral language of principles that she had been taught; she was inarticulate in conveying Mike's story, which needed to be told and heard as a narrative. When she said Mike did not want life-sustaining treatment and that she could no longer participate in the painful, futile, and unwanted dressing changes, she was transferred out of the ICU. Mike continued to receive care. When she visited, he would not communicate, avoiding any eye contact. He had multiple arrests from which he was resuscitated before he died, "in pain, frightened and alone."[69]

In another case,[69] the ICU physicians and family of an elderly, comatose stroke patient determined that care was both futile and inconsistent with what the patient would have wanted. Care, including nutrition and hydration, was documented in the chart as being withdrawn. The patient was transferred to a medical floor to die, but no one explained to the nursing staff how and why the decision to withdraw care had been made. The nurses saw a patient being "starved" and whose family never visited. The hospital had no mechanism for the nurses to get the explanations they needed. Only when the entire nursing staff told the attending physician that they would no longer provide care for the patient, in protest over withholding feedings and fluids, was there a response to facilitate communication and collaborative care.

Conclusions

The positive effects of interdisciplinary collaboration suggest that this method of decision making elicits more active involvement of nurses, patients, and families in decision making. The literature on interdisciplinary teams supports collaborative practice. Modern health care is complex; it requires the perspectives and knowledge of multiple disciplines.[70–72] Because each discipline has knowledge and skills that the others lack, teamwork is essential for good patient care. Commissions from nursing, medicine, and government have recommended collaborative care.[20,73–77] Despite a large literature that provides theoretical and empirical support for collaboration between nurses and physicians, collaboration is not the usual mode of practice. Clearly there is more work, important work, yet to be done.

References

1. *Webster's Ninth New Collegiate Dictionary*. Springfield: Merriam-Webster, Inc., 1989.

2. Webster D. Medical students' views of the role of the nurse. Nurs Res 1985;34:313–7.

3. Lewis FM. Consultation and collaboration among health care providers. In: Baird SB, McCorkle R, Grant M, eds. *Cancer Nursing: A Comprehensive Textbook*. Philadelphia: Saunders, 1991:889–935.

4. Armitage P. Joint working in primary health care. Nurs Times 1983;79(28):75–8.

5. Thomas K. Conflict and conflict management. In: Dunnette MD, ed. *Handbook of Industrial and Organizational Psychology*. Chicago: Rand McNally College Publishing Company, 1976: 889–935.

6. Weiss SJ, Davis HP. Validity and reliability of the collaborative practice scales. Nurs Res 1985;34:299–305.

7. Baggs JG, Schmitt MH. Collaboration between nurses and physicians. Image J Nurs Sch 1988;20:145–9.

8. Lamb GS, Napodano RJ. Physician-nurse practitioner interaction patterns in primary care practices. Am J Public Health, 1984;74:26–9.

9. Baggs JG, Schmitt MH. Nurses' and resident physicians' perceptions of the process of collaboration in an MICU. Res Nurs Health (in press).

10. Felton G, McLaughlin FE. The collaborative process in generating a nursing research study. Nurs Res 1976;25:115–20.

11. McMahan EM, Hoffman K, McGee GW. Physician-nurse relationships in clinical settings: a review and critique of the literature, 1966–1992. Med Care Rev 1994;51:83–112.

12. Arslanian-Engoren CM. Lived experiences of CNSs who collaborate with physicians: a phenomenological study. Clin Nurs Specialist 1995;9:68–70.

13. England DA. *Collaboration in Nursing*. Rockville: Aspen, 1986.

14. Henneman EA, Lee JL, Cohen JI. Collaboration: a concept analysis. J Adv Nurs 1995;21: 103–9.

15. McLain BR. Collaborative practice: a critical theory perspective. Res Nurs Health 1988a; 11: 391–8.

16. McLain BR. Collaborative practice: the nurse practitioner's role in its success or failure. Nurse Pract 1988b;13(5):31–8.

17. Steel JE. *Issues in Collaborative Practice*. Orlando: Grune & Stratton, 1986.

18. Styles MM. Reflections on collaboration and unification. Image J Nurs Sch 1984;16:21–2.

19. Ames A, Perrin JM. Collaborative practice: the joining of two professions. J Tenn Med Assoc 1980;73:557–60.

20. Dunbar S, Bryan-Brown C. Collaborative practice. In: J Boller, ed. *International Conference Book*. Newport Beach: American Association of Critical-Care Nurses, 1988:153–4.

21. Evans SA, Carlson R. Nurse/physician collaboration: solving the nursing shortage crisis. Am J Crit Care 1992;1(1):25–32.

22. Stein LI, Watts DT, Howell T. The doctor-nurse game revisited. N Engl J Med 1990;322: 546–9.

23. Stein LI. The doctor-nurse game. Arch Gen Psychiatry 1967;16:638–42.

24. Brown T. An historical view of health care teams. In: Agich GJ, ed. *Responsibility in Health Care*. Dordrecht, Holland: D. Reidel Publishing Company, 1982:3–21.

25. Cape LS. Collaborative practice models and structures. In: England DA, ed. *Collaboration in Nursing*. Rockville: Aspen, 1986:13–26.

26. Prescott PA, Dennis KE, Jacox AK. Clinical decision making of staff nurses. Image J Nurs Sch 1987;19:56–62.

27. Baggs JG, Ryan SA, Phelps CE, Richeson JF, Johnson JE. The association between interdisciplinary collaboration and patient outcomes in medical intensive care. Heart Lung 1992; 21:18–24.

28. Baggs JG, Schmitt MH. Intensive care decisions about level of aggressiveness of care. Res Nurs Health 1995;18:345–55.

29. Temkin-Greener H.. Interprofessional perspectives on teamwork in health care: a case study. Milbank Q 1983;61:641–58.

30. National Joint Practice Commission. *Guidelines for Establishing Joint or Collaborative Practice in Hospitals*. Chicago: NJPC, 1981.

31. Prescott PA, Bowen SA. Physician-nurse relationships. Ann Intern Med 1985; 103:127–33.

32. Knaus WA, Draper EA, Wagner DP, Zimmerman JE. An evaluation of outcome from intensive care in major medical centers. Ann Intern Med 1986;104:410–8.

33. Baggs JG, Ryan SA. Intensive care unit nurse-physician collaboration and nurse satisfaction. Nurs Economics 1990;8:386–92.

34. Feiger SM, Schmitt MH. Collegiality in interdisciplinary health teams. Soc Sci Med 1979; 13A:217–29.

35. Schmitt MH, Watson NM, Feiger SH, Williams TF. Conceptualizing and measuring outcomes of interdisciplinary team care for a group of long-term, chronically ill, institutionalized patients. In: Bachman JE, ed. *Interdisciplinary Health Care: Proceedings of the Third Annual Conference on Interdisciplinary Team Care*. Kalamazoo: Center for Human Services Western Michigan University, 1981:169–81.

36. Rubenstein LZ, Josephson KR, Wieland GD, English PA, Sayre JA, Kane RL. Effectiveness of a geriatric evaluation unit. N Engl J Med 1984;311:1664–70.

37. Koerner B, Armstrong D. Collaborative practice cuts cost of patient cares: study. Hospitals 1984;58(10):52–4.

38. Koerner BL, Cohen JR, Armstrong DM. Collaborative practice and patient satisfaction. Eval Health Prof 1985;8:299–321.

39. Mitchell PH, Armstrong S, Simpson TF, Lentz M. AACN demonstration project. Heart Lung 1989;18:219–37.

40. Fagin CM. Collaboration between nurses and physicians: no longer a choice. Acad Med 1992;67:295–303.

41. Miccolo MA, Spanier AH. Making collaborative practice work. Crit Care Clin 1993; 9: 443–53.

42. Christensen C, Larson JR. Collaborative medical decision making. Med Decis Making 1993; 13:339–46.

43. King ML, Lee JL, Henneman E. A collaborative practice model for critical care. Am J Critical Care 1993;2:444–9.

44. Daly BJ, Phelps C, Rudy E B.. A nurse-managed special care unit. J Nurs Adm 1991; 21(7/8):31–8.

45. Daly BJ, Rudy EB, Thompson KS, Happ MB.Development of a special care unit for chronically critically ill patients. Heart Lung 1991;20:45–51.

46. Alpert HB, Goldman LD, Kilroy CM, Pike AW. 7 Gryzmish: toward an understanding of collaboration. Nurs Clin North Am 1992; 27(1):47–59.

47. Anvaripour PL, Jacobson L, Schweiger J, Weissman GK. Physician-nurse collegiality in the medical school curriculum. Mt Sinai J Med 1991; 58: 91–94.

48. Larson EL. New rules for the game: interdisciplinary education for health professionals. Nurs Outlook 1995; 43: 180–5.

49. Lloyd JM. Endnote hit the right note (letter). Critical Care Nurs 1992;12(8):18.

50. Clafin N, Hayden CT.. Interdisciplinary collaboration in the special care unit. J Health Care Qual 1994;16:23–30.

51. Folcarelli PH, Kilroy CM, Tobin CM, Wheaton C.. Operationalizing collaborative quality assurance: the 7 Gryzmish experience. J Nurs Care Qual 1993; 7(3): 9–18.

52. Pike AW, McHugh M, Canney KC, Miller NE, Reiley P, Seibert CP. A new architecture for quality assurance: nurse-physician collaboration. J Nurs Care Qual 1993;7(3): 1–8.

53. Coker EB, Schreiber R.. The nurse's role in a team conference. Nurs Management 1990; 21: 46–8.

54. Dracup K, Chernow B. The team concept of patient care. In: Sivak ED, Higgins TL, Seiver A, eds. *The High Risk Patient: Management of the Critically Ill*. Baltimore: Williams & Wilkins, 1995:1555–65.

55. Ralph DD, Gleason DH. Staffing and management of the intensive care unit. In: Hall JB, Schmidt GA, Wood LDH, eds. *Principles of Critical Care*. New York: McGraw-Hill, 1992: 465–72.

56. Baggs JG. Collaborative interdisciplinary bioethical decision making in intensive care units. Nurs Outlook 1993;41:108–12.

57. Luce JM. Ethical principles in critical care. JAMA 1990;263:696–700.

58. Davis AJ, Jameton A. Nursing and medical student attitudes toward nursing disclosure of information to patients: a pilot study. J Adv Nurs 1987;12:691–8.

59. Zorb SL, Stevens JB Contemporary bioethical issues in critical care. Nurs Clin North Am 1990; 2:515–20.

60. Ouslander JG, Tymchuk AJ, Rahbar B. Health care decisions among elderly long-term care residents and their potential proxies. Arch Intern Med 1989; 149: 1367–72.

61. Haddad AM. The nurse/physician relationship and ethical decision making. AORN J 1991; 53:151–6.

62. Youngner S, Jackson DL, Allen M. Staff attitudes toward the care of the critically ill in the medical intensive care unit. Crit Care Med 1979; 7(2):35–40.

63. SUPPORT Principal Investigators.A controlled trial to improve care for seriously ill hospitalized patients. JAMA 1995;274:1592–8.

64. Pike AW. Moral outrage and moral discourse in nurse-physician collaboration. J Prof Nurs 1991;7:351–63.

65. Marsden C. Who should decide? Heart Lung 1990;19:687–9.

66. Frampton MW, Mayewski RJ. Physicians' and nurses' attitudes toward withholding treatment in a community hospital. J Gen Intern Med 1987;2:394–9.

67. Gramelspacher GP, Howell JD, Young MJ. Perceptions of ethical problems by nurses and doctors. Arch Intern Med 1986; 146: 577–8.

68. Parker RS. Nurses' stories: the search for a relational ethic of care. Adv Nurs Sci 1990; 13: 31–40.

69. Watts DT, McCaulley BL, Priefer BA. Physician-nurse conflict: lessons from a clinical experience. J Am Geriatr Soc 1990;38:1151–2.

70. Bass BM. Team productivity and individual member competence. Small Group Behavior 1980;11:431–504.

71. Kaluzny AD. Design and management of disciplinary and interdisciplinary groups in health services: review and critique. Med Care Rev 1985;42:77–112.

72. Loewy EH. *Textbook of Healthcare Ethics*. New York: Plenum Press, 1996.

73. Institute of Medicine. *Nursing and Nursing Education*. Washington, D.C.: National Academy Press, 1983.

74. Secretary's Commission on Nursing. Final report, volume I. Washington, D.C.: DHHS, 1988.

75. American Association of Critical-Care Nurses. *Summary Analysis of Critical Care Nurse Supply and Requirements*. Newport Beach: AACN, 1988.

76. National Institutes of Health. Consensus conference: Critical care medicine. JAMA 1983; 250:798–804.

77. Joint Commission on Accreditation of Hospitals. *Accreditation Manual for Hospitals*. Chicago:JCAH, 1987.

XIV Opportunities and Obstacles: Development of a True Collaborative Practice with Physicians

Colleen M. Dwyer, Patricia Hinton-Walker, Peter Coggiola, and Anthony L. Suchman

Introduction

The National Joint Practice Commission of 1972 defined joint practice as "nurses and physicians collaborating as colleagues to provide patient care." With the advent of health care reform, the topic of partnerships and collaborative practice between nurse practitioners and physicians become even more important. According to Mundinger (1994) "collaborative practice is more comprehensive, most cost effective, and, therefore more competitive than independent practice."[9] Mundinger (1994)further states that "the best model—one that reduces costs while enhancing quality and comprehensiveness—is collaborative practice."[9] The focus of this paper will be on the process of development of a true collaborative practice rather than a traditional setting where the nurse practitioner "works for" the physician. We will describe a collaborative practice that was developed as part of a community nursing center faculty practice at the University of Rochester School of Nursing.

Community Nursing Centers (CNCs) are still considered a new and innovative delivery model that provides consumers direct access to professional nursing services. These services usually include primary care, prevention, and health promotion services. Nurse-managed care is a term frequently used to describe the practices in nursing centers. However, it is in this context and through a nursing center that this collaborative practice developed. In this case, the result is effective co-management of patient care at a hospital-based primary care practice at Highland Hospital in Rochester, New York and at a nurse-managed site (the Senior Health Service) which is part of the University of Rochester CNC.

The University of Rochester CNC

The University of Rochester School of Nursing has a long history of integrating education and practice through the unification model. Early linkages for clini-

Originally published in Murphy B. (Ed.). *Nursing Centers: The Time is Now*. New York: National League for Nursing, 1995:135-55. Reprinted with permission.

cal faculty were primarily in Strong Memorial Hospital, where many practitioners could be described in some way as participating (on a daily basis) in some form of collaboration with faculty members and students from the school of medicine. However, with the development of the CNC there have been increasing opportunities for more autonomous faculty practice outside the medical center and the hospital. Moccia (1993) encourages "a faculty of the community" where loyalty shifts from traditional hospital nursing to alliances with the community.[8] The practice described in this chapter is a form of a new type of alliance with a community and reflects a model for truly collaborative practice relationships between nurse practitioners and physician groups.

The University of Rochester CNC was organized as a faculty group practice and its comprehensive services have already been described in Chapter 5. The faculty practice described in this paper evolved out of a nurse-managed practice for the elderly called the Senior Health Service. This nurse practitioner managed practice was the first practice developed under the CNC service cluster, Services for Longer Term Continuity of Care, (see Figure 1).

Development of the True Collaborative Practice

The Senior Health Service (SHS), opened in the fall of 1991. A nurse practitioner, Peter Coggolia, with a number of years of experience, began to establish an autonomous practice on the sixth floor of the Region Council on Aging building. Initially, neither the nurse practitioner, nor the physician had experience with an autonomous nurse practitioner (NP). Consequently, during the early days of the practice, the NP established a format for consultation meetings with the consulting MD. Prior to that time, the meetings had a teacher-student approach. Efforts were made for the NP to develop case studies based on a specific patient (or patients), provide background information, and present potential plans of care. Then, at the meetings with the consulting MD, the NP would present options and discuss the benefits or drawbacks of each joint plan of care. Through this approach, guidelines were established for when (and if) consultation was needed for clinical decision-making.

Even though the SHS was located within the main building for the Regional Council on Aging (RCOA), many of the seniors who attended the agency offerings were not aware the nurse-managed practice existed. A second approach was needed to increase the visibility of the SHS within the target client population. To accomplish this, a series of educational programs were presented at the nutrition center, housed on the first floor of the RCOA building. These educational programs were eventually expanded to several of the high-rise apartments in the downtown area where senior citizens lived. During that same timeframe, the public health nurse who was assigned to the RCOA building was required to cut back her overall practice time. With the change, the monthly blood pressure checks that she had performed were incorporated into the SHS practice.

In the fall of 1992, the SHS practitioner offered to participate in the Medicare

Figure 1.

demonstration project that was studying the promotion of immunizing seniors for influenza. The SHS took this program to the majority of the senior high rises in the city, as well as a few in the suburbs surrounding Rochester. Utilizing this program, the SHS staff was able to provide immunizations sessions for 350 to 450 seniors per year, while marketing the services of the CNC.

There were also opportunities to interface with other agencies who had frustrations about helping the seniors they worked with to get health care or health assessments. These agencies started referring seniors who were home bound for various reasons to SHS for house calls. Many of these visits were for patients living in federally mandated health manpower shortage areas.

As the practice began to take shape, it became a priority to work with physicians who could accept and understand the more autonomous role of the NP. Working with the traditional physician associated with a practice, the NP took on more responsibility for case management, as well as primary care. The outcome was that meetings became more streamlined and more patient focused and a more comfortable relationship with the physician developed. Consequently, the NP was able to take on a more central role in the development and delivery of patient care. After about six months, it became necessary to recruit a new physician as a consultant, since the first physicians decided to start attending a PhD program. The new consultant was selected using an interview process. This new consulting arrangement was with a group of three physicians. Initially, meetings were set up with the NP and all three physicians. However, this format became difficult due to conflicting schedules. There were many details to be worked out with the three physicians. One area of concern was which service that SHS patients would be admitted to when patients required hospitalization. After discussion with the group, it was decided that the MD consultants would rotate on a one- to three-month basis, depending on call schedules and teaching responsibilities.

The changes in physician-consultant required that trust in clinical judgement be rebuilt. This was accomplished in much the same way as it had been developed with the original consultant. As the relationship developed further, less time was needed, and meetings were less frequent. Face-to-face discussions were supplemented with telephone conversations and consultations as needed.

In the summer of 1993, the initiating NP had an opportunity to develop a rural entrepreneurial practice. Another practitioner, Colleen Dwyer, was recruited into the practice. At the same time, the primary physician-consultant expressed interest in the development of a true collaborative practice. In this new practice, the NP would continue the SHS practice on a part-time basis, and the rest of her practice would be a new collaborative practice role within the physician group practice at Highland Hospital.

Development of the New Collaborative Practice

This new innovative role between Highland Hospital and the Senior Health Service became a reality in the fall of 1993. After several meetings between the new NP, the physician group, and the CNC administrator, the preliminary details of this role were developed. The NP clinical faculty member's time would be split between the SHS and the practice at Highland Hospital. The primary advantage of this joint venture was the flexibility for role development. The mutually set goals for this position were to: (1) further develop a collaborative practice model utilizing the adult nurse practitioner; (2) initiate a patient education program that would benefit patients at both practices; (3) become involved in teaching residents and ancillary staff about the role of the NP; and (4) further enhance the role of the NP in the provision of primary, secondary, and tertiary prevention according to the Neuman Systems Model.[10]

One of the early objectives in the development of this collaborative practice was to develop mutual respect for the different abilities, knowledge, and skills that each respective discipline would bring to the practice. The building of trust would be required to make this new "collaborative practice" role a success. The focus of care provided by the NP would be nursing centered.

Nurse Practitioner and Physician Expectations

Nurse Practitioner Expectations

The application of holistic care is a critical focus of the nurse practitioner's education at the master's level, and NPs are prepared to provide care that encompasses the entire family unit. Holistic care is crucial in today's health care system. Health professionals must refocus energy not on acute care but on prevention. The Neuman Systems Model provides a basis for the focus on prevention. Neuman emphasizes that the client and family are in the center of the application of holistic care. The whole person is addressed utilizing five variables for assessment: spiritual, sociocultural, developmental, physiological, and psychoemotional. Within the Neuman model, it is the provider's responsibility to assist the patient to develop lines of resistance and lines of defense to prevent breakdown of any or all of the variables. (See Figure 2.) Examples of primary prevention include health maintenance, health promotion, and education. The case management of selected clients who were home bound was also key element in this NP practice.

The responsibility of tertiary prevention also lies within the scope of the NP. The Senior Health Service serves a population of patients that benefit from tertiary prevention services as described by Neuman. Patients seen at this site not only have multiple medical diagnoses but also have limited access to health care resources. The NP is crucial in the coordination of health care for these patients. Services provided at this site include the provision of both primary and tertiary prevention ranging from routine health maintenance to acute care visits. Case studies presented later in this chapter will describe the application of the Neuman Systems Model to patient care in this collaborative practice.

Physician Expectations

From the physician perspective, a number of key elements were required for success of the collaborative role. First, all participants (nurse practitioners and physicians alike) must be committed to shake off the more traditional assumptions about the doctor-nurse relationship. Second, all participants would need to believe that teams do better than individuals and provide better care than one discipline can do alone. Additionally, the position must be constructed in such as way as to demonstrate partnership and parity. Thus, lines of supervision are similar for physicians and for the nurse practitioners within the unit. Faculty ranks for both the nurse practitioner (as a clinical faculty member in the School of Nursing) should be comparable with the physician faculty members who are in the School of Medicine. Consequently, participation (by the nurse practitioner clinical faculty member) in teaching, administrative activities of the unit, and in Highland Hospital medical faculty meetings would not only be invited, but expected. A third element is that the nurse practitioners have their own panels (or groups) of patients for whom they are the primary care providers, just as the physicians have their own panel. Last, practitioners of both disciplines should help each other through consultation in areas pertaining to each discipline's specific professional expertise. Also, in general, both medical and nurse practitioner faculty would assist with cross-coverage during periods of absence.

Additional expectations of the role from the physician included: (1) providing quality primary care; (2) working independently within the scope of one's area of competence, asking for assistance when care moves outside of that zone; (3) offering

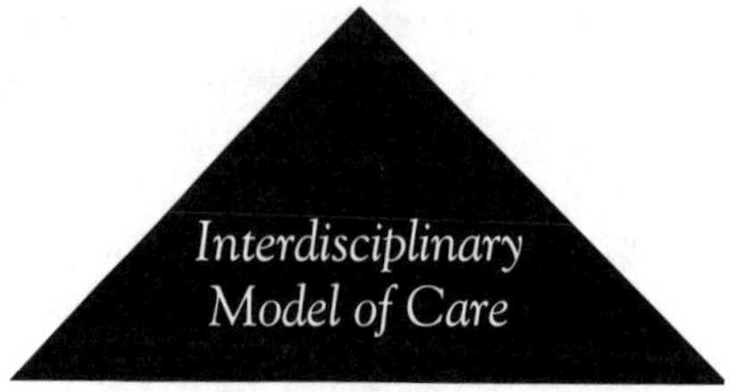

NEUMAN HEALTH SYSTEMS MODEL

- ▲ Mutual Goal Setting Between Client and Provider
- ▲ Focus of Care Consistent with Community Based Practice
  Primary and Tertiary Prevention as Intervention
- ▲ Holistic Approach With Attention To:

Stressors	Variables	
Intrapersonal	Physiological	Psychological
Interpersonal	Developmental	Spiritual
Extrapersonal	Socio-Cultural	

Figure 2.

assistance to physicians with areas that are outside their area of competence; (4) having the self-confidence to make unsolicited suggestions in the spirit of partnership and teamwork; and (5) seeking to become involved in all of the unit's activities and willingness to participate in the long-term planning of the unit.

Guidelines for Practice and Education

The guidelines established for this practice were within the context of a holistic, nursing-centered approach to care. It was agreed that the NP expectations would function as outlined in the American Academy of Nurse Practitioner's "Scope of Practice," and a collaborative practice agreement was established with the collaborating physician accordingly. The following guidelines were developed with assistance and approval of those who would play an integral part in the future success of this new practice. These guidelines were an actual part of the collaborative agreement and were available to all those who worked in the clinic:

1) The NP/MD would work collaboratively in the provision of patient care seen at Highland Hospital and the Senior Health Service.

2) The NP would have emergency access to the physician for medical backup while at the SHS.
3) The NP would orient the current staff (practice manager, nursing staff, and secretaries) to the role of the NP in primary care.
4) The NP would act as a liaison for the development of the role within the Highland Department of Medicine and the residency clinic.
5) The NP would have a faculty appointment within the residency program and conduct educational sessions for the residents.
6) The NP would develop a comprehensive educational program to promote health maintenance and wellness for patients in the practice and the community.

Another step in the development of this collaborative practice was the education of not only the residents, but the ancillary staff to the role of the NP. This would be achieved through discussion with the office staff and the practice manager. Guidelines were established to outline which patients would be appropriate for the NP to see, and those that would require referral to the MD. Written information would also be developed and was provided to reinforce this guideline. Information would then be disseminated to patients that the NP would be responsible for their first visit and this would be done prior to the patient visit.

Opportunities and Obstacles

Opportunities

This multifaceted role brings with it many opportunities. The key questions that need to be identified include whether the opportunities as supported by the literature (and the projected health care reform objectives) are relevant to this practice. This collaborative practice and the opportunity to provide care along the health illness continuum (see Figure 3), would be one of the highlights of this practice. Also, this role has the opportunity to bring the adult nurse practitioner role into the forefront of health care reform.

Cost-effectiveness is one issue that arises when looking at new opportunities for the adult nurse practitioners. Will this practice meet the challenge? Mundinger (1994) emphasizes that "teams of primary care physicians and nurse practitioners are even more cost-effective and comprehensive."[9] A number of other authors have reviewed the issue of cost-effective care provided by nurse practitioners. A meta-analysis by Brown and Grimes (1993) confirms the effectiveness of the NP in primary health care. Multiple studies have also demonstrated the effectiveness of the NP in the primary care arena.[3] Avorn, Everitt, and Baker et al., (1991) indicated that the NP was more likely than MDs to incorporate issues such as dietary, psychosocial, and cost concerns into the plan of care. These components are key elements in the objective of current health care reform.[1]

Figure 3. Continuum of care.

Cost-effectiveness of a practice can also be supported through appropriate management of primary, secondary, and tertiary prevention as described by Neuman (1982).[10] To meet the challenge of health care reform in the context of managed care, effective interventions in the primary and tertiary prevention categories can achieve a cost savings by reducing the more expensive costs of acute care, usually associated with secondary prevention (see Figure 4). Appropriateness of care, much of it enhanced by collaborative practice, can reduce costs.

From the physician's perspective, the role would be conceptualized as an opportunity to develop something between the free-standing, independent, nurse practitioner and the more traditional "physician-assistant" type of role. In this new model, the nurse practitioner and primary care physicians would work together as colleagues, providing primary care and preventive services as a team. The role was further conceptualized as an important educational opportunity. This would provide medical residents with some experience of collaborative care and, at some point in the future, a similar learning opportunity for nurse practitioner students.

Obstacles

Through the phases of development and structure of this role, several "obstacles" from the NP perspective became apparent. The expectations of the role were outlined previously, but there were still areas requiring adjustment. The presence of a nurse practitioner for the hospital-based clinic was not only new to some of the physicians but to the staff as well. Education related to the role was interjected at all staff meetings, along with written information provided on a monthly basis to reinforce this new role. However, there was still some lack of understanding of the role.

Highland Hospital had six physicians that see patients on the East 5 clinic, where this new role was to be implemented. Approximately 20 medical residents also see pa-

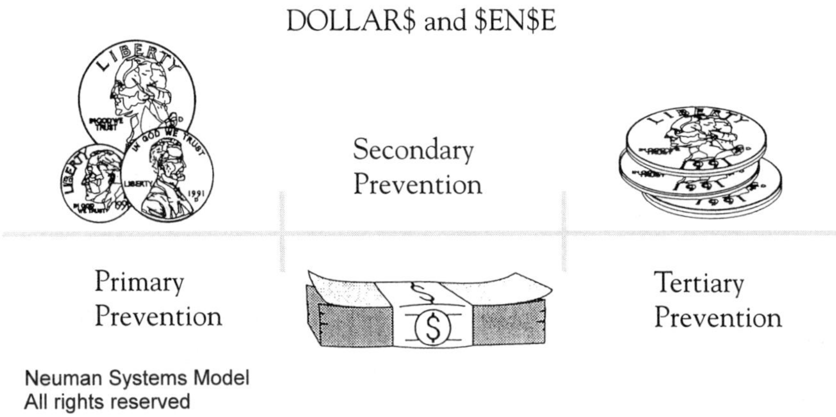

Figure 4. Cost effectiveness.

tients in this clinic. The main obstacle was role definition for this group of residents who were a culturally diverse group. None of the residents had worked with an NP in the past, and confusion regarding the role was apparent at the first meeting. Utilizing individual sessions, written information regarding the role, and responsibilities of the NP provided by the American Academy of Nurse Practitioners was beneficial.

With one exception, the six physicians that worked in the East 5 clinic also had limited experience with an NP. The fact that this NP had eight years of acute care clinical experience was a distinct advantage in gaining their trust and confidence. Over time and with much coaxing, acute care patients that were in these physicians' practices were seen by the NP in the clinic. The biggest problem was just the amount of time it took to obtain trust in clinical judgement.

Another obstacle was the chronicity of the patients referred from the SHS. A concern that was raised during the developmental phase of the role was the chronic patient population at the SHS. These patients were traditionally referred to the SHS practice either through Adult Protective or the Senior Center at the RCOA. Access to health care was extremely limited and, in most cases, absent. There was also concern about the time needed to be invested in patients from the SHS in order to make the site successful.

The first step in controlling the time involved from the physicians was to update the current files on the patients. This was done at the SHS and at Highland Hospital. A main objective was the streamlining of paper work and the continuity of charts between sites. Utilizing existing documentation forms from Highland allowed for continuity of clinical information. A database of charts was also established and updated at the hospital in case an acute care visit should require access to the information by the physician on call.

From the physician's perspective, the major obstacles are administrative and financial. There is not yet recognition of nurse practitioners as primary care providers by some HMOs and other payors. Current New York State regulations require medical affiliation in a manner which could be interpreted that the NP is subordinate. This problem continues to be further exacerbated by more flagrant regulations within some of the local HMOs in the Rochester area. These external constraints interfere with the external perception of partnership which, in turn, can perpetuate internal views and conflicts as well.

A second obstacle mentioned by the physician was financial. The question of how nurse practitioners are compensated when they do work that is similar to primary care physician's work remains quite controversial. There is a strong case for "equal pay for equal work," but this is a potential problem for some physicians. By virtue of differences in education and training, primary care provided by physicians is likely to be somewhat different than that provided by nurse practitioners. These issues will require more honest and respectful dialogue to resolve.

A third obstacle was related to the specific situation in which this collaborative practice was situated. It was difficult for one nurse practitioner to work with six attending MDs and eighteen residents, providing all with an experience of collaborative team care. As opportunities present themselves in the future, the physician group clearly see the need to increase the number of nurse practitioners to take full advantage of the opportunity to learn more about true collaboration.

Caring, Curing, and Case Studies

Clients served by the Senior Health Service are an excellent example of a population with limited access and scarce health care dollars. The provision of health care to this population can be accomplished through a paradigm shift from curing to caring. This shift in paradigm would not only allow for the allocation of health care resources, but also focus on *caring* for these patients versus the *curing* mode that has become all too prevalent.[12]

The application of the Neuman Systems Model has proven to be helpful to NPs and MDs as a way to focus on holistic family-centered care. The application of this model can best be represented through the use of a case study.

Case Study I: E.L. is a 78-year-old female patient with a past medical history of congestive heart failure, seizures, emphysema, s/p stroke, and dementia. Initial assessment of the patient occurred in May 1991. Follow-up at this time was for routine health maintenance and tertiary prevention care. She was enrolled in an adult day care program, and also has access to an emergency call system.

In March 1993 it became evident through several emergency office visits that E.L. was a subject of abuse from her family. At this point, adult protective services were utilized to monitor the situation. The client was competent to make her own decisions and elected to remain in the home. Support was given not only to the client, but also to those family members not involved with the abuse. Living with her grand-daughter and grandson proved to be an ongoing struggle in order to provide care. The location of her home was also a concern not only from Adult Protective, but also for the health care providers.

On November 1, 1993 E.L. presented with an acute onset of respiratory distress which resulted in her first acute care admission to the health care system in over six years. Referral was made to Highland Hospital, and the patient was admitted through the attending MD on call. This admission lasted one week.

During this time, concerns were raised about finances. Did E.L. have the capacity to continue to provide care in the home? What resources were available to assist in discharge planning? And, who would be the "backup" should home services be accepted? Those involved on the team included social service, attending MD/NP, family Adult Protective, home health service, and, most importantly, the patient. Her wishes to remain in her home were clear. Development of this care plan involved putting the CNC philosophy (described in Figure 5) into practice.

Barriers that were met were lack of family or emergency backup. The client refused home services. At this point, it was arranged that a daughter who lived several miles away (and worked full-time) was willing to arrange a medication schedule and assist with the housekeeping. Those family members who lived in the next apartment were not willing to participate. Requiring home oxygen and frequent medical visits, it was agreed that the NP would make home visits to follow-up and monitor her condition.

Case Study II: F.K. is a 77-year-old female patient from the SHS. The initial contact began with this patient on December 5, 1991. The diagnoses at that time included obesity, emphysema, congestive heart failure (CHF), and peripheral edema. This patient, due to morbid obesity, is confined to her home. The first visit in the home was for acute onset of shortness of breath and increased edema. At the time, it was decided to increase her diuretics, and monitor her for increased distress. Over the ensuing months, visits were primarily for acute changes in respiratory status due to CHF which responded well to an increased dose of diuretics. Home visits were made on a monthly basis and as needed. This aggressive home follow-up is an example of tertiary prevention, because it prevented costly return to the hospital in at least four instances.

On April 5 1993 F.K. experienced extreme shortness of breath and chest pain. On the initial visit, it was felt that the patient needed to be evaluated in the emergency department. On physical examination, she had marked shortness of breath, was in moderate discomfort, and had significant increase in peripheral edema. Due to the patient's fear of acute admission, the family was called to lend additional support. The providers were able to arrange for the patient to be met in the emergency department by the physician. The patient was admitted, and due to complications throughout her admission she remained in the hospital approximately two months. The patient's admission occurred at a facility where the SHS was not affiliated, and thus the patient was followed by a different physician.

On her return home, she again contacted the SHS office requesting that the NP continue with home visits and coordinate her care with her new physician. The NP at the SHS office contacted her MD who agreed to continue a co-managed care relationship in the provision of primary and tertiary prevention for this patient. Her diagnoses remained the same.

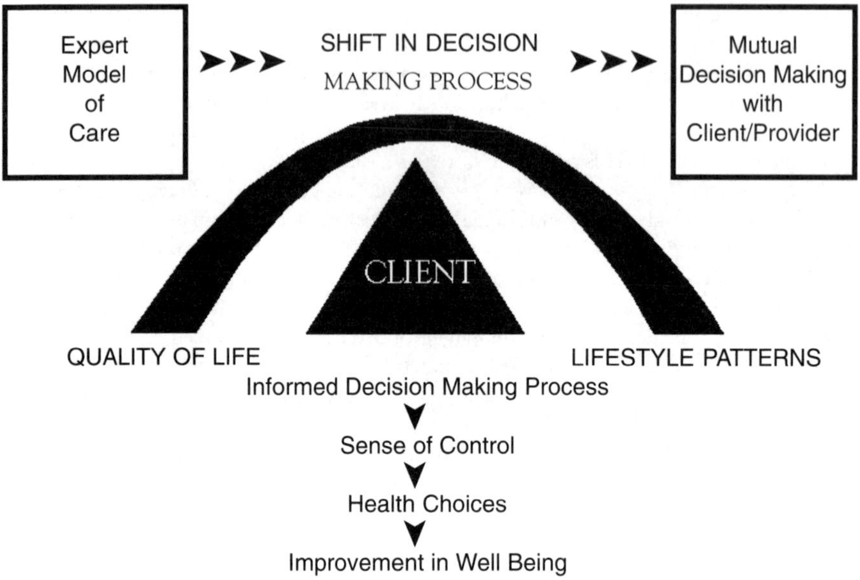

Figure 5.

Plan for home care included the following:

1. Monthly house calls by the NP.
2. Home Health Aide approximately sixteen hours per day to assist with activities of daily living (ADLs).
3. Visiting Nurse home visits every two weeks.
4. Physical therapy consult to assist with optimizing strength and mobility.
5. Dietary consult to reinforce 1500 calorie ADA diet to assist with weight loss.

This plan was developed through a case management conference held with the patient, her primary care MD and NP, Visiting Nurse, Home Health Aide, Family, Physical Therapy, Nutritionist. The main goal was to optimize physical status in order to increase ADL tolerance.

Several acute home visits have been made due to the diagnosis of cellulitis and bronchitis. Hospitalization has been avoided through an aggressive team effort. The cost savings, not only to the patient, but also the health care system is significant. A future health services research goal in the CNC is to attempt to quantify and measure the cost savings of collaborative models of care such as the one described here. The overall benefit of this tertiary prevention approach has maintained the patient in her own environment and has improved her quality of life and well being, despite acute changes in her condition. This is an excellent example of the CNC philosophy of practice as evidenced in Figure 5.

The case studies presented above, depict another advantage of having a nurse-managed clinic such as the SHS, and the benefits shard by those that utilize this practice. NPs are clearly involved in many aspects of health care across the care continuum. The advantage of the provision of tertiary prevention is well described in the SHS practice and in these case studies. Not only are health care dollars saved, but clients are able to remain an active partner in their own health care treatment and decisions.

The application of the Neuman Systems Model reiterates the importance of primary and tertiary prevention in health care. Through these prevention efforts, both NP and MD providers can decrease the cost of secondary prevention, acute care admissions, and emergency care. Collaborative efforts by the entire health care team can facilitate this paradigm shift from *curing* to a more holistic model of *caring*.

Future

The innovation and flexibility of this satellite faculty practice were key in the development of this collaborative practice. It serves as a model for both nurse practitioner students and community health students. Additionally, the joint clinical venture between a primary care residency program and the community nursing center has the opportunity to bring the role of the NP forward significantly in the context of health care reform. The focus of the adult nurse practitioner in collaborative practices such as the one in this chapter will be ongoing and continue to evoke and involve change.

Collaborative practices such as this one not only reinforce a holistic care model, but also facilitate expanding health care services to those with limited access. Mundinger (1994) also believes that patients will be less litigious "because 'patients' in those practices receive more comprehensive and satisfying care than they would be given in practices composed of physicians only or nurses only."[9] There must be attention to the paradigm shift from curing to caring, especially for the populations served by practices like the Senior Health Service.

Although there will be obstacles whenever something new is developed, the benefits of such a unique, collaborative practice cannot be overlooked. The NP has the opportunity to become the "wise hero" as described by Walker (1993), especially for the chronically ill patients requiring primary and tertiary prevention.[12] As the development of the University of Rochester's CNC continues, and the role of the NP in health care reform becomes more dominant, collaborative practices such as the one described in this chapter will become more commonplace.

References

1. Avorn J, Everitt D, & Baker M. The neglected medical history and therapeutic choices for abdominal pain: A nationwide study of 799 physicians and nurses. Archives of Internal Medicine 1991;151:694.

2. Bessman A. Comparison of medical care in nurse clinician & physician clinics in medical school affiliated hospitals. Journal of Chronic Disease 1974;27:15–125.

3. Brown S, Grimes D. Nurse Practitioners and Certified Nurse Midwives. In *A Meta-analysis of Studies on Nurses in Primary Care Roles*. Washington, DC: American Nurses Publishing, 1993.

4. Charney E, Kitzman H. The Child Health Nurse (Pediatric Nurse Practitioner) in private practice: A controlled trial. The New England Journal of Medicine 1971;285:1353–58.

5. Hirschhorn L, Biomore T. The new boundaries of the "Boundaryless" company. Harvard Business Review; 1992:104–115.

6. Lewis C, Resnjck B. Nurse clinics and progressive ambulatory patient care. The New England Journal of Medicine 1967;277:236–241.

7. McLaughlin E, Anderson E, Cesa T, Larson P, Johnson H, Gibson J, Lemans M, Delucchi K. Nurse practitioners' public health nurses and physicians performance of clinical simulation test: COPD. Western Journal of Nursing Research 1979;1:275–295.

8. Moccia P. Nursing education in the public's trust—A faculty of the community: No unreal loyalties for us. Nursing and Health Care 1993;14(9):472–474.

9. Mundinger MO. Sounding board—Advanced-practice nursing—Good medicine for physicians? New England Journal of Medicine 1994;330(3):211–214.

10. Neuman B. *The Neuman Systems Model*, 2nd Edition. Norwalk, CT: Appleton-Century-Crofts, 1982.

11. Preston RP. *The Dilemmas of Care: Social and Nursing Adaptations to the Deformed, the Disabled, and the Aged*. New York, NY: Elsevier, 1979.

12. Walker PH. Care of the chronically ill: Paradigm shifts & directions for the future. Holistic Nurse Practice, 1993;8:36–66.

XV Developing a Curriculum for Multiprofessional Teams and Community-Based Healthcare: A Partnership Challenge for Healthcare Educators

Mary R. Talen, Margaret Clark Graham, and Gordon Walbroehl

As the healthcare system changes, educators need to redesign the curriculum to train healthcare providers so that they can work in multiprofessional teams and deliver community-based, primary, and preventive healthcare. We will describe the development and outcome of one of our multiprofessional courses that are offered across curricula in the Schools of Medicine, Nursing, Professional Psychology, and Allied Health. Multiprofessional student teams visited families in their homes and consulted on treatment plans. These training experiences helped foster collaborative attitudes between allied health, nursing, medical, and professional psychology students, and expanded their perspective of community-based healthcare services. We experienced challenges at multiple levels during the creation of these courses, and our success required a committed administration, flexible faculty members, and active student participation.

The US healthcare system is shifting away from the disease model and specialty care toward prevention and primary care, from physician-centered to patient-centered practice, and from hospital-based to community-based care. Clinical practices are developing toward integrated delivery systems involving multiprofessional teams and away from fragmented, specialty practices. Much of our professional training continues to focus on a service delivery model built on specialty tracks, professional solidarity, and teaching hospital environments. Our educational curriculum for training healthcare providers lags behind these changes. As the healthcare system changes, educators need to redesign the curriculum to train healthcare providers so that they can work in multiprofessional teams and deliver community-based, primary, and preventive healthcare.[4]

Dayton, Ohio is a fertile ground for innovative health education changes because there are academic programs for physicians, clinical psychologists, nurses, allied healthcare providers (for example, dental hygienists, dietitians, mental health technicians), and a strong community neighborhood governance. Community members, healthcare administrators, providers, and educators have been acutely aware of the need and opportunity for creating models of healthcare reform in both clinical practices and healthcare education. As a result of discussions between these groups, a partnership was formed by the citizens of Dayton, the county health department, local hospitals, Wright State University, and Sinclair Community College.

The goal of this partnership has been to facilitate the process of systemic healthcare changes and to create a more coordinated, cost-effective, and comprehensive healthcare system that is responsive to the needs of the community and accessible to all community members. The philosophy of "doing with" instead of "doing for" or "doing to" underscores the partnership's mission. Overall, the partnership has been interested in enhancing the delivery of primary healthcare, reorienting health professionals' education toward primary or prevention care, and developing multiprofessional, community-responsive practice models for healthcare.

While this partnership is a multifaceted project, we will describe only its educational component, specifically the development and outcome of several multiprofessional courses that have been offered across curricula in the Schools of Medicine, Nursing, Professional Psychology, and Allied Health. We will outline not only the structure and content of these courses but also the challenges encountered in the creation of multiprofessional clinical training for healthcare providers.

Faculty Collaboration and Course Development

Prior to this project, collaborative relationships between all of these academic programs had not been formally established. We established an education committee to begin the process of building a partnership among the faculty from the Schools of Medicine, Professional Psychology, Nursing, and Allied Health, and representatives from community health programs. These committee meetings represented the first step toward developing such a partnership for collaborating on changing current educational practices. The first collective task for our committee was to identify our goals for shaping healthcare education. After much discussion, we developed a two-fold educational mission. The first goal was to create a "teaching community" by redirecting the education of health providers from the traditional institutional settings (for e.g., hospitals, ambulatory care clinics) to the community's natural environments—schools, neighborhoods, senior citizen centers, and churches. Second, we strove to educate both faculty and students about collaboration and consultation. We focused not only on our collaboration between the different disciplines, but also on the collaboration between providers, patients, family members, and the larger community, so that students would have a better understanding of diverse groups of clients, and they would learn how to be more responsive to the needs of individuals and families in the community.

After reviewing our mission statement, we decided that our first task was to introduce multiprofessional and community-based clinical courses into the curriculum of each health provider training program: allied health, medical, clinical psychology, and nursing students. Piloting a course gave our faculty the opportunity to work together, highlighted the administrative challenges in implementing courses for students across disciplines, and provided us with student feedback on multiprofessional training. A community health nursing course was selected as the basis for this pilot course because it could easily accommodate students in the other disciplines attending home visits with patients.

Course Objectives and Structure

This multiprofessional, community health course provided students in all disciplines with a hands-on opportunity to understand the healthcare needs of the community and to practice in a multiprofessional, community-based setting. The goals of the course were threefold: 1) to provide students with the knowledge and skills to work as a multiprofessional team; 2) to provide students with information on the community and community healthcare; and 3) to provide students with direct clinical experience in community-based healthcare with individuals and families.

The initial segment of the course focused on multiprofessional team development and the definition of community health concepts.[2] The first week of class was devoted to understanding the contribution of each profession to a multidisciplinary team of healthcare providers. Students were given several different opportunities to examine their preconceived ideas and stereotypic views of allied health technicians, nurses, physicians, and psychologists.[6] Then, as a contrast, they were encouraged to define their unique contribution to the team and learn about the training and expertise of other disciplines. This was a novel experience for most of the students, and it helped demystify each other's roles. Medical students for the first time became aware of the extensive training requirements of psychologists and nurses, while psychology and nursing students came to view medical students as more approachable, which thereby built bridges for future collaboration.[6] Students were also introduced to group process issues such as leadership styles, power and control issues, and conflict negotiation. We provided them with a model of multiprofessional team processes and gave them opportunities to practice multiprofessional team consultation through role-plays. This model gave the students a structured format for discussing clinical situations and ensured that each student participated in the discussion and contributed to the treatment plan. It also provided a structured format for subsequent team consultations on family cases.

Our second objective was to introduce students to the concepts of primary and community-based health care services.[4] We encouraged students to learn about the impact of culture, race, poverty, and gender on healthcare and to recognize the strengths and resources of families from culturally different backgrounds and healthcare practices.[3,5] We hoped that students would develop more sensitive attitudes and respectful behaviors toward individuals and families from different backgrounds. African-American and Appalachian health providers and community health advocates were invited to speak to the class about cultural issues and attitudes that have an impact on healthcare. The community health advocates, who are neighbors with special training in healthcare resources, provided a unique link between the students and the families in the community. These advocates lived in the community and understood the community's beliefs and attitudes about health care. In essence, these advocates acted as cultural brokers for both students and faculty. Students also watched several videotapes that pointedly captured the life stories of families as they negotiated with the healthcare system.[1,7]

With more knowledge about teams, community health resources, and cultural diversity, students were then assigned to be health managers for several families. The faculty identified three community programs—HeadStart, school-based teen pregnancy programs, and senior citizen centers—as the "teaching community" environment for the students. These agencies were chosen by the faculty because they would provide clinical experience with families from different life-cycle stages. The community agencies then identified families that needed and wanted some health-related service. As health managers, students were responsible for establishing contact and rapport with the families, assessing their healthcare needs, and coordinating any treatment plans. Students were given the opportunity to collaborate with workers from community agencies, meet with individuals in their home environment, help clarify their healthcare needs, collaborate on a healthcare goal, and develop a healthcare plan.

Students met weekly for supervised multidisciplinary team meetings to review the family's healthcare goals and treatment plans. Each team had several nursing students, a professional psychology student, a medical student, and an allied health student. Faculty members rotated between the different teams, and community health advocates occasionally participated in team meetings. These meetings provided students with the opportunity to assess families from a variety of perspectives, share resources and information about them, and decide which student could provide the most appropriate treatment. Students were encouraged to have weekly contact with the families and to collaborate with them in developing goals.

Overall, students were able to establish rapport with both their team and family members, visit regularly with families with the support of another student, identify healthcare needs, and provide some clinical treatment for clients or help link them with other providers or community resources. These healthcare plans were developed with the help of the multiprofessional team. Team work included home visits to family members, consultation with other providers on healthcare issues, referrals and collaboration with community service providers and/or advocates, and coordination of treatment and/or services. Students also experienced firsthand the frustrations of working within the community-healthcare system, recognized the obstacles in building trust between clients and providers, and witnessed the need for reform in current healthcare practices.

At the end of the course, students were responsible for terminating and/or linking families with existing healthcare programs or resources. They learned the value of collaborating with other agency workers, building trusting relationships with families, listening effectively to what clients need, developing feasible client-centered plans, and evaluating the effectiveness of these plans. This community-based, home-visiting course required constant coordination and communication with other community agencies. Students were involved in reinforcing, building on, or augmenting the goals of community workers. Students learned the importance of networking by understanding the perspective of the case worker, defining goals that were consistent with other helpers, and terminating their relationship with families while realigning them with the other providers. Students were instructed on how to establish rapport with community workers and reminded about the importance of continuity of care. Community workers were regularly informed about the contact the students had with families, and initial home visits often included agency workers. Community workers were given copies of the student's record of contact with the families and their plans, interventions, and referrals. Fostering open communication between community agencies, the students, and families has allowed the faculty to expand community-based clinical teaching opportunities.

Case Example

Mr. Higgins, a 68-year-old white male, was seen initially by a nursing student and a psychology student. Physically, he suffered from arthritis, Parkinson's disease, and hypotension. He lived in a public, senior-citizen housing complex and spent his time writing poetry, playing bridge once a week, and visiting a friend. Because of his health problems, the social service worker wanted him in assisted-living housing. Recently, he had experienced fainting spells and muscle weakness, and his sisters had found him lying on the floor of his apartment unable to move. In later meetings with him, the students discovered that his fainting was partially due to his irregular eating patterns and inconsistent use of medication. His sisters and the agency worker felt he was a danger to himself and wanted him in a nursing home. He, however, was adamant about not living in a home, and he resisted any help from the social service worker as a way to prove that he could live in his own apartment.

The students established enough of a rapport with Mr. Higgins to discuss with him how he might use *Meals On Wheels* and a community health nurse as a way to stay in his home. One morning, however, when the students came to visit, Mr. Higgins was confronted by the agency worker and supervisor about moving into a nursing home immediately. After a tense discussion, the agency workers, students, Mr. Higgins, and his sisters agreed to defer this decision, and a faculty member was called to help facilitate a decision-making process that was sensitive to everyone's concerns: Mr. Higgins stayed in his apartment with *Meals On Wheels*, but he agreed to visit several different nursing home facilities and was therefore included in the decision about assisted-living housing. This situation alerted the students to the importance of working with the entire system of helpers and collaborating with each other.

Course Evaluations

A computer-assisted program was used to evaluate the course. Students provided feedback about the structure, content, and process of the course. For the initial three courses, the home visits were, without exception, the highlight of the learning process. Nurses who had experience in hospital settings changed their views about discharge planning and recognized the importance of developing plans "with" clients instead of giving instructions "to" patients; medical students had their first experience in meeting a patient on her or his "turf"; and psychology students were awakened to the complex and systemic relationships between physical health, family interactions, and the home environment. Allied health students gained the respect of other team members and developed self-confidence in working with other professions. They had their first educational experience in working with students from another discipline. They valued learning about the education of other students and observing their clinical work. Cultural diversity training was also well received by the students, but they felt they had only scratched the surface: they wanted to know more about the attitudes and beliefs about healthcare held by different cultures.

Obstacles and Challenges

Even though we have been encouraged by the overall outcome of these multiprofessional courses, we continually revised them because of the obstacles and challenges we experienced. We evaluated these courses from three perspectives: administration, faculty, and student.

Administration

From an administrative perspective, the implementation of multiprofessional programs required intensive top-down and bottom-up administrative support. From the top-down, four Deans, four Academic Deans, and one Program Director from the professional healthcare schools were consulted and informed about the development of this multiprofessional curriculum. They were involved for several reasons. They needed to understand and agree with the principles of the program, approve its goals and objectives, appoint faculty to develop and teach courses, and recruit students to register. Some of these obstacles were addressed when we first started, but the implementation of the program became cumbersome when Deans had to hire, reassign, or provide faculty with release time to develop courses.

The bottom-up details of coordinating schedules of faculty and students from four different graduate and undergraduate programs proved to be *the* major obstacle. Deans and faculty were either reluctant to release students from certain courses and clinical practice, or they were hesitant about adding one more thing to the students' already full schedule. Negotiating a 4–hour block of time for 10 weeks challenged each participant's investment in the partnership, their inclination to protect professional territories, and their commitment to multiprofessional team work.

Consequently, the course schedule was initially driven by student and faculty availability rather than by student and faculty interest and expertise.

Faculty

These pilot courses have been faculty intensive and expensive—approximately 1 faculty member to 5–7 students. Because of this, administrators were slow in assigning faculty. During the initial course, faculty from professional psychology had no preparation for the course, and the medical school rotated three physicians to cover the supervisory responsibilities. The nursing faculty member became the fulcrum of the team, providing structure and the teaching experience for a community-based healthcare course. Having faculty members with different investments and responsibilities did not create a cohesive working team. Instead of modeling a new approach to multiprofessional education, the first course seemed to reflect more traditional multiprofessional interactions and stereotypic professional roles. The nursing faculty took charge of the details and organization of the course; the medical faculty rotated in and out and were called out for emergencies; and the psychology faculty observed and commented on the group dynamics.

After a chaotic beginning, the faculty learned better ways to work together because a consistent team of experienced faculty members was selected to revise subsequent courses. They learned to take time for faculty development by planning and reviewing course objectives, sharing responsibilities for teaching, coordinating clinical experiences and clinical supervision, and understanding and respecting each other's perspective and clinical contribution. Both medical and psychology faculty went on home visits, developed the syllabus, and organized clinical experiences. The revised courses modeled a more equitable, multiprofessional faculty team for students and created better working relationships between the professional schools.

Students

From the student's perspective, the initial course structure and expectations for students created barriers between students and accentuated their differences. Because of scheduling complications, students who were at different stages in their clinical training were admitted into the course. Consequently, advanced, third-year medical students were "invited" to volunteer for the first course while first-year psychology students with little clinical training were recruited. These six graduate doctoral students were then placed in multiprofessional teams with five or six well-seasoned, registered hospital nurses who were finishing their BS degrees. For the nursing students, this course was a requirement, and they were expected to keep case records, write a paper, and attend all sessions. The "volunteer" medical and psychology students, however, did not have these requirements, and their level of participation reflected this status. A later course had the reverse situation—advanced psychology students matched with first-year medical students and senior-level nursing students.

Mixing students who were at different stages in their clinical training and with different expectations for course credit created difficulties. Because they were at different stages in their training, experience, and knowledge about their professional contribution, they had different levels of clinical competence. Consequently, the development of their professional identity was significantly different, and their team case consultation could easily become derailed and digress in different directions, which would then require more attention to team process issues. Expectations of the different team members had to be clarified and reviewed more frequently. In addition, students who were taking the course for more credit had more health management responsibilities. This situation created tensions among team members and an unequal investment of time with client families. Their clinical involvement with families was also significantly different because of variations in the students' level of training, comfort in their professional roles, and clinical skills. Having first-year medical students make a health-risk assessment for an elderly patient was too advanced for their training level, just as assigning first-year psychology students to assess sexual abuse allegations concerning a 5-year-old was beyond their level of competency.

The revised courses now have students at the foundation stage of professional training, and the course structure and content can be more easily tailored to match the knowledge base and skill level of these students. In addition, all students now have the same requirements and earn the same credit hours for the course. Equalizing these course requirements and expectations lays the foundation for equal participation and mutual respect. Balancing the students' level of training has become a salient factor in determining when the course is offered. Including students who are at similar stages in their professional training now resolves scheduling questions because it significantly affects the course content and clinical expectations of the students.

Conclusions and Recommendations

Multiprofessional and community-based clinical practices are emerging as new models for future healthcare services. These alternative approaches for healthcare delivery require educating the next generation of healthcare providers about multiprofessional consultation and community-based primary care. Multiprofessional clinical courses help to shape collaborative attitudes between healthcare providers, expand their perspective on how, when, and where health care is to be delivered, and enhance their understanding and skills in working with individuals, families, and communities.

The necessary groundwork for implementing these courses, however, is complex and time-consuming. Administrators must be committed to the mission of redirecting the healthcare education of their students and be willing to invest their time, resources, and money to make it possible. Flexible faculty members from the different health programs are also essential for developing a multiprofessional curriculum. The multiprofessional faculty team must grapple with power and control issues, challenge professional roles and territories, and forge new roles and clinical partnerships. This requires frequent and open communication along with compromise and negotiation among the faculty team members. And last, equal student participation is an important area for developing positive multiprofessional learning experiences. Students gain more when they enter the course from a similar stage of professional development, when they are given equivalent course requirements, and when they work together in a hands-on clinical practice experience.

References

1. American Cancer Society. *Cancer and the Poor. Report to the Nation.* Atlanta GA: American Cancer Society, 1989.

2. Baldwin, D.C., & Baldwin, M.A. Education for teamwork in primary health care. In H. Knopke & N. Dickelman, eds. *Approaches to Teaching Primary Health Care*. St. Louis MO: C.V. Mosby, 1981.

3. Gonzalez, V.M., Gonzalez, J.T., Freeman, V., & Howard-Pitney. B. *Health Promotion in Diverse Cultural Communities*. Palo Alto, CA: Stanford Center for Research in Disease Prevention, 1991.

4. *Healthy People 2000* (Videotape). National Health Promotion and Disease Prevention Objectives, U. S. Department of Health and Human Services Public Health Service (DHHS Publication No. PHS-9150212), 1990.

5. Martin, M., & Henry, M. Cultural relativity and poverty. Public Health Nursing 6: 28–34, 1989.

6. McDaniel, S.H., Hepworth, J., & Doherty, W.J. *Medical Family Therapy: A Biopsychosocial Approach to Families with Health Problems*. New York: Basic Books, 1992.

7. Public Broadcasting System. *Our Children at Risk* (videotape). New York, 1991.

PART 4

Community Partnerships

Health care delivery systems, educational institutions, and businesses are experiencing revolutionary changes. One major change is that organizations must attend in a very real way to the needs, preferences, and values of the consumer. Previously, this was only addressed in the context of consumer satisfaction with individual services. However, there is now an increased recognition that institutions serve populations of individuals, usually expressed in the context of community. Community health, education, and well-being are now considered important markers of success; to change the status of the community in these areas requires community partnerships (see the Community Partnership chapter in the Perspectives on Partnership Process section).

Community partnerships are particularly challenging to develop and depend upon the core principles presented in this book—relationship, understanding, and empowerment. In many communities, it is difficult for institutions to shed their hierarchical image and reputation. Stereotypes of certain communities must also be addressed. For members of some communities, there is a distinct distrust of institutions that translates into distrust of the individuals who are responsible for building the partnerships. Conversely, there is occasionally a devaluing of the people in certain communities because diversity has not been valued in the past by society as a whole. Institutions must first realize that they too are "communities" with their own values, perceptions, and stereotypes. Institutions must undertake a self examination in each of these areas if they hope to build effective partnerships with other communities, be they organizations, populations, or groups of individuals from a geographic area.

The process of building community partnerships is more complicated and takes more time than the dyadic relationships described previously in this book. This process frequently begins with dyadic communication between two individuals, then groups of persons from two communities (frequently an institution and a selected population) begin to develop a relationship. If initial dialogue and shared experiences between dyads and small groups are successful, then key members (stakeholders) of the two larger communities begin to come together, find common purpose, and move forward from a plan to implementation, hopefully culminating in the desired outcome as reflected in a formal evaluation.

Three papers in this section address the value, challenges, and benefits of the development of community partnerships. *Natalie Baturka, Margie Siegel, Julienne*

Roach, and *John Schorling* write about the challenges of improving health care services to a predominantly African-American population. This chapter highlights the challenges of bridging not only a cultural gap created by racial differences, but the gap between two diverse institutions—a university and rural churches. The value of the success of the initial dyadic relationships is evident in this chapter, as well as the willingness to address the "unfavorable image" of the university in the community. Finally, the persistence of the university in attending meetings, addressing rumors and other evidence of mistrust, and focusing on the process as well as the outcome resulted in a successful community partnership.

In order to improve the health of students in a New Jersey community, a partnership was created between the local school system and the department of pediatrics of the regional medical center. *Gary Eddey* and *Anthony Suchman* describe outcomes of the partnership, and highlight both the benefits and barriers. Consistent with successful community partnerships, there were reciprocal benefits to both the school system and to the department of pediatrics. Improvement in access to consultation, knowledge and skills of school health professionals were valuable to the school system, while education of residents and medical students in community settings was facilitated by this partnership. Barriers were consistent with the challenges of building community partnerships. Physicians had to learn to work collaboratively with other professionals and family members, and learn to work as participants on teams. Again, the value of persistence and consistency in relationships throughout the process was a key to the success of this project.

Finally, *John Wheat* and *William Curry* describe the challenge of developing a community partnership in rural Alabama. The purpose of the partnership development was to increase the number of primary care providers in an underserved rural setting. This chapter also highlights the challenge of bringing diverse groups into a partnership with a university. A common thread in these chapters, which is critical to the development and success of community partnerships, is the time it takes to build the relationships, develop trust, achieve common goals and purposes, and to implement a program. Finally, this chapter describes particularly well the challenge of developing partnerships with subgroups within the "community" partnership in a geographically large rural setting.

XVI The Alliance of Black Churches Health Project: A Successful Partnership for Health Promotion in Rural African-American Communities

Natalie L. Baturka, Margie Siegel, Julienne Roach, and John B. Schorling

Introduction

African-Americans suffer a disproportionate burden from many potentially preventable illnesses, including heart disease, diabetes, stroke, and cancer. In 1989, the rate of potential years of life lost due to preventable conditions was 141.8 years per 1000 persons among black men, compared to 65.3 for white men. For black women, the rate was 74.3 years, compared to 34.0 for white women.[1] In response to such disparities, the Alliance of Black Churches Health Project (ABCHP) was begun at the University of Virginia Health Sciences Center in 1991. The project's purpose has been to work with coalitions of black churches to help improve the health of the African-American population in two rural counties in central Virginia.

The use of community coalitions to address the health problems of populations has become more common recently.[2] This is a result of the realization that behavior change at the individual level can be fostered by changes at the institutional and social levels as well. Universities, through their material and intellectual resources, can aid the development of these coalitions.[3] However, coalition building requires the empowerment of others in the community, a role that is not always easily accomplished by academic institutions.

In this chapter, we will: 1) describe the genesis of the Alliance of Black Churches Health Project, and the initial organizational efforts; 2) provide a brief overview of interventions which have been undertaken; 3) discuss the partnership benefits to the community and to the University of Virginia; 4) describe several of the challenges faced by the ABCHP staff; and 5) highlight some of the lessons that we have learned from the partnership.

The authors would like to thank the officers, board members, and volunteers of the Louisa Alliance of Black Churches and the Buckingham Health Education Board, without whose trust, interest, and effort this project would have never succeeded. Although we started out to teach you, it is certainly we who have learned the most.

Program Background and Description

The stimulus for this project was a research grant from the National Heart Lung and Blood Institute. The purpose of this research was to develop and evaluate a smoking cessation program for rural African-Americans. In deciding how best to do this in central Virginia's relatively sparsely populated rural counties, we adapted a model developed by Dr. John Hatch of the University of North Carolina.[4,5] Dr. Hatch pioneered the use of black churches as sites for disease prevention and health promotion programs, and the ABCHP was designed around interventions to be delivered through church coalitions. The coalitions were established not only for the research study, but also to create health promotion organizations that would become institutionalized and continue their efforts beyond the funding period.

Two communities, defined by the geographic boundaries of two counties, were chosen as the study sites. These counties are demographically similar, adjacent to Albemarle County (where the University of Virginia is located) but not contiguous to each other, and are in different health districts. Each county has about 35 churches that are identified as primarily serving black residents. Over 90% are Baptist.

Coalition development. The development of the coalitions was directed by two coordinators in each county, who devoted 50% of their time to these activities. They began by organizing the two coalitions using community organization strategies similar to those promoted by Dr. Ron Braithwaite.[6,7] By contacting local agencies and associations, they identified community leaders to serve on ad hoc nominating committees. They then developed community resource inventories, conducted community health needs assessments, and worked with the committees to organize the coalition boards. Within six months, elected board members were in place, regular coalition meetings had begun, and board members and University staff had identified initial target health conditions.

The coalitions in each county were composed of volunteer lay persons and clergy; the "core" group in each county numbered about 15 individuals, but training and services reached a much wider population. The project began with a philosophy that community members knew best how to deliver services to their friends and neighbors. Specific guidelines for proceeding were not prescribed, other than that smoking cessation needed to be a priority in Buckingham County. As a result, the coalitions developed in somewhat different ways. In one county, the coalition board included no ministers and was called the Louisa Alliance of Black Churches. In the other, the president of the board was a minister, but the members did not want the coalition to be identified specifically as black or church focused, so they named themselves the Buckingham Health Education Board.

Risk factor survey. To monitor the effect of the smoking cessation program and to establish the prevalence of other cardiovascular risk factors in the African-American populations of these counties, we conducted an extensive two-phase survey. Phase I was a face-to-face survey conducted by the University of Virginia Center for

Survey Research. It included questions on smoking beliefs and attitudes, as well as items adapted from the 1990 *Behavioral Risk Factor Surveillance Survey* developed by the Centers for Disease Control and Prevention. Over 3,744 (52%) of the 7,183 adult African-Americans were contacted, of whom 1,679 individuals were randomly selected to receive the survey, and 1,148 (68%) actually completed it.[8]

In Phase II, health professionals affiliated with the ABCHP contacted a sample of 403 Phase I survey respondents to offer free health screenings.[9] Phase II measurements included blood pressure, height, weight, waist-to-hip ratio, serum glucose, total cholesterol, HDL-cholesterol, and glycosolated hemoglobin.

Cardiovascular risk factors were found to be extremely prevalent among the populations in both counties. Over half the adult population was obese, 38% had hypertension, 27% smoked cigarettes, 18% had high cholesterol, and 15% had diabetes.

These ABCHP data, combined with other needs assessments, were presented to the coalitions to help members plan health and prevention activities. Together with the ABCHP staff, they developed a strategy of training lay health educators from the churches and providing educational programs and health screenings to their communities.

Coalition health promotion activities. Within the African-American community, the church serves not only as a place of worship, but also as a location for educational and social activities. The coalitions took advantage of these mixed roles of the church in order to promote healthy behaviors in the target communities. In Buckingham County, health educators were trained as smoking cessation counselors, using culturally appropriate materials developed by the ABCHP. Their smoking cessation efforts also included several "Gospel Quit Nights." These events were open to the public and featured entertainment by gospel groups complete with information on smoking cessation programs. These gatherings reflected the existing spiritual and social patterns within the churches.

Health educators in Louisa County were trained in blood pressure screening, self-breast exam techniques, and CPR/first aid. The Louisa coalition also focused on educating its members and health educators in areas that would improve not only the health, but also the quality of life in their community. In addition to health education programs, their topics included how to access hospice services, write living wills, and develop a county comprehensive plan. The Louisa coalition also sponsored an extremely successful Certified Nursing Assistants course and an ongoing low-impact exercise and weight control program.

Both coalitions strived to reach beyond their church populations to offer services to the entire community, both white and African-American. The coalitions have organized free community-wide screenings at various church locations. The services provided include, among others, blood pressure and cholesterol measurement, mammography, and screening for glaucoma and diabetes. All educational programs were open to the public; special programs were developed to attract a significant cross-section of the community. Civil rights activist Julian Bond spoke in both counties, AIDS workshops have been offered, and the Buckingham coalition has conducted two community-wide smoking cessation contests, and sponsored anti-smoking poster and essay contests in the schools.

Changing behaviors through the church. Throughout the project, the focus has remained on interventions delivered through the churches. The church not only serves as a center of social interaction, but the existing social networks also provide a means of disseminating information and providing support when behavior change is attempted. Some churches have been difficult to reach, for a variety of reasons. These include: infrequent services at some small congregations; unreliable communication, especially if no one active in the coalition knows someone at the church; and varying support from the ministers.

The African-American churches in both counties hold meetings, revivals, homecomings, and reunions that often last for hours. Food is usually served, and it is often high in fat, calories and salt; a healthy lifestyle is not regularly promoted. To address these issues, educational programs have been developed to address dietary factors. Within the church communities, changes have started to occur. Church groups have begun requesting "heart-healthy" cooking demonstrations for special events, and coalition members have begun preparing low-fat foods at their revival and reunion offerings.

The Role of the University in Coalition Development

From the outset, the University, as represented by the ABCHP staff, has served three roles, providing facilitation for coalition development, access to resources, and liaison to other parts of the University and the community. The ABCHP staff initiated formation of the coalitions and supported implementation of their projects. The Project Coordinator provided expertise in community organizing; other staff members provided research, medical, evaluation and grant writing expertise. The staff also helped with such tasks as publicity, publication of a newsletter and fundraising.

The ABCHP has been able to tap the vast resources within the University Health Sciences Center to benefit the coalitions. It has provided access to consultants and speakers, including the then Surgeon General, Dr. Jocelyn Elders, and legal expertise for incorporating and obtaining tax exempt status. When coalitions want to offer training, screening, and medical services, ABCHP staff help to identify and recruit qualified professionals to help.

The staff has also helped the coalitions gain access to resources outside the Medical Center. These resources included both contacts and grants at the state and federal levels. For example, the Project Coordinator helped to form a link between the coalitions and the Virginia Department of Health; this relationship led to grant-writing workshops and ultimately resulted in a State Minority Health grant. ABCHP contacts also resulted in an alcohol and drug abuse prevention workshop which, in turn, led to the development of a county-wide drug abuse prevention program involving a diverse cross-section of the community.

ABCHP staff have fostered collaborative projects in the community. One very successful example is the Certified Nursing Assistant Program which has trained 18 graduates to date. This course, sponsored by the Louisa Alliance of Black Churches, was the result of a collaboration between Medical Center nurses, a local nursing home, the local agency on aging, the local community action agency, and two home health care agencies.

The ABCHP staff also provided access to resources outside the Medical Center that were important to the coalitions' ongoing viability. While the coalition members worked to elicit support and donations within their respective communities, the Project Coordinator made countless contacts to identify financial support for existing and future programs.

Benefits to the University from the Partnership

The ABCHP is one of the University's first major research efforts which has included community organizing and empowerment as one of its primary goals. Conducting a valid and useful study has been important; however, providing long-term benefits to the community members, who in return assisted us in doing our research, has been equally essential to the project's success. While the community has, in fact, benefitted from services and programs provided by the ABCHP, the University has received numerous benefits as well.

Educational benefits. Due to its partnership with the coalitions, the University has had the opportunity to provide educational experiences for students and medical staff. These have enhanced many existing academic programs. Medical students, particularly those involved with the National Student Medical Association, have participated in many training services for health educators and in all of the free community screenings. The ABCHP has also facilitated participation by nursing and medical students, residents, and staff physicians from a variety of specialties.

Opportunities to work in the community have been particularly beneficial to minority students and residents. They often experience frustration attempting to change patient behaviors in clinical settings. Working in the community has allowed them to gain a better appreciation for the complex influences in their patients' lives.

Clinical practice benefits. The educational programs which have taken place outside the Medical Center have enhanced clinical practice. All the problems addressed by the coalitions are common among the rural patients seen at the Medical Center. Not only has the ABCHP increased the medical staff's understanding of cultural and environmental influences on behavior, it also has increased knowledge about the availability, or lack of, community resources. Patients seen at the Medical Center have been referred to the coalitions' smoking cessation programs and exercise and weight control classes. These resources were not available to physicians and their patients prior to the implementation of the project.

Research opportunities. The ABCHP has not only increased Medical Center access for a hard-to-reach population from a clinical perspective, but it has also fostered the development of collaborative research projects. By serving as a liaison between researchers at the University and the community coalitions, the ABCHP staff has been able to help design research proposals to benefit both the academic community and these minority populations.

These grant-writing projects have served as an excellent vehicle for ABCHP staff to educate colleagues about community dynamics and influences, aspects of rural African-American culture specific to the two counties, the importance of coalition autonomy, and appropriate roles for the University in community-based research. Previous traditional research tended toward conducting the research, collecting data, and then returning to the University to analyze it. The ABCHP approach asks, "What can we put in place within the community that will remain after the research is completed"? These "research residuals" might include training and education of community members, improved or new local health care programs, or increased access to Medical Center services. Collaborative grant proposals have been written with the University's Diabetes Center, the local mental health research center, and a health policy institute.

Acceptance of the University by the community. Finally, the ABCHP has had a positive impact on the Health Sciences Center's image in these communities. Initial acceptance of the project was a major challenge, in large part because of the communities' previous negative experiences with Medical Center research. In addition to rural-University distrust, there was also suspicion among African-Americans in these communities. As described above, University-based medical studies typically entailed researchers coming to them to collect data and then leaving, never to be heard from again. At one point early in the project, rumors spread that the Medical Center was injecting African-American men with the HIV virus as part of another research study.

In Louisa County, after listening to our initial proposal, one woman said, "That's all fine, but you must show us that you're going to do something, that you're committed. The talk's fine, but what are you really going to do for us?" ABCHP staff recognized that attending to this woman's question as we implemented the project would be critical to our success.

Eventually this perception did diminish, and the personalities and character of project staff were important in breaking down barriers. The staff members' ability to establish rapport with coalition and community members was invaluable. At a panel discussion held at the Medical Center, one of the Buckingham coalition members spoke to this issue. He said that one reason the relationship had been successful was due to the respect the Buckingham organizer showed him and others. "She always called me 'Mr. Earl' [not his actual name]. Most people don't do that. I finally had to tell her to call me 'Joseph'."

Staff also addressed the acceptance problem with time, flexibility, and persistent community contact. Most importantly, ABCHP staff made certain to follow-up on every request and promise made. During the first year of the project, the Louisa County supervisors asked the coalition to participate in developing the comprehensive plan for the county. When members wanted more information about the process, the ABCHP Coordinator arranged for two African-American University faculty in the Planning Department to conduct workshops for coalition members. Interestingly, the woman who questioned our intentions in the beginning served on the county planning board. She then enrolled in the Masters Degree program in Planning at the University. Three years later, at the annual meeting, her comment was, "This project changed my life."

For the University and Medical Center, the goodwill fostered by the ABCH Project has opened doors in these rural African-American communities. The coalitions are now seen as a way to contact previously hard-to-reach populations. Whether University members need information from the community or want to publicize available medical services, there is now a more efficient means to do it.

Challenges Faced by ABCHP Staff

The staff have faced other challenges besides initial community mistrust. These centered around three issues: cultural differences between staff and the communities; role definitions; and limited resources. In this section, we outline these three areas and our responses to them.

Cultural differences. Initially, none of the permanent ABCHP staff were African-American, and none had any previous significant experience with rural African-American churches. In order to address this potential barrier, the staff attempted to approach the coalitions without preconceived ideas about the best ways to implement interventions in the communities. In addition, we have remained open to the need to address both spiritual health and physical health in coalition activities.

We have attempted to reduce any barriers presented by our different cultural backgrounds by interacting frequently and also by trying to listen and observe. Staff members have regularly attended church services and other community events. Developing a collaborative rather than directive relationship with community coalition members has been key, as has been our emphasis on fostering trust and respect in all our interactions.

Defining our roles. The fluid roles of ABCHP staff have been an ongoing struggle. Because our primary objective has been to establish coalition autonomy and create permanent community institutions, walking the fine line between being supportive and being directive has often been difficult. How do staff provide support while still fostering and maintaining coalition autonomy?

In hindsight, this problem was resolved by our overall philosophy and mission regarding our work with the coalitions. While it was rarely articulated, the existence of a shared set of values among staff facilitated our ongoing role definition. In essence, we have seen our role as facilitators whose prime directive was to empower and foster autonomous and sustainable coalitions. This shared belief simplified ongoing decision-making in numerous areas. Given a choice, if an activity would benefit the coalitions, then we pursued it. If it would not enhance their autonomy, then we did not pursue it.

This guiding philosophy also clarified the roles of coalition members in relation to ABCHP and Medical Center staff. Coalition members were always asked to participate in presentations about the project to the University community and others. Credit for success has always been given to the coalition volunteers who have made the project possible.

It seems that one reason the coalitions have been so successful is their similar shared philosophy. While ABCHP staff worked to serve the community, the coalitions worked to serve God and to improve the health of their neighbors and families. This clarity of purpose has served them well throughout the course of the project.

Limited personnel and financial support. Because the National Heart Lung and Blood Institute funds were primarily for research, there were limited resources available for other coalition activities and programs. ABCHP staff spent considerable amounts of time identifying potential support from the University and communities. While these contacts on behalf of the coalitions increased their visibility and created roles for non-coalition members, the fund-raising or (donation-raising) time might better have been spent on other activities. These donations included everything from money to pay prayer breakfast caterers to free bagels for training workshops to barbecue grills and radios as prizes for community contests.

Because of limited funds, nearly all training services required volunteers from the Health Sciences Center or organizations such as the American Lung Association or American Cancer Society. We were fortunate in that there was considerable interest at the Medical Center in the ABCHP and its work. Many nurses, nursing students, medical students, and physicians expressed support and willingness to donate time for specific projects. However, the demands on their time within the University often prohibited their participation. Institutional support for medical staff to provide services within their community, which we have been lacking, is key for the development of long-term successful relationships. The time demands for this project have been relatively large because of the long driving distances to the counties from the Medical Center.

The project has succeeded in leveraging funding by collaborating with many other groups who are also interested in serving this population. In addition to providing the needed resources for programs, the collaborations also forged relationships that will improve the sustainability of the coalitions.

Conclusion: Lessons Learned

Nearly five years have passed since the initial formation of the ABCHP, and both coalitions continue to do well. Initially, the coordinator for each coalition was a staff member at the University. These roles have now been assumed by individuals from the communities, and each coalition currently has a totally independent local office and small paid staff funded by contributions and new grants. This transition to greater local ownership has been very important for the maturation of the coalitions, and will hopefully lead to long-term success. At a recent ABCHP Annual Advisory Board meeting, volunteers reported on new programs that had been added to their outreach: a Stroke Survivors program, a Preacher-to-Preacher initiative, an exercise program for senior citizens, and a Gospel Fest for breast and testicular cancer awareness. The Buckingham Health Education Board has received funding from Project ASSIST to expand its smoking program to target youth prevention.

Perhaps the most lasting result of this particular partnership between these communities and the University has been the collective lessons we as a staff have learned. We hope that these can be integrated into future programs and relationships. The following is a summary of these lessons, which have evolved over the three-year partnership between the University and the two coalitions.

Acknowledge and address preconceptions concerning the University's agenda. The best way to identify these preconceptions is to ask coalition members what their past experiences with the University have been and to explore with them community perceptions. This often cannot be done until a good relationship has developed between University staff and the coalitions. Once the dialogue begins, however, the concerns need to be addressed openly and directly.

Early visibility in the community is important to validate individual commitment. Staff members must interact frequently with community and coalition members early in the project in order to demonstrate a serious commitment to coalition development. Again, positive interaction facilitates discussion about community concerns and affords staff the opportunity to address them. Early visibility also enhances the coalitions' ability to recruit and retain members.

Keep promises—big and small—made to the coalitions. Staff members have made efforts to follow through on any commitments made to the coalition members, and this has been critical to establishing our credibility with them. By always doing what we said we would do, we have also increased trust between the coalitions and the community and churches.

Flexibility and responsiveness are essential when working with coalitions. Staff must be responsive to the changing needs of the coalitions in designing and implementing programs. The agreed-upon goals of the coalition should take precedence, even at the expense of the needs of the outside partners. Preconceived protocols are often unworkable, and wasted efforts can be avoided by eliciting coalition input.

We also learned that we could not limit activities to health care, that we needed to include other coalition interests such as community planning and small business workshops.

Cultural sensitivity is necessary in all planning and interactions. This includes accepting the realities of the community's "calendar." Summer revivals, holidays, other church and community activities, and the busy schedules of active coalition participants limit the number of meetings per year and the speed of tangible progress. Staff expectations changed often as we became more knowledgeable about coalition and community needs.

Sensitivity to the expectations of coalition members also meant designing projects that incorporated spiritual and religious elements into the health interventions and coalition procedures. Some examples were inclusion of a devotional booklet as part of the smoking cessation package, prayers at meetings, churches as meeting locations, and ministers as speakers at special events.

Community control is the key to long-term success. From the outset of the project, community members need to be actively involved in the decision-making process. As trained health professionals, we are used to telling others what to do. Yet if the goal of coalition building is local empowerment, individuals within the communities need to develop the skills and be given the responsibility to make the decisions that will affect them. These decisions may then not be in the best interest of the University, and developing the discipline to accept this can be difficult.

Design projects that create high visibility for the coalitions in their communities. High visibility events, such as county-wide health screenings, school contests, and national figures as speakers, produce public recognition for the coalitions and members who generously donate their time and talents. For these events, invitations to participate should be extended to a broad range of community members, helping to breakdown racial and religious barriers. University staff can play an important role by helping to organize high visibility activities and by publicizing them in a variety of media.

Realize that the rural, African-American community is not homogeneous. First, arbitrarily chosen populations or communities may not have meaning to the individuals involved. Second, there are significant differences between churches, socioeconomic groups, and other subcultures. Heterogeneity abounds, and this necessitates, among other things, a wide variety of programs and services.

Develop a strong working relationship with local public and private agencies. Working closely with existing community groups eventually provides increased opportunities and resources for the coalitions. This begins to happen once the coalitions establish credibility in the community, and other groups view them as gatekeepers to a target population which is often underserved.

Overall, this project has been quite successful, as judged by the benefits enjoyed by both the coalitions and the University. This has largely resulted from the development of a shared sense of purpose,[10] and the gradual shift of power from the University to the communities. This shift has occurred as the coalitions have gained financial and administrative independence—the ultimate measure of success.

References

1. CDC. Trends in years of potential life lost before age 65 among whites and blacks in the United States, 1979–1989. MMWR 1992;47:889–891.

2. Butterfoss FD, Goodman RM, Wandersman A. Community coalitions for prevention and health promotion. Health Ed Research 1993;8:315–330.

3. Levine DM, Becker DM, Bone LR. Narrowing the gap in health status of minority populations: a community-academic medical center partnership. Am J Prev Med 1992;8:319–323.

4. Eng E, Hatch J, Callan A. Institutionalizing social support through the church and into the community. Health Ed Q 1984;12:81–92.

5. Hatch JW, Voorhorst S. The church as a resource for health promotion activities in the black community. In: Health Behavior Research in Minority Populations: Access, Design, and Implementation. Becker DM, Hill DR, Jackson JS, Levine DM, Stillman FA, Weiss SM, eds. USDHHS, NIH Publication No. 92–2965, 1992.

6. Braithwaite RL, Murphy F, Lythcott N, Blumenthal DS. Community organization and development for health promotion within an urban black community: a conceptual model. Health Ed 1989;20:56–60.

7. Braithwaite RL. Coalition partnerships for health promotion and empowerment. In Braithwaite RL, Taylor SE, eds., *Health Issues in the Black Community*. San Francisco: Jossey Bass Publishers, 1992.

8. Schorling JB. Stages of change of rural African American smokers. Am J Prev Med 1995;11:170–177.

9. Willems JP, Hunt DE, Schorling JB. Coronary heart disease risk factors and cigarette smoking among rural African Americans. J Nat Med Assoc, in press.

10. Allen RF, Allen J. A sense of community, a shared vision and a positive culture: core enabling factors in successful culture based health promotion. Am J Health Promotion 1987 (Winter);40–47.

XVII Partnership between a School District and a Department of Pediatrics

Gary E. Eddey and Anthony L. Suchman

Introduction

Although the field of school health dates back to the early part of this century, educational institutions and health care delivery systems have tended to remain isolated from one another. Health care providers are often the first professionals that parents consult regarding poor performance of students, yet these providers seldom become involved in educational systems. More recently, the school-based health center movement has been developed to provide continuity of care of school-aged children, assure follow-up of care, and prevent gaps in health care delivery by improving the link between existing school health services and the community health care delivery system.[1] These school-based health centers are often sponsored, organized, and managed by school boards, hospitals, public health departments, nonprofit organizations, community health centers, and family planning clinics.[2]

Previously, many educators were unaware of the potential contributions health care providers could make to understanding and responding to students' school performance problems, particularly those related to the "new morbidities" of childhood and adolescence: emotional and behavioral problems, learning disorders, attention deficits, developmental delays, and problems such as alcohol and drug use as well as violence among peers.[3] Such "social morbidities" are further complicated by the fact that 20% of teenagers in the US have at least one serious medical problem ranging from heart disease and epilepsy to severe psychological difficulties.[4,5]

Physicians are now required to become more involved with schools by recent federal legislation, specifically Public Law 94–142 and its subsequent amendments of 1986 and 1990, which address the educational concerns of children from infancy through age 21 who are developmentally disabled or are otherwise at risk for significant academic, social, and/or medical difficulty in a regular classroom setting. This legislation gives physicians and other health care providers such as nurse practitioners,

---

The authors wish to acknowledge the help of Drs. R. Schonhorn and G. Barabas in understanding the team process and for editorial assistance; and Drs. J. Truman, S. Wang, and J. Mattingly for help in the maintenance of the program.

physician assistants, social workers, and psychologists who care for children responsibility for optimizing school functioning for this large group of educationally disadvantaged students.[6]

This chapter describes how a partnership between a New Jersey public school system and a Department of Pediatrics succeeded in addressing a wide range of school performance problems in children and adolescents. It describes how multidisciplinary collaboration can be singularly successful in resolving complicated, longstanding, systemic problems, and how traditional institutional structures, behaviors, and attitudes may thwart collaboration even without anyone's direct intention to do so. Finally, it offers an interesting story of how a critical and serendipitous bridge between two poorly interacting institutional cultures permitted creative and effective solutions to emerge.

How and Why the Partnership Was Created

Educational regulations in New Jersey mandate a variety of physician services in public schools.[7] The school physician may be called upon to perform routine physicals for students in certain grade levels, to provide pre-participation sports examinations for middle and high school students,[8] or in some instances to function as a consultant.[9,10,11] Most public school districts contract for these limited services with one or more physicians working in conjunction with the school nurses. Typically, very little communication occurs or is expected to occur between the school physicians and school district staff. Because this type of interaction has historically offered limited educational opportunities, school health programs have been viewed as generally unsuitable for medical education.

In the Morristown School District, which consists of ten schools serving approximately 6000 students in three large, demographically diverse towns in New Jersey, the role of school physician was traditionally filled by primary care physicians from the community who would devote approximately two hours per week to this role. When, in the late 1970s, the District had difficulty recruiting physicians, it contracted with the Department of Pediatrics at Morristown Memorial Hospital to provide school health services. Attending physician coverage was to be provided by the hospital with additional coverage by pediatric residents and medical students (all under an attending pediatrician's supervision). This relationship provided additional manpower for the school district and at the same time provided residents and medical students with an excellent opportunity to learn how to perform routine physical examinations and an introduction to school health. Over time, this simple contractual relationship evolved into a more comprehensive partnership that could best be characterized as a school-linked service effort.[12]

Early in the partnership the pediatricians concentrated only on the basic obligations of the contract, primarily performing routine physical examinations. By this time, the school district required six hours per week of scheduled time from the physicians, medical students and residents. There was little time to interact with school district staff and little attention was paid to the broader range of issues involving school health. In the mid 1980s, the Department of Pediatrics at Morristown Memorial Hospital hired a pediatrician with previous experience as a primary school teacher to help coordinate the school health program. This "accidental factor" ultimately played a pivotal role in the history of the program; the pediatrician's ability to feel at home in both the school and hospital environments allowed him to serve as a bridge between the educational and medical communities, leading to progressive growth in their interaction.

Collaborative Projects

The development of partnership between educators and clinicians is best illustrated by the evolution of the Child Study Team (CST)[13] which is charged with performing a comprehensive evaluation of children who are performing or functioning below their expected level, and then developing an individualized educational plan to help them reach their potential. The CST is comprised of educators, social workers and psychologists. Initially, its only interaction with the health care system consisted of referrals of children to the school health physician for routine physical examinations in fulfillment of regulatory requirements; communication between the CST and the physician was rare. The CST did not routinely provide its assessments of the children's problems to the examining physician thinking that the physician would have little need for that information. Similarly, the physician provided brief reports that only rarely included findings or hypotheses pertaining to the children's learning difficulties. The physician's exam was geared to identify health problems that might affect the child's ability to participate in physical education or sports. In the absence of effective communication, collaboration was not possible.

Two factors led to a striking reversal of this situation. The first was the establishment of a policy by the school district mandating more communication: at least one CST member was to meet with the examining physician just before or after each medical examination. When this was not possible, the school nurse was to convey detailed information between the CST and the physician. The second was the establishment of a dialog between medical and educational professionals leading to a better understanding of each others' disciplines. Neither group adequately appreciated the "blind spots" of its own approach or the potential contributions of the other (especially their unique observational vantage points and analytic frameworks) in evaluating students' problems. This dialogue was enormously facilitated by the presence of the physician with prior experience as an educator. He was able to identify, and help others see, the synergistic potential of a truly collaborative approach. For example, teachers, by virtue of the amount of time they spend with students, have an overview of behavior and performance that is unequaled. They witness every day what physicians struggle to learn second hand, namely, a "history" of behavior. Yet, more often than not, the observations of teachers are not sought by the medical profession and when they are, the teachers do not have an adequate clinical language to structure their observations and convey them to the physicians. In the Morristown program, interdisciplinary workshops were organized to give teachers better tools for observing and communicating about student behavior.

With these improvements in both the quantity and quality of communication, the physicians began to function as team members; phone contact between physicians and school personnel has increased steadily and the school physicians are then able to communicate the educators' perspectives to other members of the medical team. Sometimes physicians observe children in the classroom setting, and meet afterwards with teachers and principals. They also hold informal teaching rounds with CST members and school nurses. The CST has come to value the medical contribution to its evaluation process and to expect constant communication rather than episodic consultation.

The role of school nurses on the CST has changed as well. Previously, their involvement was limited to scheduling medical examinations. Now their clinical observations and recommendations are often sought and they have operational responsibilities for such key functions as neurodevelopmental assessment, the monitoring of medication effects in children with attention deficits, and the on-site evaluation of somatic complaints that can sometimes accompany learning problems.[14]

The changes in the CST process go beyond increased respect and recognition of team members' contributions; there is a fluidity of function and roles that allows the team to deploy its resources differently in each case to obtain maximal benefit. A rigid and compartmentalized distribution of responsibilities and authority has been replaced by a more flexible pattern of shared decision-making and leadership. One team member or another will rise to leadership in formulating the team's assessment, designing and implementing solutions and communicating with others according to the specific needs of a given case and the abilities of the team member.

The teamwork of the CSTs led to other creative collaborations between educational and medical professionals. A case in point is the formation of interdisciplinary absenteeism committees. Excess absenteeism is an area of school health that has been traditionally neglected both in clinical practice and in the literature.[15] In the Morristown district, the traditional evaluation of excess absenteeism consisted of a discussion among the parent, teacher and principal of a child's absenteeism. The typical response was retention in grade, a strategy that had both remediative and punitive intent. However, despite its widespread use and strong public support, retention does not help children to "catch up" and may increase their risk for future failure.[16]

Building on the success of the interdisciplinary CSTs, interdisciplinary groups now address absenteeism at each school in the district. As the physicians became involved in this program, they became concerned about the consequences of retention and advocated for treatment of excess absenteeism as another form of school performance problem.[17,18,19,20] They suggested timely intervention by the CST to prevent the need for retention.

As a result of collaborative planning, a new process was developed. Each school's principal now chairs its absenteeism committee and contacts the school physician through the school nurse to perform the initial review of cases of excess absenteeism. With the parent's permission, the school physician contacts the student's health care provider(s). If a medical issue appears not to be the primary reason for the excessive absences, the child is referred to the Child Study Team, again, with parental approval. The school physician serves as the liaison between the absenteeism committee, the family and the primary care physician. Through this collaborative team approach, students are now more likely to receive timely help for the real causes of poor attendance and to avoid a secondary crisis precipitated by retention.

The Benefits of Collaboration

Collaboration between educational and health care institutions has had a number of benefits for the Morristown School District. First, there has been improvement in the quality of evaluations, particularly with regard to comprehensiveness, of the evaluation of school performance problems and absenteeism. Second, there has been substantial growth in the knowledge and skills of educational professionals with regard to attention deficit disorder and other "new morbidities." As a result, the expectations and performance standards of educators in this area have been raised. For example, in-service courses on these topics which used to be presented by the school physician are now given by school personnel. Third, there is now a network for informal consultation that was not previously present. Educators have access to interdisciplinary consultation at a much earlier stage in the evolution and recognition of a problem than they did previously.

The collaborative School Health Program has given Morristown Memorial Hospital the opportunity to deliver on its commitment to provide and support community-based services to complement its tertiary care programs. The hospital and community-based pediatric continuity clinics have become better able to address the needs of their school-aged patients. Pediatric residents and coordinators of nursing service in the hospital continuity clinics now communicate more regularly and more effectively with school personnel. The school district is now linked to the full service public health pediatric clinic in its region. The close relationships that the school nurses have with the nursing service departments in the various pediatric clinics ensures prompt and appropriate access to the health care system for all children seen through school health programs. The Department of Pediatrics has gained an opportunity to introduce its residents and medical students to the field of school health and to teach them about the role of physicians in community organizations. The school district's partnership with the School Health section of the Department of Pediatrics has encouraged new relationships with other divisions in the department. For example, with new lines of communication open, the Adolescent Service was able to create a forum for educators and physicians to interact and collaborate.

As a result, the Adolescent Service is now working with the secondary schools to develop one program for rapid evaluation of suspected substance abuse and another to evaluate the students exhibiting violent behavior, and has also developed conferences for educators addressing many of the needs of the regional secondary schools.

Barriers to Partnership

There were many potential barriers to the process of partnership in the Morristown school health program. The inability of educational and health care professionals to appreciate the strengths and limitations of each other's fields was causing misunderstandings and potentially adversarial situations. Each group eventually learned about each other's roles in improving students' school performance through collaborative interdisciplinary staff development and continuing education.

Reimbursement for health care provider services was costly. Fortunately, the Board of Education recognized the value of physician services when utilized in a comprehensive school program, and that these services might result in savings through the more rapid resolution of chronic and costly student problems. More recently in many states, mid-level providers such as nurse practitioners and physician assistants are serving student populations for minor illness, reproductive health care, treatment of chronic illnesses, immunizations, health promotion and psychological counseling.[21]

Physicians were needed who could work collaboratively with other professionals and with families in a community setting. In general, physicians have not been trained to function as participants on teams. In this case, the fact that one of the school physicians had previous experience as an educator facilitated the team building process.

Educators were initially reluctant to supervise the performance of physicians in the school health program. Over time, however, responsibility for the supervision of physician services has shifted from the school nurses (who functioned more as coordinators than supervisors) to the senior district administrators, who define specific responsibilities, performance standards and guidelines for continuing professional education to assure that students' needs will be met.

Conclusion

This partnership between the health care and educational communities in Morristown is a good example of interinstitutional collaboration and offers several lessons. As is true of most successful partnerships, this one was based on both deliberate, skillful effort and serendipity. The previous teaching experience of the school health physician gave him a greater understanding of the needs of students and schools from the perspective of the educators. He was able to communicate this to other medical professionals, and likewise helped educators understand the medical perspective. As the two professions gained a better understanding of one another and the problems they were facing, communication improved and new collaborative solutions became possible.

School professionals came to recognize the usefulness of physicians in addressing school performance problems and other issues and learned how to utilize them as a resource. These educators also learned to stipulate training needs for health care providers to prepare them for working with students. School administrators learned that they could and should direct (or at least co-direct) the collaborative school health programs.[22] Their leadership and active involvement are essential for setting policies and overcoming barriers.

The physicians and other health care providers learned to work effectively as team members within the educational system, departing from a historical pattern of limited, contractually specified services and poor communication. The obstacles encountered by public school systems when developing policies to integrate health care professionals into their programs are not insurmountable.[23] Medical education and residency training programs should recognize the importance of collaboration skills and include them in their curricula. Collaborative school health programs should be recognized by pediatric residency programs as excellent sites for exposing pediatric residents to community experiences.

References

1. Falsetti D, Kovel A. How one school-based clinic is meeting the challenge of adolescent health care. Journal of the American Academy of Nurse Practitioners 1994;6(8):363–368.

2. Lear JG. Building a health-education partnership: The role of school-based health centers. Pediatric Nursing 1992;18(2):172–173.

3. Haggerty RJ, Roghmann KJ. The stress model for illness behavior. In RJ Haggerty, KJ Roghmann, IB Pless, eds., *Child Health and the Community*. New York, NY: John Wiley & Sons; 1975.

4. Velsor-Friedrich B. Schools and health, Part II: School-based clinics. Journal of Pediatric Nursing 1995;10(1):284–289.

5. Lavin AT, Shapiro GR, Weill, KS. Creating an agenda for school-based health promotion: A review of 25 selected reports. Journal of School Health 1992;62(6): 212–228.

6. Purvis P. The public laws for education of the disabled—The pediatrician's role. Dev & Behav Ped. 1991;12:327–339.

7. New Jersey State Department of Education Administrative Code and Statutes; Chapter 1. Trenton, NJ: State Department of Education, 1990:5–6.

8. Hoekelman RA. The preparticipation sports physical examination (editorial). Ped.Ann. 1992;21:145–146.

9. Nader PR, Caldwell BS, Parcel GS. Overview of school health. In RA Hoekelman,

S. Blatman, SB Friedman, NM Nelson, HM Seidel, eds., *Primary Pediatirc Care*. St. Louis, MO: CV Mosby Co, 1987:657–661.

10. Committee on School Health, AAP. The Role of the physician in schools: Policy statement. AAP News, April 1992.

11. Newton J, and Committee on School Health. *School Health: A Guide for Health Professionals*. Elk Grove, IL: American Academy of Pediatrics; 1987.

12. Behrman RE. School linked services. In RE Behrman, ed. *The Future of Children*. Los Altos, CA: Center for the Future of Children, David and Lucile Packard Foundation, 1992;2:4–5.

13. Purvis P. The public laws for education of the disabled: The pediatrician's role. Dev Behav Pediatrics 1991;12;5:327–339.

14. Levine MD. Developmental variation and learning disorders. Cambridge, MA: Educ Pub Svc; 1987:505–506.

15. Nader PR. School absenteeism. In RA Hoekelman, S Blatman, SB Friedman, NM Nelson, HM Seidel, eds., *Primary Pediatric Care*. St. Louis, MO: CV Mosby, 1981:670.

16. Shepard L, Smith ML. Synthesis of research on grade retention. Educ. Leadership. 1990;84–88.

17. Rogers KD, Reese G. Health studies: Presumably normal high school students. Amer J Dis Child, 1965; 109:9–27.

18. Weitzman M, Klerman LV, Alpert JJ, et al. Factors associated with excessive school absence. Pediatrician 1986;13:74–80.

19. Weitzman M, Alpert JJ, Klerman LV, et al. High-risk youth and health: The case of excessive school absence. Pediatrics 1986;78:313–322.

20. Klerman LV, Weitzman M, Alpert JJ, et al. Why adolescents do not attend school. The views of students and parents. J Adol Health Care 1987;8:425–430.

21. Kaplan, DW. School-based health centers: Primary care in high school. Pediatric Annals. 1995;24(4):192–200.

22. Behrman RE. Concerns about school-linked services: Institution-based versus community-based models. In Behrman RE, ed. *School Linked Services*. Los Altos, CA: Center for the Future of Children, David and Lucile Packard Foundation, 1992:107–117.

23. Jehl J, Kirst M. Getting ready to provide school-linked services: What schools must do. In Behrman RE, ed. *School Linked Services*. Los Altos, CA: Center for the Future of Children, David and Lucile Packard Foundation, 1992:2.

XVIII Rural Alabama Health Alliance: A Strategic Partnership for Rural Health Development

John R. Wheat and William A. Curry

Introduction

Alabama's rural health crisis[1] reflects a greater crisis in all sectors of rural communities and in the quality of rural life.[2] Urban centers provide opportunities that are missing from small towns. Even children of prosperous farmers, who may be assured a local livelihood, are finding the attraction of distant opportunities too great to ignore.[3]

In the 1970s, Alabama's response to the lack of rural physicians was to expand its existing medical school and to establish a new one. Although a few towns gained new family physicians,[4] the strategy of increased production was generally unsuccessful; for the problem was not that too few physicians were being trained, but that too few trained physicians wanted to live in rural Alabama. Policy makers, who were predominantly of an urban perspective, did not recognize what most rural Alabamians knew: that it was fallacious to regard all graduates as potential rural doctors. Young doctors with urban mindsets were not likely to find rural living consistent with their expectations regarding lifestyle. Instead, it was young people who grew up in rural areas who were most likely to return there to practice.[5–9] Alabama's rural voices, which advocated for the recruitment of rural school children and the integration of rural medical facilities into the medical education system, were too weak to be influential in the development of the state's medical education policy. The result was an "urban undertow"[10] in which resources that were intended to support rural development were captured instead by urban institutions, exacerbating rather than reversing the decline of rural communities.

To gain a stronger, more effective voice in the political struggle over health policy and resources, representatives from three rural counties in West Alabama organized the Rural Alabama Health Alliance (RAHA). The purpose of this non-profit corporation was to:

1. promote medical and other health-related education in rural communities;
2. enhance rural health care;
3. stimulate rural community development; and
4. engage in fund raising related to such activities.

RAHA communities wanted to create local opportunity via medical and health education, improved professional resources, and associated economic development.[11] They also wanted to influence policy in the areas of education by increasing the number of rural-reared physicians and service to underserved populations (for example, authorizing the practice of mid-level practitioners and home health nurses in a way that allows health care businesses in rural communities to compete more effectively with urban hospitals, universities, and public health for these personnel). Prior to forming the alliance, individual RAHA communities had no hope of competing with two state medical schools for state funds to support innovative medical education in the rural community. Afterwards, they competed successfully.

The remainder of this paper expands on methods used in RAHA's development and reports on its current progress, including RAHA recommendations to other rural communities with similar concerns. We begin with personal statements to set the stage for the discussion.

Contextual Statements

J.W. statement: Bill (W.C.) and I became involved in deliberations that eventuated in the Rural Alabama Health Alliance in 1990. Our colleagues and we found a rural voice lacking in policies being developed to deal with Alabama's rural health care crisis. For example, the Alabama Department of Public Health initiated an Office of Rural Health, appointing a director without openly consulting any rural constituents; the Office of Rural Health incorporated the Alabama Rural Health Association with a board dominated by ex officio members whose official positions of employment were urban-based institutions; and the Governor, at odds with the State Health Officer, decreed a separate Governor's Office of Rural Health and appointed a director who was guided by a board of the Governor's cabinet members. Nowhere, it seemed, could an authentic, rural Alabama voice find leverage to impact policies of these offices and organizations.

We took somewhat different routes in coming to a common interest in seeing the rural voice strengthened. Both of us were from small towns in rural West Alabama, from contiguous counties actually, and attended the University of Alabama at Birmingham (UAB) as pre-med students in the early 1970s when Governor George Wallace was finding political support for expanding the state's medical education capacity, expressly to meet rural health needs. We vaguely remember having seen each other in some classes and having had some mutual acquaintances. From the University in Tuscaloosa our paths diverged as we went to medical school, Bill (W.C.) to Vanderbilt supported by a National Health Service Corps Scholarship and I (J.W.) to UAB supported by a U.S. Navy Health Professions Scholarship.

While we were immersed in medical education and service obligations, significant changes in medical education occurred in West Alabama. The University of Alabama campus in Tuscaloosa became the recipient of a branch campus of the School of Medicine whose main campus was in Birmingham at UAB. The Tuscaloosa branch campus became home to 40 medical students on clinical rotations and 36 Family Medicine residents who were required to take community rotations with rural preceptors, many of whom practiced in West Alabama.

Bill completed medical school and residency in Internal Medicine at Vanderbilt, then served in the National Health Service Corps in his hometown, Carrollton, for two years before returning to Vanderbilt for a year as Chief Medical Resident. While in Nashville, he was an early member of the Vanderbilt Student Health Coalition (which became a model for other such student-run organizations around the country). He then returned home to establish a practice in Carrollton, where he became a preceptor for students and residents of the Tuscaloosa branch campus. In the 1980s, Bill became deeply involved in rural health policy as his community struggled and succeeded in saving its small hospital. Later in the decade, he joined with the Dean at the Tuscaloosa branch campus to propose a rural geriatric center which did not materialize.

I interrupted medical education after an internship with a stint in the Navy as a primary care medical officer in rural, Western Australia and in Virginia, then studied rural health services in a Masters of Public Health program at the University of North Carolina. Then I completed residencies in Internal Medicine at the Mayo Clinic and Preventive Medicine at North Carolina, where I was a Robert Wood Johnson Clinical Scholar and Chief Resident of Preventive Medicine. From 1986–1990, I was a junior faculty member in Preventive Medicine and Internal Medicine at the University of Tennessee in Memphis, before returning to Tuscaloosa as a member of the Department of Community Medicine to replace a recently retired faculty member.

Our teaching roles for the Tuscaloosa branch campus brought Bill and I together in 1990. A trip to Chapel Hill, North Carolina for a conference on rural research provided us time to share ideas. Our similar backgrounds and aspirations for rural Alabama were probably critical to the initiation of the Rural Alabama Health Alliance.

W.C. statement: The three founding counties of Bibb, Fayette, and Pickens include physicians who have known each other to some degree for years. But because the counties are not all contiguous, we do not see each other frequently. In the mid to late 1980s we realized that we had several things in common. For one thing, we were among the survivors. The late 1970s and the 1980s in Alabama, as elsewhere, were a real shake-out time for rural hospitals. Alabama had at one time the highest number per capita of rural hospital closures in the nation. Eleven rural hospitals in Alabama closed in one year. This was second only to Texas in absolute numbers, but was the highest per capita across the United States. It was through strong bonding between the local community, its medical community, and the mutual support so generated, that these three communities each managed to sustain a strong medical community, hospital, and local health care system through the hostile economic environment that existed through most of the country, certainly in Alabama.

There were other commonalities. Our location in West Alabama made us somewhat similar demographically and in our economic base. Though we had differences, we were comparable. We had in common the fact that our local health care systems had become, in slightly different ways, local economic centers both as an industry and as a stimulus for other health related economic development. They were making large economic impacts in our rural counties.

Another thing we had in common was that we each had a good relationship with the University of Alabama School of Medicine branch campus in Tuscaloosa. Many of our communities' physicians were graduates of either the medical school or the residency program there. Many of us, like myself, teach part-time with the medical school, both going into Tuscaloosa to teach and hosting students locally who come out for

> preceptorships in family medicine, community/rural medicine, and internal medicine. Thus, through our teaching role we already had a network partially established.
>
> In the late 1980s and early 1990s we began to strengthen our links with each other, in part, because of our teaching roles in Community/Rural Medicine. Faculty in the Tuscaloosa branch campus' Department of Community Medicine, particularly John (J.W.), encouraged us to get together to discuss educational issues that we saw were important for the future of health care both in our communities and other rural communities in Alabama. In each community we saw needs that we were not really meeting which heightened our desire to affiliate with each other and other partners. We were spending all day long doing the everyday business of physicians, nurses, hospitals, and other parts of the healthcare system. We were not really addressing the broader issues of public health, patient education, community education, occupational medicine, and a number of other things. We knew we needed help and we knew that we needed to be community based. We needed to come together and share our experiences, and we needed the expertise of people like the faculty of the Department of Community Medicine to stimulate us to do so. Our mutual interests in rural health developments and supportive medical education were so strong that we continued meeting and, consequently, developed the Rural Alabama Health Alliance. RAHA first included only our three counties. Since then, we have expanded to include another county and to involve institutional affiliates including the medical school branch campus and community health center organizations.

In retrospect, the commonalties noted above probably were important factors in the ability of the three communities to work collaboratively in RAHA. Other communities that had been invited to become early participants in developing RAHA cited dissimilarities among these factors as reasons for holding off. Participants from Monroe County were more distant from the Tuscaloosa branch campus and looked to Mobile as their regional center. Those from Greene County had already developed a legacy of successful health programming around a community health center as opposed to a hospital. However, as RAHA has progressed both of these counties have remained in contact, and Greene County has begun to participate as a member.

Methods in the Development of RAHA

Role of Branch Campus-Community Liaison Relationships. The three communities have had a variety of ties with the Tuscaloosa branch campus over the years, promoting a sense of trust for continuing partnership. A physician in one of the communities had been the campus' first residency program director while maintaining his practice at home. Another two had been the first preceptors to maintain students and residents in their local practices and homes for three-month stints of rural community medicine. These physicians and others supervised weekly attendance in the branch campus clinic or attended on ward rounds. These activities created for them a sense of credibility among branch campus faculty.

One of the most constant relationships between RAHA communities and the branch campus involves the community physicians' role as adjunct clinical faculty, serving as preceptors for students on Family and Community Medicine rotations. These rotations require campus-based faculty to visit the community during the course of each student's rotation, thereby promoting face-to-face visits between campus and community faculty, cultivating relationships, and enhancing the interchange of ideas and values. Thus, the teaching activities of the rural practitioners generated a steady stream of students and concomitant visitations from campus faculty.

In 1990, the Department of Community Medicine recruited a physician faculty member (J.W.) with a career interest in rural health development in Alabama to replace a recently retired colleague, the late Robert Gloor who had led the branch campus in extending medical school relationships into communities.[12] In the course of his assignments to visit community-based preceptors, J.W. initiated a series of unstructured interviews to learn about the community physicians' perceptions of rural health needs in Alabama. At the top of their concerns was the need to attract more physicians to practice in their communities and help maintain their local hospitals. Close behind was the concern that urban hospitals, universities, and public health organizations were drawing primary health care and its accompanying financial support out of rural communities.

With the participation of the interviewees, the Community Medicine (CM) faculty member drafted a white paper[13] to document their concerns and opinions, propose a plan to develop, and use model rural practice sites for teaching primary care and family medicine. These sites would assist in the transition of medical education from a tertiary center base to a community base. In the rural communities the teaching would depend primarily on experienced, adjunct clinical faculty and, secondarily, on faculty from branch campuses or the primary care departments of central campuses.

Model sites would be linked with nearby medical school campuses at which adjunct clinical faculty in the model site community held appointments. Programs centered around the model sites would increase the numbers of rural school children pursuing medicine and health careers, increase the numbers of medical students and residents from rural Alabama seeking educational experiences in the model sites, and develop support and service programs to make the model communities exemplary in sustaining satisfied health professionals and their families. The white paper outlined a process for developing the model clinics, campus, and administrative support.

The Role of Rural Physicians and Hospital Administrators. Subsequently, the CM faculty member met jointly with a group of these physicians and hospital administrators from five rural communities to confirm in conference that there was a consensus about these concerns. The community physicians' concerns were uniform and were endorsed consistently by their hospital administrators. These physician-administrator pairs agreed that developing rural oriented health professionals and maximizing the use of local health care were their key concerns for developing the potential of their communities. They decided to work with the CM faculty as a Rural Health Working Group, choosing one of the physicians (W.C.) as the group's chairman and another as vice-chairman. The CM faculty member (J.W.) was designated as staff to the group. Furthermore, they agreed that the issues involved were so important that a joint effort among the communities might be feasible.

In conference among physicians and hospital administrators, the CM faculty suggested that one test of feasibility might be to determine if the group's interest extended beyond the health care sector. Each of the physician-administrator pairs agreed to consider a three-community effort, with each pair identifying a key community leader from a non-health sector who would accompany them to another consensus conference and the three resulting trios attempting to form a joint venture.

The Role of Community Consensus Conferences. Greene and Monroe County representatives did not convince administrators or other community leaders to attend these conferences. The three remaining communities that eventually developed RAHA did convene a consensus conference that expanded the physician-administrator pairs to include an industrialist from Bibb County, a retired accountant from Pickens County, and a community college administrator from Fayette County. The interest in strengthening these communities' health care systems was all the more fortified by the inclusion of these individuals, who brought fresh perspectives on the economic impact of the local health care industry in rural communities. Almost immediately this enlarged group decided that the issues in the white paper were of such importance to the life of their communities that a formal organization tying themselves together to address rural health care concerns should be accomplished.

The Rural Health Working Group set about the task of determining the nature of its organization to promote a rural community perspective in all aspects of its work. The group decided through discussions covering several meetings that a full representation of the interests of people in each community should be maintained with local community committees. The local committees should elect messengers to a central board. Each county committee would then also designate one of the three officers of the central board with the positions rotating among the three counties on a regular basis. Thus, the local community would have a continual check on the direction that the organization would take.

The Role of Community Committees. The working group members returned to their respective counties and organized community committees with representatives from all sectors in the community. There were 15 members on Bibb County's community committee, 12 on Fayette's, and 21 on Pickens'. These committees discussed the merits of the rural health plans being pursued and selected their messengers, including officers to attend organizational meetings of the three-county group.

The Role of a Non-Profit Organization. The messengers and officers designated from the counties met to formalize an organizational structure that would ensure accountability to the wishes of the local communities. A non-profit organization was planned with a governing board comprised of three to five messengers from each county. Among a county's messengers would be the individual that the county committee designated as the RAHA officer from that county. Policy decisions would be decided by a vote by county, with each county's group of messengers being required to caucus to decide their county vote. Two out of three votes would decide a policy. There would be a Chairman, Vice Chairman, and Secretary-Treasurer rotating counterclockwise among counties every two years.

Articles of incorporation and by-laws were formulated and endorsed. Rural Alabama Health Alliance was the name taken by the organization. Robert's Rules of Order[14] was taken as the procedural guide. One of the first items of business was to move the board meetings away from the urban setting in Tuscaloosa and rotate meetings among the three communities' hospitals.

Activities In Priority Areas

Table 1 lists activities of RAHA according to priority area. The following comments highlight some of the activities.

Promoting Medical and Health Education in Rural Communities. This was the area in which RAHA invested most of its energies initially. Early on, RAHA engaged in advising the University of Alabama School of Medicine on rural medical education by reviewing the strategic plan (white paper),[13] noted above, that was presented to a subcommittee of the Dean's Family/Rural Practice Advisory Committee. Subsequently, RAHA worked to develop individual elements of the plan. As an example, the RAHA Education Committee worked with the Family Medicine Department of the branch campus to develop a one month rural rotation for Family Practice residents that has since become a required rotation for all residents at the branch campus.

The Family Medicine Department secured non-salary grant funding to set up the residents' rural rotation. One RAHA community reported a problem in the lack of residency program funding to pay the residents' salary during the rotation. The funds necessary could not be generated by the residents' service in the community. After consultation with DCH Regional Medical Center, the teaching hospital associated with the residency program, the three RAHA community hospitals sought to recover graduate medical education funds directly from their federal source as the residents rotated through the community hospitals.

In developing various elements of the strategic plan, the RAHA Education Committee developed working relationships with the Family Medicine and Community Medicine Departments at the branch campus, University of Alabama School of Medicine (UASOM), University of South Alabama College of Medicine, the branch campus Occupational and Preventive Medicine Unit, Bevill State Community College, Women Involved in Farm Economics (WIFE), Alabama Farmers Federa-

Table 1. Rural Alabama Health Alliance Activities in Priority Areas.

Priority area	Goal	Partners*	Outcomes
Rural Medical Education	Influence UASOM	Deans Family/Rural Practice Advisory Committee	Strategic white paper
	Rural rotations	Branch campus FM Dept.	FM resident rotation
		Branch campus CM Dept.	Community/Rural Medicine student rotation
			Pre-med preceptorships
		UA College of A&S	IM resident elective
		USACOM IM Dept.	Rural Medicine selective
		UASOM	Stipends and honoraria
		FPRHB	
	More rural pre-med students	UA, branch campus, WIFE, CES, AFF, MASA, AL-AAFP, FPRHB	Rural Health Scholars Program
	More rural medical students	UASOM	Two members on UASOM Admissions Committee
	Longitudinal track of rural medical education	UA, UASOM, branch campus, AL-AAFP, ASIM, AAP	Rural Medical Scholars Program
	Rural preventive/occupational medicine course	UA, branch campus, CES, AFF, FPRHB, PHD	Agrimedicine Program grant
	Nurse/technician training	Bevill State Community College	Unfunded plan
	Model rural training sites	Branch campus, USACOM FM Dept.	AHEC Program proposal
Rural Health Services	Recruit family physicians	Branch campus FM residency program	Six FM physicians recruited to RAHA communities

	Referral network	Branch campus faculty, DCH Regional Medical Ctr.	Physician Access Line plan initiated
	Telemedicine services	UA, branch campus, CHC, DCH Regional Med. Ctr., USDA	Telemedicine grant funded
	Influence rural health policy	ORH, MASA, AL-AAFP, AHA, ARHA, APHCA, AEA, AFF, ASIM, AAP	Consultations to ORH, physician extender/mid-level practitioner laws passed, rural student loans law passed
	Primary care access for all children	Local school system, BCBS, branch campus CM Dept.	Child Caring Foundation program
	Maternity care for low income women	Branch campus OB Dept.	Medicaid waiver program proposed for region
Rural Community Development	Community organization	Local institutions, branch campus CM Dept., UA New College, UA graduate students	Needs assessment projects in 2 counties
	Health-related economic growth	Local institutions	New jobs associated with recruited physicians' practices
Fund Raising	Grants	UA, branch campus departments, FPRHB, BCBS, USDA	Agrimedicine ($80,000), Telemedicine ($300,000), FP residents, RHSP, Child Caring Project, Medicaid Waiver project.

*Abbreviations are as follows: AAP (Alabama Academy of Pediatrics), AEA (Alabama Education Association), AFF (Alabama Farmers Federation), AHA (Alabama Hospital Association), AHEC (Area Health Education Center), AL-AAFP (Alabama Chapter-American Academy of Family Physicians), APHCA (Alabama Primary Health Care Association), ARHA (Alabama Rural Health Association), ASIM (Alabama Society of Internal Medicine), A&S (Arts and Sciences), BCBS (Blue Cross/Blue Shield of Alabama), CES (Cooperative Extension Service), CM (Community Medicine), FHC (Family HealthCare Corporation, a community health centers program), FM (Family Medicine), FPRHB (Family Practice Rural Health Board, a state funding agency), IM (Internal Medicine), MASA (Medical Association of the State of Alabama), OB (Obstetrics), ORH (Office of Rural Health), PHD (Public Health Department), UA (University of Alabama), UASOM (University of Alabama School of Medicine), USACOM (University of South Alabama College of Medicine), USDA (United States Department of Agriculture), WIFE (Women Involved in Farm Economics).

tion (AFF), Family Practice Rural Health Board (FPRHB), Cooperative Extension Service (CES), University of Alabama (UA), Medical Association of the State of Alabama (MASA), Alabama Chapter-American Academy of Family Physicians (AL-AAFP), and several small hospitals. Particularly notable was the success at helping start the Rural Health Scholars Program, Rural Medical Scholars Program, and Agrimedicine Program in addition to the required resident rotation. Taken together, these programs offer rural high school students the opportunity to enter a sequence of educational experiences that lead through pre-med, medical, and residency experiences with a rural orientation.

The branch campus has been a good partner, perhaps because of commonalities noted above and the concordant missions focusing on rural health care. RAHA was able to supply access to scarce grant funding via the FPRHB, a fact which heightened the branch campus' interest. Most of the other partners joined because of similar missions (WIFE, AFF, CES), RAHA influence on their boards (MASA, AL-AAFP), and others (UASOM) because of RAHA's strength with potent political lobby or policy groups (MASA, AL-AAFP, University of Alabama Board of Trustees).

Enhancing Rural Health Care. RAHA has been active in efforts to improve rural health care locally and state-wide. The organization has been helpful for recruiting rural physicians and for linking service components of the rural communities with needed referral and consultation services at the branch campus and DCH Regional Medical Center. RAHA's Education Committee took the lead role in proposing a telecommunications linkage among the RAHA hospitals, branch campus, DCH Medical Center, and the University of Alabama System and in developing a telephone access line for reaching branch campus faculty and DCH Regional Medical Center specialists for referral care. The branch campus activated the Physician Access Line (PAL) to provide RAHA physicians direct contact with branch campus physicians on service in the regional hospital. The telemedicine proposal was submitted in 1994 and funded by the US Department of Agriculture in 1996.

State-wide, RAHA's influence is being felt in state policies regarding rural health. With its credibility as a grass roots organization, RAHA has been sought for consultative advice by the Alabama Office of Rural Health on issues of importance to rural Alabama. Individual members of RAHA have become effective spokespersons on various committees and boards that are formulating health policies for the state, including the Board of Medical Examiners, committees advising the medical schools and legislature, and the Alabama Chapter-American Academy of Family Physicians. One RAHA physician (W.C.) was elected to the Board of Medical Examiners, an influential position with respect to the state health department, medical practice, and state medical association. His rural credibility was recognized, and he was charged with chairing a committee to inform the medical association on policies to assist rural Alabama. Since this committee has been in action, the medical association has sponsored or co-sponsored bills to: a) repay loans of medical students agreeing to practice in rural Alabama, b) authorize the practice of physician assistants, and c) authorize the practice of nurse practitioners. All three of these bills were passed and signed into law. This is especially notable because the practice of physician extenders had been opposed by the medical association, a powerful lobby in Alabama, for years.

Stimulating Rural Community Development. Rural community development is the long-range plan of RAHA. RAHA has laid the foundation for economic development related to health care by bringing together in each community a spectrum of leaders in health, agriculture, education, industry, and government. Strategic planning that makes use of the expertise and resources available to these various leaders has been instigated. A prime example is the effort behind the proposal for a telecommunications network. One physician and his administrator took the lead role in developing the idea, which the physicians on the Education Committee endorsed and expanded. An industrialist member of one community committee provided his airplane and pilot so that the committee could study a successful network in another state. RAHA officers then made liaison with Bevill State Community College and the University of Alabama to get expert assistance on packaging the proposal, which was submitted and funded.

Engaging in Fund Raising for RAHA Activities. To date RAHA has been the principle recipient of funds for the Agrimedicine Program ($80,000), has been a partner in helping the branch campus gain funding for Family Practice residents to do rural rotations ($2000 per resident), and for the Rural Health Scholars Program ($200,000) for high school students. The successful telemedicine proposal made RAHA the principle recipient of $300,000 for a telecommunications network.

While RAHA has accomplished a great deal, a 1993 survey of the RAHA Board also identified several areas of disappointment. The state's largest medical school continued to be unresponsive to rural health issues; the nearby state university was also hesitant to respond to RAHA; and the state's smaller medical school was too distant for RAHA to influence effectively.

Four board members wrote in special concerns:

- "The RAHA groups need to persevere. Certainly, many things we have been involved in seem to progress slowly, but we need to hang together because I feel there have been improvements."
- "We need to be more focused and set priorities."
- "Model clinics seem to have lost their priority somewhat, [though they] were a favorable focus in the beginning."
- "The University of Alabama and University of Alabama in Birmingham seem to have only a perfunctory interest and involvement [in RAHA's efforts]."

There were five board members who had recommendations for other communities wishing to duplicate the successes and avoid any pitfalls associated with RAHA. These recommendations were:

- "Do something; don't simply decide to not be involved."
- "Develop identifiable grassroots leadership willing to spearhead project development. This will require an individual or small group to step forward and assume control, but be backed with broad-based community support."
- "Incorporate other organized groups, if possible, to gain a broader base of support."
- "Develop educational sessions to help community members and organizations become prepared to develop their own directions based on needs and resources."
- "Study RAHA and what it has been through; use that which seems applicable and make adjustments to fit local situations."

There were ten recommendations that the board members made for future directions of RAHA. These were:

- "Become a policy leader [for rural Alabama]."
- "Increase influence on appropriate political and academic interests."
- "Become a greater voice of rural medicine at the state level and show what can be done."
- "Inform the state of how urgent the need is for [support of] rural hospitals."
- "Concentrate on building a [more] solid grassroots support to influence health policy and reform issues."
- "Consider networking to have a greater voice in the state's ultimate health care reform package."
- "Develop a better communication system among the three communities to move RAHA along faster."
- "Push for medical education to provide more prolonged exposure to rural medicine; do the same for nursing and allied professions."
- "Focus RAHA activities; refocus on model clinic development."
- "Expand RAHA's directions to other types of rural development beyond health care."

Continuing Developments

In the 30 months since this survey, it is of interest to note some of RAHA's accomplishments with respect to areas of concern and recommendations listed above. RAHA has become a well recognized entity in many professional and policy circles around the state. One RAHA board member (W.C.) has risen in rank in organized medicine and was elected in 1996 without opposition to the highly influential position of President of the Medical Association of the State of Alabama. Another was elected President, and still another President-Elect, of the Alabama Chapter-American Academy of Family Physicians. Thus, RAHA's influence on the medical profession's contribution to health policy in Alabama has been virtually assured.

RAHA has proceeded to expand and network. A community health center corporation with clinics in seven counties, two of which already were represented in RAHA, joined as an affiliate member. Greene County, contiguous with Pickens, also joined. Two other counties are still considering invitations to become involved. A RAHA board member has been nominated to the Alabama Rural Health Association board and another sits on the board of the Alabama Primary Health Care Association. Finally, the board is well represented in the Small and Rural Hospital Constituency of the Alabama Hospital Association.

With respect to the state's medical schools RAHA has made sizable inroads. Two board members sit on the Admissions Committee at the larger medical school. Representing the Alabama Chapter-AAFP, they also sit on the same school's recently formed Primary Care Advisory Committee. It was no coincidence that this school recently endorsed a plan for a longitudinal rural medicine track, the Rural Medical Scholars Program, that admits 10 rural students per year and provides special programs involving rural preceptors, RAHA communities, and the branch campus in Tuscaloosa. RAHA has hosted a think tank involving the state's smaller medical school, and one board member has become a clinical instructor precepting internal medicine residents for that school. As noted earlier in the paper, RAHA members were central to the successful effort to enact legislation to sanction the use of physician extenders in Alabama, an almost unthinkable happening because of the state medical association's long history of staunch opposition. RAHA's major telemedicine grant has created even greater interest among regional health care institutions and educational centers.

Finally, RAHA has tentatively begun to expand its scope in rural developments. First, it has become involved in the study of managed care options for health care providers in West Alabama. Second, it has begun to lend its local organizational structure to developing, in partnership with Blue Cross/Blue Shield^{SM} of Alabama, a program that assures access to primary medical care to all school children in its service area. In summary, RAHA has the promise of maturing into another successful partnership that involves community organizations and academic medical centers.[15]

Conclusions

RAHA is a maturing organization generating from three rural communities in Alabama because of a perceived need to strengthen the rural voice in policy-making for medical education and health care. RAHA's unique organizational structure captures each community's need for independent input to the board's decision process. The communities have benefitted from their combined strengths and have begun to address their concerns for educating more rural health care providers in rural settings, improving rural health services, stimulating rural development, and fundraising. RAHA could become a vehicle for developing a regional network of health development in the coming years of healthcare reform. As such, it might become a model for other regions in Alabama and the rural South.

References

1. Fox CE and the Rural Health Task Force. Alabama's rural health care crisis. Rural Health Task Force Report to the Alabama Legislature. January, 1990.

2. Beaulieu LJ (ed.). *The Rural South in Crisis: Challenges for the Future*. Boulder, CO: Westview Press, 1988.

3. Garkovich LE, Bokemeier JL. Factors associated with women's attitudes toward farming. In Beaulieu LJ (ed.). *The Rural South in Crisis: Challenges for the Future*. Boulder, CO: Westview Press, 1988: 120–140.

4. Willard WR. Medical education and medical care: Some inadequacies, some solutions. University of Alabama, 1983.

5. Rosenblatt PA, Whitcomb ME, Cullen TJ, Lishner DM, Hart LG. Which medical schools produce rural physicians? JAMA 1992; 268:1559–1565.

6. Rabinowitz HK. Evaluation of a selective medical school admissions policy to increase the number of family physicians in rural and underserved areas. NEJM 1988;319:480–486.

7. Brazeau NK, Potts MJ, Hichner JM. The Upper Peninsula Program: A successful model for increasing primary care physicians in rural areas. Fam Med 1990;22:350–355.

8. Verby JE. The Minnesota Rural Physician Associate Program for medical students. J Med Educ 1988;63:427–437.

9. Head RE, Harris DL. Characteristics of medical school applicants and implications for rural health care. Fam Med 1989;21:187–190.

10. Wheat JR. Strategy to overcome urban undertow: An obstacle to rural health development. J Agromed 1994;1:59–67.

11. Doeksen GA, Miller DA. Rural physicians make good economic sense. J Okla Med Asso 1988;81:568–573.

12. Gloor RF, Tapp J.W. The medical extension agent. Arch Environ Health 1968;17:832–835.

13. Wheat JR and the Sub-committee of University of Alabama School of Medicine Family/Rural Practice Advisory Committee. Strategic Plan for addressing Alabama's rural health needs: Medical Schools' role. Birmingham, AL: UAB, March, 1991.

14. Robert HA (ed.). *The Scott Foresman Roberts Rules of Order*, Newly Revised, 9th ed. Glen View, IL: Harper Collins, 1990.

15. Maurana CA, Goldenberg K. A successful academic-community partnership to improve the public's health. Academic Med 1966;71:425–431.

Part 5

Educational Partnerships

In this last section, we turn our attention to education. There are clear parallels between educational and clinical contexts and processes. Both involve assessment, goal setting, and behavior change. The student-teacher dyad mirrors that of the patient and clinician with respect to traditional hierarchical roles and the improved outcomes that can result from collaborative decision making. Also, education, like clinical care, is a team sport; it requires good working relationships between professionals and institutions. So it should not be surprising to find the same principles of partnership at work in these two domains. Moreover, educational partnerships have additional significance in their potential to provide experiential learning and role modeling to students. After all, it is in professional education programs that young clinicians establish the ideals of practice to which they aspire. Here they acquire attitudes towards their colleagues and expectations as to their roles. The tone of their educational experience will color the rest of their professional career: will it be one of respect and personal presence or one of abuse, humiliation, and withdrawal?

Jane Westburg and *Hilliard Jason* address the need for collaborative education that prepares healthcare professionals to work as partners with patients and colleagues. They contrast authoritarian and collaborative partnerships in educational and clinical settings. They also call for a shift away from authoritarian toward collaborative ways of working together in both the traditional practitioner-patient and teacher-learner relationships (a similar theme to that in Suchman's comparison of the control and relation paradigms in the Perspectives on Partnership Process section). They describe factors that support the development of such collaborative partnerships and specific strategies for overcoming some of the barriers that impede movement toward collaborative education.

Diane Schwartz and *Richard Sarkin* describe the administrative and educational partnerships that were necessary to develop a Teaching Effectiveness Program (TEP) as part of the University of Buffalo's Primary Care Resource Center. The TEP prepares community faculty to teach medical students and residents about primary care in their practices. This program assisted the University in achieving an ambitious goal: to have at least 50% of its graduating medical students entering primary care disciplines.

Alberto Cardelle and *Lynn Bickley* describe the infrastructure support for organizing international collaboration for students to gain training in another healthcare system: in particular, training in community-oriented primary care. The program they describe is part of the International Health Partnerships for Community Based Medical Education (IHPP). They describe the process of selecting and setting up partnerships for this collaborative educational program. Case vignettes provide information about the success, challenges, and difficulties of providing worthwhile learning opportunities for students. The curriculum they devised for participating students includes a strong focus on cross cultural work—learning to understand the perspective of the "other."

Margaret Blandin-Clark's paper exemplifies the deliberate use of the learner-teacher relationship to model the partnership skills to be used by the clinician, and to help students reflect on their own experience of a potentially one-down situation to better understand the experience of their patients. She has developed an educational consultation approach for helping managed care companies assist physicians who had either low rating on patient satisfaction surveys or frequent patient complaints. Managed care companies are attentive to this quality improvement issue, in part because of HEDIS reporting and the accreditation process organized by the National Council Quality Assurance. The Chief of the Family Medicine Department at a local HMO was concerned about a physician on his staff who was well respected for his biomedical expertise, but who had recurring complaints from patients about how he related to them. The chief asked a behavioral health professional and educator to help this physician deal with the issues identified in these complaints. She met with this physician over a period of time (six patient care sessions over a five month period). The consultation process improved the physician's interactions with patients and reduced patient complaints. This service is now available on as needed basis for this HMO.

Geoffrey Swain and *Eric Weiner* describe a collaborative educational model: the family medicine/family systems case consultation. Interdisciplinary faculty collaboration has long been touted as an ideal way to model to learners about how to collaborate together. They have developed such a collaborative endeavor in a Family Practice Residency setting. They organized a case-driven, consultation method for residents. The residents use these consultations to help themselves get "unstuck" in dealing with complex, psychosocial issues. They provide an example of how they work together in helping residents deal with challenging situations. They evaluated the impact of these consultations on their residents; for example, they appreciated the supportive atmosphere to explore troublesome cases.

Readers with an interest in education may also want to see the chapter by *Mary Talen* and her colleagues in the Health Care Team section. This chapter describes a most innovative effort to teach interdisciplinary collaboration. Students of several professions (medicine, nursing, psychology, and the allied health professions) were formed into teams. After receiving an orientation to group process and to each others' disciplines, they were assigned to patients in the community with the task of conducting a multidisciplinary assessment and creating and implementing a treatment plan. In keeping with the theme of parallel process, the educators involved in this project had to attend to their own partnership process before this program could succeed.

XIX Collaborative Education: Preparing Health Professionals for Functioning as Partners with Patients and Colleagues

Jane Westberg and Hilliard Jason

Authoritarian, hierarchical models have dominated much of healthcare practice and health professions education. These models tend to disempower patients and learners and are often inadequate in meeting patient needs. In hierarchical, patriarchal systems the skills of "lower status" providers can be devalued and underused, resulting not only in the loss of potentially valuable contributions but also in the loss of providers who either burn out or drop out when treated in these ways. To counter these negative outcomes, in both healthcare and health professions education, we need to accelerate our movement toward more responsive, collaborative[1] models, in which healthcare professionals function as partners both with patients and clients,[2] their families, and with each other (colleagues in their profession, as well as colleagues in other health professions).To prepare students to be collaborative professionals, we need to foster partnerships between teachers and students and among students. We also need to ensure that we help students develop and maintain collaborative attitudes and capabilities by, for example, involving them in cooperative, interdisciplinary learning activities. In this chapter, we present the rationale for promoting partnerships between healthcare professionals[3] and their patients and among health professionals. We describe behaviors that are characteristic of collaborative relationships by contrasting them with behaviors in authoritarian[4] relationships. Then we identify attitudes, knowledge, and skills our students need for being collaborative practitioners. In the balance of the chapter we describe strategies for helping ensure that learners develop these attitudes and capabilities.

Need for Partnerships between Health Professionals and Patients

In contrast to the situation earlier in this century, when the leading causes of premature death were infectious, much of the preventable morbidity and premature mortality in the industrialized world is now closely linked to modifiable human behavior. Today, most of the leading causes of premature death (much of cancer, heart attacks, strokes, chronic obstructive pulmonary disease, suicide, violence, and unintentional injuries) can be delayed or fully prevented through changes in the ways people conduct their lives. Few people, however, make long-lasting behavioral changes just because their healthcare provider tells them to do so. Health professionals can raise issues and provide information and support, but unless people take "ownership" of their own health and health goals they are unlikely to sustain whatever plans they make for changing the ways they live.[5]

The successful management of most conditions requires that patients be active partners in their care. Yet, between one-third and one-half of physicians' orders are not followed. Most patients do not adhere fully to long-term treatment plans.[6] (This is especially important because of the rise in the numbers of people of all ages with chronic problems.) As with prevention, the likelihood of patients adhering to plans is greatest if patients develop ownership of and participate with practitioners in developing and monitoring their management plans. When the plan involves the support or direct help of other significant people in the patient's life (e.g., family members), these people may also need to be included in some phases of the planning and/or monitoring process.

Increasingly, consumers of healthcare are requesting, even demanding, that they be active participants in setting priorities and making decisions that affect their lives.[7,8,9] And medicine is no longer the mysterious, exclusive province of health professionals. With daily health-related stories in the public media, it is no longer uncommon for patients to learn about new developments in medicine even before their healthcare providers do.

Health Professional-Patient/Client Relationships

The characteristics of helpful relationships between health professionals and patients/clients vary widely. On one end of a continuum are fully collaborative relationships in which the persons seeking help are actively encouraged to be partners in their care. On the other end of this continuum are authoritarian relationships in which people seeking help are kept in a passive role. (See Table 1.) Traditionally, health professional-patient relationships have been closer to the authoritarian end of the continuum, but we, and many others, strongly recommend preparing students to be more collaborative.[10,11]

In collaborative healthcare, health professionals and patients are partners, with each person having unique contributions to offer. Health professionals have skills, resources, technical knowledge, and experiences that can be useful to their patients. Patients have special knowledge of their symptoms, their resources and support systems, their living and work circumstances, their abilities and willingness to make lifestyle changes, their beliefs, and their personal values—all of which are important elements in their health status, health prospects, and approaches to care. In authoritarian relationships, patients are seen primarily as recipients of care and exert little influence on key decisions.

Table 1. Contrasting Behaviors in Collaborative and Authoritarian Health Professional-Patient Relationships[11]

Collaborative (Partnership, Cooperative)	Authoritarian (Hierarchical, Patriarchal)
Patients are treated as valuable contributors to their care.	Patients are treated primarily as recipients of care.
The patient and health professional jointly set the agenda.	The health professional sets the agenda.
The patient participates in assessing his or her needs.	The health professional determines the patient's needs.
The patient and health professional jointly establish the goals of care.	The health professional determines the goals of care.
The patient and health professional jointly establish the management plan.	The health professional tells the patient what to do.
The patient is involved in monitoring his or her progress.	The health professional monitors the patient's progress.
The health professional modifies his or her approach according to the patient's needs.	The health professional is always directive.
Independence is fostered.	Dependence is fostered.
The patient is empowered.	The patient is disempowered.
Care is patient-centered.	Care is provider-centered.

Health professionals who function collaboratively do not demand that patients function collaboratively from the start. Rather, they begin where their patients are, even if that means accommodating people who are accustomed to being highly dependent. They encourage their patients to assume increasing control of, and responsibility for, their care but do not push them faster than they can go. Collaborative health professionals are also flexible with patients who strive to be active in their care. They are aware that serious illness, such as major surgery or depression, can cause even strong, assertive patients to want (or need) to be more dependent for a time. Authoritarian health professionals, however, tend to be rigid about their relationships with patients. They are likely to be bothered, even threatened, by people who want some control over their care.

Relationships among Health Professionals

When caring for patients, the vast majority of health professionals in the industrialized world work with other health professionals. Most so-called "solo practices" are not solo at all. Typically there are at least two health professionals (e.g., a physician and nurse) in a practice. Even the few clinicians who are truly alone at a clinical site or other setting typically interact regularly with other practitioners.

Many health professionals work in teams. For decades hospital-based teams have provided care in highly technical, crisis-oriented, life-threatening situations (e.g., the care of people undergoing surgery, and the care of trauma and burn victims). Teams in a variety of settings care for people with complex, long-term problems (e.g., the frail elderly with multiple problems, persons undergoing physical rehabilitation). Primary care teams have been around for decades, but they are being given increasing attention by those who feel that teamwork is essential to high quality primary care,[12,13] and by cost-conscious healthcare corporations trying to save money by having lower-paid health professionals assume some of the tasks traditionally done by physicians. Some healthcare teams work together on a regular basis (though some professionals may be rotated on and off teams). Other teams are brought together to address specific tasks.

Clearly, health professionals do work together. Our focus here though is the *way* they work together. To describe how they work together, we again use a continuum ranging from collaborative to authoritarian. As with practitioner-patient relationships, we realize that most relationships among colleagues fall somewhere between the two extremes. We also realize that we are oversimplifying the issues to make some points briefly. Our focus is on teams and on groups of professionals who potentially could operate as a team.

Authoritarian Model. Even though they may be working in the same space and caring for the same patients, health professionals functioning in this model have minimal communications with each other about their work. In general, they function hierarchically, with the leader (typically the highest paid professional with the most status and professional education, but not necessarily the most competent in all the tasks faced by the group) unilaterally making the important decisions and delegating assignments. The professionals who are at the top of the hierarchy are often not fully aware of the other professionals' ranges of capabilities (especially their higher-level skills), and so these other professionals are often underutilized or called on to do work that is inappropriate for them. Although the professionals at the top of the hierarchy are usually dependent upon the skills of the other professionals, this dependency is not always acknowledged. In addition, as already indicated, authoritarian health professionals tend not to see patients as partners in their care or as members of the health team.

Collaborative Model. Health professionals working in this model genuinely share responsibility for patient care. The division of labor is organized around common goals with members contributing their expertise as needed. Outcomes and goals are regularly re-evaluated.[13] Collaborative professionals respect each other and each other's autonomy. They see one another as partners. Developing and maintaining a partnership relationship is viewed as a dynamic, ongoing process, and so they invest time in getting to know each other and in developing ways of working together. They also strive to make the best possible use of each member's unique experiences, skills, and perspectives. Most of the important decisions are made jointly, although in some situations (e.g., some emergencies), decisions may be made unilaterally. Patients and family members are actively involved in the discussion of care and are regarded as important members of the health team. Collaborative professionals tend to work together flexibly, even shifting leadership responsibilities accordingly as the nature of the task changes. (Such shifting of leadership responsibilities tends to be more common among members of primary care teams, who jointly focus on complex multi-dimensional issues, than among surgical teams, who focus on more delineated, clear-cut issues, so to speak.)

Need for Collegial Partnerships in Providing Healthcare

With the growing costs of, and diminishing resources for, healthcare, it is more important than ever that we not underuse or misuse health professionals. There is also some evidence that the burn out and drop out of some health professionals has been linked to their dissatisfaction with aspects of the hierarchical model.[14] Burn out and drop out are costly, both in human and economic terms. When people burn out there is likely to be more tension in the work place, and burned out people may be less productive or even absent from work. If professionals drop out altogether, finding and training replacements is a costly undertaking. Collaboration can lead to more professional and personal satisfaction, and it can result in the learning of new skills and reduced burn out and drop out.[15,16] When people feel that they are listened to and respected, and when they feel they can use their expertise in meaningful ways, they are more likely to feel satisfied. Costs can also be saved in other ways. For example, when professionals work closely with each other and know what each team member is doing, duplications of effort are less likely.

Making maximum use of the expertise and perspectives of people from multiple professions can increase the likelihood of being able to provide high quality, comprehensive healthcare. This is particularly true when patients have complex, interlocking problems that go beyond the focus of any one profession, as in the care of some older people. Also, the likelihood of creating fresh, innovative approaches to problems is heightened when attempts are made to integrate rather than simply coordinate different perspectives.[13,17]

For decades, both providers and consumers have complained about fragmented care. Although we give lip service to the holistic nature of individuals, we typically divide them, and their care, into parts. As healthcare becomes ever more complex and specialized, integration and coordination of care is all the more critical. When health professionals work as partners, leadership and coordination tasks are likely to be assumed by appropriate persons. Professionals accustomed to working with colleagues from other health professions tend to be concerned about and grasp the big picture. They are less likely to focus only on their own narrow areas of interest. When patients sense that their caregivers are communicating with each other, they are more likely to be satisfied with their care.

As we try to solve the problems within existing healthcare settings, and, even more important, as we search for new, more responsive ways to provide healthcare, we need multiple perspectives and ways of thinking. For example, using Deming's[18] notion of quality circles, teams of people in healthcare settings have devised fresh, creative solutions to complex problems.

As society takes steps toward greater equality for women, minorities, and others, growing numbers of patients and health professionals want to participate in making decisions that affect their lives and the goals they are trying to achieve. Collaborative models help them have such a voice.

Collaboration and collegial partnerships can also be key to health professionals' ongoing competency. There is evidence that peers who practice alone are at greater risk (over time) of experiencing difficulty in maintaining their professional competence.[19] Colleagues who work collaboratively can be important sources of learning for practicing professionals.[20]

Attitudes, Characteristics, Knowledge, and Skills Needed by Collaborative Professionals

To help our students become collaborative health professionals, we must ensure that they respect their patients' knowledge and perspectives and see them as vital partners in their own care. Also, we need to ensure that our learners genuinely value and are curious about the perspectives and contributions of other health professionals. They need to be open to, even eager for, learning from others, tolerant of different perspectives and ways of functioning, and sufficiently secure so they do not feel diminished by the contributions of others. We also don't want them to have the kind of "professional ethnocentrism" that causes some professionals to feel separate from, perhaps superior to, others.[21]

To be effective partners with patients, learners need to be active listeners and clear communicators. They need to be able to assess their patients' "readiness" to be partners in their care and, as appropriate, help their patients participate in defining and clarifying their healthcare goals, developing contracts for achieving these goals, and more.

To be effective members of healthcare teams, learners need both leadership and team membership skills and an understanding of group dynamics. They need to be able to work with colleagues in creating trust-based relationships, building consensus, making complex decisions, developing goals, identifying strategies for achieving those goals, evaluating the team's work, negotiating disagreements and conflicts, and mediating disputes. Team members need to respect, understand the roles, and recognize the contributions of other members.[13]

And to work effectively together, learners from different healthcare disciplines need shared values, language, and conceptual frameworks.[22]

Providing Learners with Role Models

If learners are to become collaborative professionals, they need to be exposed to practicing professionals who function collaboratively with patients, families, and colleagues, and who can also articulate what they are thinking and doing. And if learners are to learn to collaborate with other health professionals, they need to see their teachers being collaborative in both classrooms and clinical settings.

To serve as successful role models for learners, some clinician-educators (particularly those whose formal education took place in environments that were more authoritarian than collaborative) may need to make a special effort to be collaborative.

Finding clinical settings in which learners can observe collaboration between health professionals and patients and among health professionals (e.g., in multidisciplinary teams) can be challenging. Sometimes it's necessary to create model practices.[23]

The Need to Work Collaboratively with Learners

People tend to treat others as they have been treated. There is good reason to believe that the way we treat learners can help shape the way they in turn will treat their patients and colleagues. If we treat learners in collaborative ways, they are more likely to treat others collaboratively. If we treat them in authoritarian ways, they are more likely to treat others in authoritarian ways.[11]

In recent decades, research has demonstrated that significant relationship patterns in families are repeated across generations. The most dramatic and saddest illustration of this phenomenon is the perpetuation of child abuse. People who were abused as children are more likely to abuse their own children than are their counterparts who were not abused.[24] Using this process as a metaphor, McKegney[25] compares medical education to a neglectful, abusive family system. When the "children" (medical students and interns) and "adolescents" (residents) are abused, they are likely to perpetuate this hurtful system. They, in turn, abuse more junior members of the "family."

If negative patterns can be transmitted from generation to generation, it is reasonable to assume that positive patterns can be passed down as well. Pellegrino[26] described such a phenomenon: "When the teacher helps the student in a compassionate and understanding way, he illustrates how the student can, in turn, give the same understanding to the patient, who is dependent on his humanness as the student is dependent on the teacher's (page 1290)."

Teacher and Learner Relationships

The ways learners and teachers interact in health professions education range from collaborative to authoritarian. At the collaborative end of the continuum, students and residents are active partners in their learning. At the authoritarian end of the continuum, learners are passive. Most instruction in the health professions falls between the extremes but closer to the authoritarian end of the continuum. The contrasting behaviors in collaborative and authoritarian teacher-learner and learner-learner relationships are summarized in Table 2.

Collaborative and authoritarian teachers make different explicit and implicit assumptions about learners and learning. Collaborative teachers respect each student's autonomy. They assume that students have understandings, experiences, skills, feelings, and interests that are relevant to their learning experiences, and that they can teach each other. In fact, collaborative teachers find they often learn from their students. Collaborative teachers assume that learning is something that learners have to do for themselves and that their job as teachers is to facilitate this process. Learning is also seen as a shared process in which students and teachers work together in building knowledge.

Given these assumptions, when observing a collaborative group you are likely to see learners actively engaged with each other and with the tasks at hand. They raise questions, express opinions, and more. The teacher may not use much air time but will, instead, be listening actively and observing. When he or she speaks, it is likely to be in the form of a question or feedback that helps learners examine what they are thinking or doing. When appropriate, the teacher also encourages learners to do their own thinking, to develop and express their own points of view. Students are not asked to blindly accept each other's opinions and ways of doing things. Rather, the leader encourages them to challenge and critique each other constructively and to share their knowledge and skills.

Teachers functioning in an authoritarian model tend to assume that their job is to *tell* and that the student's job is to absorb what they hear. Even when working with small groups of students, these teachers often turn the sessions into lectures or a series of mini-lectures. They do most of the talking, only occasionally turning to

Table 2. Contrasting Behaviors in Collaborative and Authoritarian Teacher-Learner Relationships[1]

Collaborative	Authoritarian
Learners are treated as valuable contributors to their own and to each other's learning.	Learners are treated primarily as recipients of teaching.
The teacher and learners jointly set the agenda.	The teacher sets the agenda.
Learners participate in assessing their learning needs.	The teacher presumes to know the learners' learning needs.
The teacher and learners establish individual and shared goals of learning.	The teacher determines the goals of learning.
The teacher and learners develop individual and group learning plans.	The teacher may develop a learning plan.
Learners monitor their own progress and provide feedback to each other.	The teacher monitors the learners' progress.
The teacher provides "situational leadership."	The teacher is always directive.
Independence and collaboration are fostered.	Dependence and competition are fostered.
Instruction is learner-centered.	Instruction is teacher-centered.

the group with questions, many of which are perfunctory or rhetorical. When there is interaction, it is chiefly between the teacher and individual students rather than among groups of learners.[27,28,11]

In collaborative education, learners are helped to be self-directed while also being encouraged to work with and teach each other. In authoritarian education, either intentionally or as a side-effect, dependence and competitiveness are fostered. Learners who accept and adopt an authoritarian approach often do not trust their capacity to direct their own learning successfully, and they tend to compete with their peers for their teacher's favor. Collaborative teachers typically give learners increasing responsibility for their own and each other's learning. Authoritarian teachers often cling to their responsibilities and their exercise of control over others.

Collaborative teachers provide "situational leadership."[29] They do not demand that learners function collaboratively from the start or at all times. Rather, they begin wherever their learners are — even if initially the learners prefer being dependent and passive. As soon as possible, though, these teachers encourage and help learners assume more control of and responsibility for their own and each other's learning. When working for the first time with individual learners or (groups of learners) or when introducing new tasks within established relationships, teachers tend to go through the following four stages: (1) assessing; (2) modeling; (3) coaching; and (4) fading. First the leaders assess the leaders' needs. Second, they provide any direction that is needed (perhaps by modeling what the learners need to do). Third, as students are able to take more responsibility for their learning, they serve as coaches. Fourth, as learners are able to work more independently, the leaders fade into the background.[30,31]

Authoritarian instructors, in contrast, tend to be rigid about the pattern of their relationships with learners. Some authoritarian instructors are threatened by, and demeaning toward, students who want some control over their learning. Collaborative teachers encourage students and residents to work with each other toward common goals. Authoritarian teachers often pit learners against each other.

Collaborative Education Can Foster Collaborative Healthcare

Students who feel valued as contributors to their own and each other's learning are likely to find that it is not a huge leap to value the contributions that their patients can make to their own care. Students who actively participate in identifying their learning needs and learning goals, in developing plans for achieving their goals, and in developing plans for monitoring their progress, are likely to gain a sense of ownership of their learning tasks and understand the need for parallel activities in healthcare. They are likely to understand that if patients are to feel responsible for sustaining their health and have ownership of the management plans, they need to be involved in all phases of their care: identifying their healthcare needs; developing goals for their behavior change or care; and making plans for achieving their goals and for monitoring their progress. Also, the skills students develop, as they assess their learning needs and plan their learning, are parallel to the skills they will need for helping patients participate in assessing their healthcare needs, and in planning and monitoring their healthcare. In addition, if students develop written learning contracts (containing their goals and plans), they gain experience with an approach to contracting that can be useful with patients/clients in healthcare settings.[11]

Interdisciplinary Education

Currently, many faculty and students in the health professions have little significant contact with each other, even though they might be on the same campus.[13] However, some students from different health professions do attend some of the same classes, and some students have even shared an integrated core curriculum taught by interdisciplinary teams.[23] Sharing courses, particularly when they are designed and taught by interdisciplinary teams, can be a form of collaborative education, but some of these courses are taught in authoritarian rather than collaborative ways and do little to promote future collaborative work. For example, students who sit passively together in large halls listening to the same lectures are not gaining any skills toward being collaborative. In optimal models of classroom-based collaborative education, interdisciplinary groups of students and faculty engage in lively, interactive, small group sessions usually focusing on shared healthcare challenges.

During clinical learning experiences, such as clerkships and practicums, some students from the different professions do work and learn in the same settings. In the least optimal situations, faculty compete to secure clinical sites for their students. If students end up being at the same site at the same time, they may interact only a little and even openly compete for patients. In contrast, in truly collaborative clinical learning experiences, students work as partners on student health teams or work on existing healthcare teams. (See below.)

Most interdisciplinary educational experiences have been isolated, elective, extra-curricular offerings.[32] A significant attempt to develop an integrated didactic and clinical interdisciplinary educational program throughout the curriculum took place in the 1970s at the University of Nevada.[73]

Providing Learners with Opportunities They Need for Developing and Valuing the Capabilities Required for Being Collaborative Professionals

If learners are to develop and retain the attitudes, characteristics, understanding, and skills needed for being collaborative health professionals, they need the following opportunities.

Opportunities to Clarify the Capabilities They Are to Develop. Learners need to see written descriptions of the capabilities they are to develop (e.g., in the form of learning goals or objectives). Perhaps even more important, as we described above, they need role models. They need to see living examples of who they are to become and what they will be able to do.

Opportunities to Understand the Importance of Being Collaborative. If learners are to value and use these capabilities throughout their careers, not just while they are being supervised, they need to value these capabilities. Being exposed to credible, competent role models who value and work collaboratively with other health professionals is an important start in preparing learners for sustaining their collaborative skills into their careers. Learners also need to witness different models of practice (e.g., by directly observing practitioners or viewing videotapes of their work with patients) and reflect on the implications, including likely outcomes, of these different models. Learners can reflect aloud on their own experiences as patients, caregivers, and members of teams. They can also role-play interactions with both patients and colleagues, enabling them to experience the process and likely outcomes of different ways of relating to patients and colleagues. In addition, they can read and discuss the growing literature on collaborative models of practice.

Health professionals can learn to value collaboration with patients and colleagues by being helped to appreciate all they can learn from others and how these people can help them be more effective in their work. In support of this process, students can be guided toward relevant experiences, such as asking people with chronic health problems what it's like to live with their condition and how health professionals can be most helpful to them. Most students do have opportunities to work with and learn from professionals in their intended discipline. They can also observe other health professionals in action and talk with them about what they do.

Opportunities to Develop Ownership of the Learning Process. As we discussed above, students are much more likely to feel ownership of the learning process if they are partners in identifying their learning goals and in developing and monitoring their learning plans.

Opportunities to Acquire Needed Knowledge and the Tools for Seeking Knowledge. Learners need to understand the principles and concepts of group dynamics, collaborative decision-making, consensus-building, and negotiation. They need to understand the perspectives and areas of expertise of other health professionals and the role of health teams in providing quality care. Learners need to begin building this and other knowledge. But even more importantly, they need to learn how to learn about these areas, and they need to *want* to learn more, so they will continue learning, even after graduation.

The knowledge that is most likely to be deeply internalized by students is the knowledge they seek as the result of being exposed to real world challenges. For example, students are most likely to want to learn the principles and skills of negotiation if they are confronted with the need for this knowledge as they try to work through simulated or real challenges that require negotiation skills.

Opportunities to Engage in Staged, Supervised Practice of New Capabilities. As when learning any capabilities, students who are learning to be collaborative need repeated practice in being collaborative. In the relative safety of a classroom they can practice collaborating with patients by role-playing with simulated patients or each other. During CLEs (clinical learning experiences), such as clerkships and practicums, they can try using their collaborative skills with patients.

Students can prepare for working as partners with colleagues in their own discipline by working collaboratively with each other, during classroom-based courses and CLEs. A helpful model for learning collaborative attitudes and skills is problem-based learning in which small groups of learners jointly address a problem (typically a patient problem) by sharing current relevant knowledge, identifying learning issues, seeking information, building knowledge, and solving problems.[33,34] Teams of learners in the same discipline can jointly work on projects (e.g., providing care to the homeless). During CLEs they can share some clinical tasks, and they can share and jointly build knowledge in patient-centered case conferences.

Students can also practice working collaboratively with other health professionals. For example, in the classroom, pairs of medical and nursing educators can facilitate sessions in which mixed groups of medical and nursing students work through patient care challenges (cases), using strategies that parallel those used by interdisciplinary health teams.[35] In classrooms, students from different disciplines can role-play interdisciplinary team issues.[36] Interdisciplinary student teams can jointly work on projects. As they select an idea, set goals, assign responsibilities, resolve conflicts, and make decisions, they can learn about team dynamics and strategies.[37] Multidisciplinary student teams can also learn about creating partnerships with communities when they work with members of the community in identifying neighborhood needs and helping health centers design programs to address these needs.[38] And, student health teams under the supervision of faculty members (e.g., faculty teams whose disciplines match those of the students) can provide healthcare in a variety of settings (e.g., family practice centers, patients' homes, nursing homes), addressing challenges that are congruent with their current capabilities.[23,36,37,39,40,41,42] During CLEs, students can also work as members of existing interdisciplinary health teams. Although their main preceptor should probably be a member of their intended discipline, who is on the team, they can also receive helpful supervision from professionals from other disciplines.

Opportunities to Reflect on and Assess Their Practice of New Capabilities. To extract the most possible from learning experiences, students need to reflect on their practice of new skills. This includes identifying what they are doing competently and what still needs work. Reviewing videotapes of their role-plays or their real interactions with patients and colleagues can be an effective aid to this process.[43] Also helpful are open-ended discussions that enable students to reflect on how they feel and think about partnerships with patients and colleagues. Students who are jointly caring for patients can reflect on their efforts in case conferences.[42]

Opportunities to Receive Feedback from Teachers, Peers, and Others. Timely, constructive feedback is vital to learning. Sources of feedback include teachers, peers, other health professionals, other health professions students, and simulated and real patients. Before peers give feedback to each other, they should be helped to understand the characteristics of constructive feedback (e.g., it is concrete; it focuses on the person's behavior not on the person himself or herself; subjective comments are labeled as such). Feedback from patients can be particularly powerful when patients and learners jointly review videotapes of their encounters.

Opportunities to Learn from and with Teachers and Learners Whom They Trust. If learners are to take the risks typically involved in making changes and developing new capabilities and if they are to be open in their self-assessments, they need to trust their teachers (including peers who give them feedback and teach them in other ways). Mutual trust implies that partners in a relationship feel confident that whatever each reveals or is revealed to them will be used to help, not hurt, the other person, and that confidentiality will be maintained, if needed.

Opportunities to Develop a Professional Identity. Currently, most learners do have opportunities to develop some aspects of their professional identity. Unfortunately, particularly in some health professions and certain settings, that identity does not include an expectation of working collaboratively with others.

Multiple Opportunities to Learn and Work Collaboratively with Other Health Professionals. If learners are to become and remain professionals who work as partners with other practitioners, they need far more than occasional or episodic opportunities for collaborative, interdisciplinary learning. These opportunities should be an integral, continuing part of the curriculum. Collaboration, like other core attributes of their identity, needs to be routinely modeled, practiced, and reinforced. It needs to be so well integrated that it is part of the ever-present "landscape."

Factors That Can Be Barriers to or Support for Collaborative Education

In general, in the US, we have tended to place higher value on competition and individual achievement than on cooperative, group accomplishments. We are a "star"-oriented culture. This cultural heritage, together with our substantial traditions of using authoritarian approaches in healthcare and education, pose considerable barriers to the evolution of collaborative education. We have a paucity of role models (especially in medicine) of professionals working collaboratively with either patients or colleagues, and, in many institutions, it is difficult to find truly collaborative healthcare teams for students to observe and join.

Because of our long tradition of authoritarian education, some teachers who decide to function in collaborative ways must first learn how to work collaboratively with patients and colleagues before they can teach collaboration to their students. Educators desiring collaboration may find that other faculty members, administrators, and students don't want to take the risks or undergo the discomfort that can be involved in changing their approaches. Collaborative teachers may also have to contend with institutionally imposed practices (e.g., competitive grading systems) that foster distrust and competition. Educators who want to engage in collaborative, interdisciplinary teaching often have additional hurdles, such as incompatible calendars among schools, the reluctance to add more to students' already busy learning agendas, attempts to protect perceived professional territory and curriculum time, the logistics of matching maturity or experience levels of different student groups, the lack of institution recognition and reward, and clinical sites that are not structured for interdisciplinary learning.

A key factor in support of collaborative education, including collaborative, interdisciplinary education, is the general shift in many areas of society from authoritarian to collaborative paradigms. An increasing number of healthcare consumers, administrators, and students are demanding more collaborative approaches. As a growing number of schools, especially prestigious institutions such as Harvard, are shifting to more collaborative approaches (e.g., problem-based learning in collaborative tutorial settings), it becomes easier for other schools to move toward greater collaboration, particularly with reports that teachers are finding collaborative models of education to be stimulating, even exhilarating.[44] Foundations and the federal government are funding interdisciplinary collaboration in healthcare and health professions education, particularly in primary care in underserved communities.[12,36,45]

Strategies for Moving toward Collaborative Education

Cohen and colleagues[46] identify three factors that can facilitate institutional change: (1) committed leaders, (2) education-specific funding, and (3) centralized curriculum governance. Maurana and Goldenberg[47] recommend building trust and consensus, letting go of old ideas, thinking creatively, creating win-win situations, being persistent, and keeping a sense of humor. When planning an interdisciplinary course, Grant[13] suggests: (1) allowing ample preparation time for organizing and training diverse faculty; (2) establishing strong institutional support; (3) identifying visionary leaders within the institution; (4) seeking external funding; (5) emphasizing patient-centered teaching; and (6) accepting the "non-ideal." When developing interdisciplinary clinical experiences, Kappelman and colleagues[42] recommend developing well-functioning health teams before involving students, and having regular meetings of the preceptors from the various disciplines to ensure common goals. Talen and colleagues[36] found that in developing an interdisciplinary course it was important for faculty to be equally committed and to avoid stereotypical roles (e.g., nursing faculty always taking charge of organization and details of the course, medical faculty rotating in and out and being called for emergencies, and psychology faculty observing and commenting on group dynamics). They also found it helpful for students to be at similar levels in their clinical training, to have the same expectations for course credit, and to have "hands-on" clinical practice experience together. We also recommend systematic faculty development to help ensure that faculty members value collaboration, function collaboratively with patients, colleagues, and learners, and have the skills required for doing so. Particularly during the development of new programs, there needs to be frequent and open communication among faculty members.

Conclusion

To prepare health professionals who want and are able to work as partners with patients and families as well as with colleagues (in their own and other disciplines), we educators need to foster partnerships with learners and faculty (including faculty from other health professions) and use collaborative models of education. We need to provide learners with opportunities to witness collaborative role models, practice collaborative skills, critique their efforts, and receive feedback on their efforts from others. And, as we all face the inescapable awkwardness and discomforts that accompany all fundamental changes in outlook and behavior, we need to be patient with ourselves and others.

References

1. Historically, especially during wartime, people who cooperated with an enemy were referred to as "collaborators." Increasingly, in the health professions literature, as here, this word has a far different, positive meaning. Throughout this chapter, "collaboration" is meant to convey an approach to teamwork that is highly constructive, in which genuine partnerships are formed, where people are valued for their perspectives and skills, and where the focus is on the pursuit of shared goals. Differences in educational backgrounds, titles, and experiences are relevant only to the extent that they pertain to the tasks needing to be done. They are not barriers to communication or cooperation.

2. In this chapter we primarily use the word "patient", although the word "client" could also be used.

3. In this chapter, we used the words "professional", "clinician," "practitioner," and "provider" interchangeably.

4. We use the term "authoritarian" to describe a way of relating to others based on obedience to authority.

5. Westberg J, Jason H. Fostering healthy behavior: the process. In SH Woolf, S Jonas, RS Lawrence (eds.) *Health Promotion and Disease Prevention in Clinical Practice*. New York: Williams & Wilkins, 1996:145–162.

6. Sackett DL, Haynes RB, Tugwell P. *Clinical Epidemiology: A Basic Science for Clinical Medicine*. Boston: Little Brown, 1985.

7. Cousins N. Anatomy of an Illness as Perceived by the Patient. New York: W.W. Norton, 1979.

8. Haug MR, Labin B. *Consumerism in Medicine*. Beverly Hills, CA: Sage, 1983.

9. Strull WM, Lo B, Charles G. Do patients want to participate in medical decision making? *JAMA* 1984;252(21):2990–2994.

10. Jason H, Westberg J. *Teachers and Teaching in U.S. Medical Schools*. Norwalk, CT: Appleton Century-Crofts, 1982.

11. Westberg J, Jason H. *Collaborative Clinical Education: The Foundation of Effective Healthcare*. New York: Springer Publishing Company, 1993.

12. *Community Partnerships: Redirecting Health Professions Education toward Primary Healthcare*. Battle Creek, MI: W.K. Kellogg Foundation, 1992.

13. Grant R. *Interdisciplinary Collaborative Teams in Primary Care: A Model Curriculum and Resource Guide*. San Francisco: The Pew Health Professions Commission, 1995.

14. Fagin CM. Collaboration between nurses and physicians: no longer a choice. *Acad Med.* 1992; 67(5):295–303.

15. Mechanic D, Aiken LH. Sounding board: a cooperative agenda for medicine and nursing. *N Eng J Med.* 1982;307:747–50.

16. Steele JE. *Issues in Collaborative Practice*. Orlando, Florida: Grune & Stratton, Inc., Harcourt Brace Jovanovich, 1986.

17. Kitzman H. Education in nursing. In JE Steele (eds.). *Issues in Collaborative Practice*. Orlando, FL: Grune & Stratton, Inc., Harcourt, Brace, Jovanovich, 1986:139–159.

18. Deming WE. *Out of Crisis*. Cambridge, MA: Massachusetts Institute of Technology, 1986.

19. McAuley RG, Paul WM, Morrison GH, Beckett RF, Goldsmith CH. Five-year results of the peer assessment program of the college of physicians and surgeons of Ontario. *Canadian Medical Association Journal* 1990;143:1193–1199.

20. Mann KV. Educating medical students: lessons from research in continuing education. *Acad Med* 1994;69:41–47.

21. Ducanis AJ, Golin AK. *The Interdisciplinary Health Team: A Handbook*. Germantown, MD: Aspen Systems Corporation, 1979.

22. Sand RG, Stafford J., McClelland M. I beg to differ: conflict in interdisciplinary team. *Social Work in Health Care* 1990;14(3):55–72.

23. Baldwin DC, Baldwin MA. Interdisciplinary education and health team training: a model for learning and service. In AD Hunt, LE Weeks (eds.) *Medical Education since 1960: Marching to a Different Drummer.* East Lansing, MI: Michigan State Foundation, 1979:190–221.

24. Helfer RE, Kemp CH, (eds.) *Child Abuse and Neglect: The Family and Community*. Cambridge, MA: Ballinger, 1976.

25. McKegney CP. Medical education: a neglectful and abusive family system. *Fam Med.* 1989;21(6):452–457.

26. Pellegrino Ed. Educating the humanist physician: An ancient ideal reconsidered. JAMA 1974;227(11):1288–1294.

27. Jason H. A study of medical teaching practices. *J Med Educ.* 1962;37:1258–84.

28. Jason H. A study of the teaching of medicine and surgery in a Canadian medical school. *Canadian Medical Association Journal* 1964;90:813–819.

29. For a description of "situational leadership" that is relevant to education, see Blanchard KH Situational leadership. In RA Ritvo, AH Litwin, L Butler, (eds.). *Managing in the Age of Change*. New York: Irwin Professional Publishing, 1995:14–33.

30. Collins A, Brown JS, Newman SE. Cognitive apprenticeship: teaching the craft of reading, writing and mathematics. In LB Resnick (ed.) *Knowing, Learning, and Instruction: Essays in Honor of Robert Glaser*. Hillsdale, NJ: Lawrence Erlbaum Associates, 1989:453–494.

31. Westberg J., Jason H. *Fostering Learning in Small Groups: A Practical Guide*. New York: Springer Publishing Company, 1996.

32. Beloff JS, Willett M. Yale studies in family health care III: the health care team. JAMA 1968;205:73–79.

33. Barrows HS, Tamblyn R. *Problem-Based Learning: An Approach to Medical Education*. New York: Springer Publishing Company, 1980.

34. Barrows H. *How to Design a Problem-Based Curriculum for the Preclinical Years*. New York: Springer Publishing Company, 1985.

35. For a description of interdisciplinary, case-based learning in the continuing education of mixed groups of physicians, nurses, dietitians, pharmacists, social workers, and health educations. See Mann KV, Langille DB, Viscount PW, et al. Multidisciplinary learning in continuing professional education: the heart health Nova Scotia experience. *Journal of Continuing Education in the Health Professions* 1996;15:50–60.

36. Talen MR, Graham MC, Walbroehl G. Introducing multiprofessional team practice and community-based health care services into the curriculum: a challenge for health care educators. *Family Systems Medicine* 1994;12(4):353–360.

37. Mazur H. Beeston JJ, Yerxa EJ. Clinical interdisciplinary health team care: an educational experiment. *J Med Educ.* 1979;54(9):703–713.

38. Center for Community Health Education, Research, and Service (CCHERS) Boston, Massachusetts. In *Second National Primary Care Conference: Case Studies.* Dallas, TX, 1949.

39. Baldwin DC, Edinberg MA. Models for faculty teams. In *Interdisciplinary Health Team Training. Proceedings of a Workshop Held at Snowbird, Utah, September 6–8 1976.* Washington, DC: DHEW Publications, 1978.

40. Croen LG, Hamerman D, Goetzel RZ. Interdisciplinary training for medical and nursing students: learning to collaborate in the care of geriatric patients. *Journal of American Geriatrics Society* 1984;32(1):56–61.

41. Itano JK, Williams J, Deaton, MD, Oishi, N. Impact of a student interdisciplinary oncology team project. *Journal of Cancer Education* 1991;6(4):219–226.

42. Kappelman MM, Bartnick LA, Cahn B, Rapoport, MI. A nontraditional geriatric teaching model: interprofessional service/education sits. *J Med Educ.* 1981;56:467–477.

43. Westberg J, Jason H. *Teaching Creatively with Video: Fostering Reflection, Communication, and Other Clinical Skills.* New York: Springer Publishing Company, 1994.

44. Wilkerson L, Maxwell JAAQ. A qualitative study of initial faculty tutors in a problem-based curriculum. *J Med Educ.* 1988;63:892–899.

45. Adams LJ, Miller ME. *Interdisciplinary Rural Teams Training Project: Resource Manual.* Denver, CO: University of Colorado Health Sciences Center SEARCH/AHEC Program, 1994.

46. Cohen J., Dannefer EF, Seidel HM, et al. Medical education change: a detailed study of six medical schools. (Submitted for publication: cited by Grant, #13.)

47. Maurana CA, Goldenberg K. A successful academic community partnership to improve the public's health. *Acad Med.* 1996;71:425–431.

XX Teaching Effectiveness Program for Primary Care: An Educational and Administrative Partnership

Diane G. Schwartz and Richard T. Sarkin

Introduction

In 1992, the Council on Graduate Medical Education and the Association of American Medical Colleges recommended that 50% of practicing physicians be generalists, that funding for Graduate Medical Education (GME) be based on workforce needs, and that financial incentives be created to educate more primary care physicians.[1,2] President Bill Clinton's Health Security Plan further reinforced the need to provide incentives to train more generalist physicians.[3]

To meet this goal, the School of Medicine and Biomedical Sciences, State University of New York at Buffalo (UB) and the Graduate Medical Dental Education Consortium of Buffalo (GMDECB)—the teaching hospitals affiliated with UB—entered into an agreement with the New York State Department of Health, and local insurance carriers to return a percentage of the Indirect Medical Education (IME) funds to the consortium. Between 1992 and 1995, the GMDECB committed approximately $5,000,000 of hospital training funds to this effort.

The Primary Care Resource Center (PCRC) was created in September 1992 to focus on three fundamental areas of program development: 1) fostering resident and medical student interest in primary care careers; 2) recruiting medical students to primary care residency programs in the GMDECB; and 3) providing faculty development opportunities in teaching effectiveness, research, and medical informatics. In this chapter, we will first provide some information about the development and administration of the PCRC, and then describe two aspects of the center:

- Community-based training sites for residents and medical students, and
- Teaching Effectiveness Program.

The Primary Care Resource Center

The PCRC's staff consists of a Director, Executive Director (Medical Education Specialist), Associate Directors for Recruitment and Administration, and several support staff. The work of the Center is carried out by several committees, whose membership is composed of full-time as well as voluntary faculty from the community. An executive committee provides advice and counsel to the staff and is composed of the chairs of the primary care departments and the coordinators of each of the committees.

The goals and objectives of the center are: 1) to formulate a recruitment plan suitable for all generalist programs; 2) to devise strategies to encourage half of UB graduates to select GME in primary care; 3) to attract a greater number of American medical school graduates to the GMDECB primary care residencies; and 4) to share information and experiences relative to reaching these goals. The recruitment effort was considered top priority and a variety of strategies were used to achieve these goals. This chapter will focus on two components of the PCRC: the Community Academic Practice (CAP) sites and the Teaching Effectiveness Program. One of the most important achievements of the Center is the collaboration that has been developed between the Departments of Family Medicine, Medicine, Pediatrics, and Social and Preventive Medicine to coordinate the recruitment of primary care residents to the GMDECB.

Community Academic Practice Sites

The CAP program was designed to expose residents and medical students to outstanding community-based primary care role models by offering learners clinical training in doctors' offices as well as hospitals. CAP site selection is based on stringent peer-reviewed criteria. As a result, CAP sites offer students and residents the opportunity to develop a mentoring relationship with some of the best primary care physicians in Western New York. Between 1992 and 1997 twelve CAP sites became operational.

CAP sites receive an initial start-up grant and in subsequent years a maintenance stipend. Each selected site also receives a microcomputer, modem, and printer. Physicians and medical learners are expected to use the computers to access Hospitals and University at Buffalo Library Resource Network (HUBNET), a computerized medical information system; the Internet; World Wide Web; and electronic mail. The objective in placing computers in the CAP sites is to integrate the use of these resources into clinical practice and stimulate the development of life-long learning skills.

CAP physicians have stated that their involvement in the program has many benefits, including an increased academic environment in which to practice, greater motivation to learn, enhanced recognition from peers and patients, and expanded teaching responsibilities. At CAP sites, learners are able to work with several physicians to observe a variety of different practice styles, which helps to develop a unique style of their own. Residents assigned to CAP sites have evaluated the program as outstanding. One CAP site director summed up his experience by saying,

"It is simply more fun now that my practice is a CAP site." The CAP committee meets bimonthly to share experiences, both positive and negative, and learn from one another.

Teaching Effectiveness Program

The driving force behind the development of the Teaching Effectiveness Program (TEP) has been the belief that the career decisions of medical students and residents are greatly influenced by their clinical teachers. Primary care faculty and residents with excellent teaching skills will be better preceptors, mentors, and role models to medical students.[4–6] Thus, we have defined faculty development as a core component of the School of Medicine's long-term strategy to enhance the status of the primary care disciplines on the UB campus.

Mission, Goals, and Objectives

Most medical faculty have not received formal training in effective teaching techniques. As clinical education shifts from the inpatient hospital ward to the physician's office and clinic, the range of physicians becoming involved in clinical teaching is broadening significantly. As a result, these clinical preceptors need to be trained to teach effectively.[7] Since residents also have significant teaching responsibilities to medical students, they were included in the training initiative. As primary care faculty and residents become more effective teachers in primary care, we believe that students will view the generalist disciplines as more intellectually challenging, rewarding, and fun.

The TEP is a collaborative effort of full-time and volunteer, community-based, primary care physicians dedicated to improving their own teaching skills and those of their colleagues. The TEP mission is achieved by presenting instructional programs that enhance the teaching skills of primary care physicians, creating an environment in which quality teaching is of major importance to the School of Medicine, enlisting the participation of volunteer and full-time faculty as well as residents and medical students, ensuring that faculty and residents develop life-long learning skills, and providing residents and faculty with access to and the ability to retrieve current medical information.

The activities of the TEP are carried out by a committee of twenty-five primary care faculty, residents, and medical students, as well as educators from the Graduate School of Education and the Office of Medical Education. Approximately half of the Committee members are full-time faculty. Generally, one or two residents and several medical students attend monthly meetings. As a result, a unique partnership was created at the outset. The TEP has become a visible and productive demonstration of how colleagues from a range of medical disciplines and professional settings can pool resources, work towards a common goal, and create successful partnerships in a very short period of time.

Setting Priorities

One of the TEP Committee priorities has been to develop innovative ways to enhance the teaching skills of faculty and residents. A team effort involving faculty from each of the primary care departments develops and presents workshops on teaching effectiveness. To foster mentoring, we paired junior faculty, residents, and medical students with senior faculty. Committee members thus enhanced their managerial, organizational, and instructional skills while creating a product that many faculty and residents could utilize.

Another important priority for the TEP is to teach primary care physicians how to use computers to access medical information databases such as MEDLINE, search the Internet and World Wide Web, and use electronic mail (e-mail). The decentralized clinical environment at UB makes communicating via e-mail an essential skill, and the PCRC has provided one of the few opportunities, either on campus or at the Consortium hospitals, for developing these skills. To promote such computer use, the medical informatics workshops are offered several times per year.

TEP Staff

The Committee Coordinator is a full-time, primary care faculty member who devotes fifteen percent time to the program goals and is responsible for the activities of the TEP. His responsibilities include recruiting new members, conducting monthly committee meetings (a business meeting followed by an educational presentation), and facilitating collaboration with other units on campus. He serves as a member of the Executive Committee of the PCRC and participates on School of Medicine committees to maintain an institutional commitment for faculty development.

The Medical Education Specialist assists with developing, conducting, administering, and evaluating all TEP activities. One of her primary responsibilities is to establish partnerships with physicians in each of the primary care departments and to offer individualized and group instruction on a variety of educational topics.

Workshop Development

The TEP, in collaboration with faculty who have expertise in teaching, has organized workshops to help faculty understand the theoretical framework of adult learning that is relevant to ambulatory settings. Initially, the TEP Committee chose four workshop topics: communication, feedback, learning styles, and accessing information. Since then, the range of TEP workshops has expanded significantly (a list is available on request).

To publicize workshops, promotional brochures are mailed to approximately 1,000 primary care physicians located throughout Western New York. Most workshop attendees are university hospital-based physicians from the primary care disciplines, but a significant representation of community physicians, residents, medical students, specialists, and nurses also regularly attend sessions.

Workshop evaluation

Workshops have been shown to produce changes in physician teaching behavior[8–10] if they accommodate the busy schedules of practicing clinicians. Participants assess the TEP workshops using a standardized evaluation form and have rated them very well. In addition, the Center mailed a survey questionnaire to clinical preceptors who had attended at least two TEP workshops (n=75). The questionnaire, a modified version of the Self-Assessment Inventory created by Irby,[11] uses "before" and "after" self-reports of eight teacher behaviors: organization/clarity, enthusiasm/stimulation, instructor/knowledge, rapport, instructional skill, clinical supervision, clinical competence, and professional characteristics. Thirty-five usable responses were received. Overall, the perceived gain in preceptor characteristics was impressive ranging from 22 to 47 percent improvement in performance on the eight teacher behaviors listed above. This is preliminary data assessing the impact to date of TEP workshops.

Master's Degree in Education for the Health Professions

The Master's Degree is a 32 credit-hour program being offered in conjunction with the UB Graduate School of Education. Candidates will complete eight formal courses, i.e., twenty-four credits, propose and conduct an applied research project; and participate in a series of professional seminars, jointly planned and administered by the Graduate School of Education and the PCRC. The program's goals are to enable physicians to understand how people learn, how teaching facilitates learning, and how to develop the skills needed to design and create effective curricula. Another objective is to create a cadre of physician educators who can provide leadership and direction in curricula reform and innovation on campus. This new program helps to further increase the status, respect, and recognition of primary care physicians among medical students and faculty in other disciplines.

Progress Toward Goals

The TEP fostered collaboration among the primary care clinical departments, family medicine, medicine, social and preventive medicine, and pediatrics to improve the teaching effectiveness of medical school and community-based faculty, as well as residents. The interactions among educators, residents, and medical students has created an exciting and creative atmosphere. In addition, the interaction between the Center and faculty from the Graduate School of Education has led to the development of a Master's Degree in Health Professions Education. Working with faculty educators has given an expanded dimension to the program by providing a foundation of knowledge and skills from which to develop and experiment with new curricula, and to implement and practice new instructional techniques. Through TEP workshops, primary care physicians have developed the ability to use computers to access medical information and communicate via e-mail, an essential skill given UB's clinical environment. Another important achievement is that many non-primary care faculty also have enrolled in TEP workshops.

Since the inception of the program, many TEP members have engaged in a variety of scholarly activities, including presentations at local, regional, and national meetings. In addition, manuscripts describing program achievements, experiences, and ideas for enhancing teaching effectiveness also have been prepared. In the four years since the program began, many faculty have successfully presented a range of continuing medical education workshops, revised the first and second year curriculum to focus on the practice of primary care medicine, and developed a solid understanding of the principles of education.

TEP workshops have provided physicians with opportunities to learn new precepting skills; fine-tune their abilities as mentors and role models; develop their medical information literacy techniques;[12] use evidence-based medicine to support clinical decision-making; and create new avenues of communication with other primary care physicians. CAP physicians, in particular, have become a network of primary care educators with enhanced linkages to the School Medicine and Consortium of hospitals.

Conclusion

The PCRC is dedicated to changing the institutional climate of the UB School of Medicine from one that focused on specialty practice, to a more even distribution of positions in which at least 50% of GME trainees are in primary care. Table 1 summarizes UB School of Medicine graduates' career choices and the number of students who have decided to stay in Buffalo for their residency.

Table 1. UB Graduates Primary Care Match Analysis

% UB Graduates Entering Generalist Residencies

	1991	1992	1993	1994	1995	1996
Total Primary Care	40	37	38	44	42	50

% Distribution Within Generalist Residencies

	1991	1992	1993	1994	1995	1996
Family Medicine	14	13	27	15	18	30
Internal Medicine	48	44	36	48	28	28
Pediatrics	34	38	29	27	43	32
Medicine/Pediatrics	3	5	8	10	10	10

Number of UB Students Staying in Buffalo for Generalist Residencies

	1991	1992	1993	1994	1995	1996
Family Medicine	2	2	1	2	5	8
Internal Medicine*	7	3	3	8	3	3
Pediatrics	5	5	8	8	11	10
Medicine/Pediatrics	0	2	2	3	2	4
Total Staying in Buffalo	**14**	**12**	**14**	**21**	**21**	**25**

To achieve this goal, we embarked upon a recruitment drive and organized faculty development activities. This process has brought together faculty from different departments to work collaboratively. UB has demonstrated how colleagues from a range of departments and professional settings can pool resources and create successful partnerships to achieve a common goal.

References

1. Council on Graduate Medical Education. Improving Access to Health Care Through Physician Workforce Reform: Directions for the 21st Century (third report). Washington, D.C., U.S. Department of Health and Human Services, 1992: viii-ix.

2. AAMC Policy on the Generalist Physician; as adopted October 8, 1992. Acad Med 1993; 68:1–5.

3. Health Care Update: The Need for Health Care Reform, 1993:29

4. Hitchcock MA, Stritter FT, Bland CJ. Faculty development in the health professions: conclusions and recommendations. Med Teach 1993;14(4):295–309.

5. Irby DM. How attending physicians make instructional decisions when conducting teaching rounds. Acad Med 1992;67(10):630–8.

6. Nobel J, Friedman RH, Starfield B, et al. Career differences between primary care and traditional trainees in internal medicine. Ann Int Med 1992;116(6):482–7.

7. Jennett PA, Edworthy SM, Rosenal TW, et al. Preparing doctors for tomorrow: information management as a theme in undergraduate medical education. Med Educ 1991; 25:135–9.

8. Skeff KM. Enhancing teaching effectiveness and vitality in the ambulatory setting. J Gen Int Med 1988;3(2 suppl.):926–33.

9. Skeff KM, Stratos GA, Berman J, Bergen MR. Improving clinical teaching: evaluation of a national dissemination program. Arch Int Med 1992;152:1156–61.

10. Irby DM. Teaching and learning in ambulatory care settings: a thematic review of the literature. Acad Med 1995;70(10):898–931.

11. Irby DM. Evaluating teaching skills. Diab Educ 1986;11(special issue):37–46.

12. Schwartz DG, Nolan LA, Schlicht MP. Teaching medical computing skills to internal medicine residents. Med Ref Serv Quart, in press, 1997.

XXI International Partnerships for Community Health Education: Development and Maintenance

Alberto J. F. Cardelle and Lynn Bickley

The preparation of future physicians for Community-Oriented Primary Care (COPC) remains one of the most crucial and largely unmet challenges of the health sector. A myriad of reports and conferences conclude that community-based, problem-solving curriculum reforms within medical education are vital; particularly reforms emphasizing skills needed to better respond to the needs of communities and promote healthy lifestyles.[1] In the United States, these recommendations arise as a response to the low level of attention and funding granted for preventive and community health care services. With a medical specialists to generalists ratio of 7:3 and interest among graduating medical students in generalism dropping from 30% in 1985 to less than 15% in 1992, the United States lacks the necessary training opportunities in COPC. Trends indicate that the medical education system has been unable to encourage interest in primary care among future physicians, in part due to a lack of positive experiences and role models in primary care. In contrast to the United States however, conditions in developing countries have promoted community-oriented, medical education reforms. According to Dr. Arthur Kaufman, director of community medicine at the University of New Mexico, "the incentive for reform—the gross inequities between rich and poor, urban and rural areas—are so blatant in developing countries that the impulse for medical education reform is more palpable there than in industrialized countries."[2] Dr. Gabriel Smilkstein, of the University of California-Davis, expands this concept and maintains that when it comes to medical education reforms, the industrialized countries should examine and learn from international experiences, especially those from the developing world.

The need for improved collaboration with international schools is supported by the increasingly blurred line between international and domestic health concerns. The world's entire population is steadily transforming into one global entity with shared ecological and health concerns. If current political, social, and economic trends continue, the resulting US populace will be comprised of a mosaic of cultures, ethnicities, languages, and economic classes. Health personnel will need socially innovative and clinically sound skills to operate efficiently in this new heterogenous and interdependent environment.

Unfortunately, past international collaboration and assistance programs have operated on the principle that *we* (the developed countries of the world) know how, and that we have the answers for *their* (the developing countries) problems.[3,4]

Medical schools and institutions in the industrialized world have employed a very narrow and ethnocentric international relations paradigm that disregards local realities and the need for common learning and development. In short, international programs historically implemented by medical schools have not focused on the growing commonality between the health conditions of all of the world's countries.

Given historical trends in international programs and the critical need for modification, the American Medical Student Association Foundation designed and implemented the *International Health Partnerships for Community-Based Medical Education* (IHPP). The IHPP, which links medical schools in four different countries, has the goal of increasing the interest, understanding, and commitment to community-oriented primary care among medical schools, faculty, and students. This is accomplished by creating a system of collaboration and exchange between an assortment of unique and diverse COPC service and training models. The long-term objective is for these partnerships to develop into the infrastructure of a self-sustaining international support and communication system among medical schools and faculty interested in educational COPC programs.

This chapter discusses the process behind the creation and maintenance of the partnerships which involve annual and bilateral faculty and student exchanges. The IHPP, with a three-year grant from The Pew Charitable Trusts, created partnerships linking the *University of Rochester School of Medicine* and the *Universidad Autonoma de Chiapas*, in Mexico; the *University of Colorado School of Medicine* and the *Universidad del Valle*, in Cali, Colombia; and the *University of Massachusetts Medical School* and the *Universidad Autonoma Nacional de Honduras*. Participating students and faculty travel to their affiliated international partner school, participate in ongoing community health and primary health care services, and research projects that expose them to the realities of community health care in Latin America and the United States. Student training programs include exposure to community-based health promotion and disease prevention programs, primary care clinical services, and to a myriad local health professional committed to primary care. As part of faculty exchanges, medical educators collaborate with their counterparts in primary care research projects, curriculum development, faculty development, and COPC teaching experiences.

Selection of Partners and Creation of Partnerships

The creation of the partnerships entails a two step process of: 1) identifying an initial assortment of institutions that compliment the guidelines set by the partnership; and 2) selecting the most appropriate institutions. Guidelines used for this process must be based on a set of general goals and mechanisms developed before initiating the creation process. Since the creators of the IHPP decided that partnerships should strengthen community-oriented medical education through the sharing of experiences in diverse international COPC experiences, the search for suitable partners was based on identifying institutions that would compliment this goal. Generally speaking, when working with international partnerships it is important to recognize that the partnership will not only be affected by conditions internal to the partner institutions, but by circumstances in the international arena. Thus it is necessary to develop one set of guidelines that examines the geopolitical aspects and another set which surveys the internal aspects of the institutions.

Geopolitical Guidelines

The purpose of geopolitical guidelines is to set general parameters for a baseline satisfactory level of political and economic stability. If this is not present, the environment is not conducive to the partnership implementation, development, and sustainability. The guidelines are presented below with examples:

A. Determine what countries have identified the health field on which the partnership will be based (in the case of the IHPP it was COPC) as an area of major health importance. For example:

> The IHPP considered Mexico, Central America, and the Caribbean because the countries of the region had indicated a need for strengthening their primary health care delivery systems, as a consequence of having to improve quality of life with more people and fewer resources.

B. Analyze the countries' health care and medical education systems and determine whether the partnership would positively influence or detract from local medical education and service delivery programs. For example:

> Officials from a medical school and Ministry of Health in a Caribbean country hesitantly removed themselves from consideration for participation in the IHPP because officials were concerned that their participation would divert resources away from locally initiated programs which they had already committed to developing.

C. Evaluate a country's current political situation and determine if its stability can ensure appropriate impact, and guarantee participants' safety. For example:

> While recruiting a medical school in Central America, it became apparent that factional student organizations were objecting to working with North American groups and were denouncing the IHPP. It was ultimately determined that any eventual program would raise political tensions creating a dangerous and unstable environment for the partnerships.

D. Evaluate the country's current international status and how partnerships will affect that status. For example:

> The IHPP concluded that the proximity of Mexico, Central America, and the Caribbean to the United States, the permeability of their borders, and the growing interdependence of the countries' economies and societies, reinforced IHPP's goal of linking domestic and international health concerns.

Internal Guidelines

A set of internal guidelines was also created to establish the parameters defining suitability as a partnership institution. Internal guidelines delineated the necessary conditions *within* the institution at the institutional, administrative, and faculty level. Academic, administrative, and logistical guidelines with examples are presented below:

Academic and Administrative Guidelines

A. Assess the school and faculty's commitment to developing, implementing, and maintaining community-based medical education programs, in target settings and with target populations (e.g., underserved, migrant populations, rural vs. urban).

This criteria is critically related to the existence of a national health care delivery infrastructure and the availability of appropriate community-based services. It is vital to have an understanding of the services offered by the ministry of health (MOH) of the country and the involvement of the school within those services. If the MOH programs are inadequate, determine how much effort has been made by the school to develop alternative services through non-governmental entities. For example, the Aga Khan Medical School in Karachi, Pakistan, in the absence of adequate MOH programs, established seven mini-health-system prototypes in Karachi that provide its students with appropriate COPC experience.

B. Determine the faculty's philosophy and experience as preceptors.

Preceptorship protocol in other countries can be quite different from the protocol employed in the United States. The partnership will suffer if these differences are not understood before implementation. Although very rare, there are examples of overworked physicians overestimating the capacity of the student and using the presence of a foreign medical student as a good opportunity for vacation, thus leaving the student unsupervised and solely responsible for delivering clinical services.

C. Measure the level of enthusiasm, commitment, and time available on the part of the faculty.

This criteria requires balancing the advantages of all three factors. For example, one has to balance recruiting a faculty person who has experience in implementing and sustaining COPC and/or international programs with the fact that faculty with this amount of experience may already be over-extended. The partnership will suffer if it becomes one more project for an already overworked professor, but will gain from the experience of an active faculty member.

D. Confirm what the school, as an institution, will provide in terms of financial and in-kind support to the partnership.

The initial contact with a school is typically with a motivated faculty person. Unfortunately, faculty persons do not sustain a partnership by themselves. There is a need for institutional support, and, although preferably in the form of financial support, there is also a great need for in-kind support, especially in administrative services. Faculty persons are vital in the development and implementation of the academic program, but without the institutional support much of the partnership potential will be unexploited.

E. Determine the level of experience of the school and faculty with international programs.

The program should determine how it desires to invest its resources. Although schools which possess a long history of international programs are a benefit and offer a better chance of sustainability, a school without experience will have more to gain from the international partnerships and will be willing to invest more in the partnership.

Logistical Guidelines

A. Determine if the area in which the school is located has adequate and affordable housing for students.

B. Ascertain if the area offers a reliable local transportation system which allows participants to move about adequately and experience diverse regions of the country.

C. Assess whether or not the country possesses an efficient communication infrastructure which will facilitate communication between partner institutions.

D. Appraise whether the transportation to and from the country is sufficiently developed to assure a reasonable travel budget.

Application of Guidelines

The selection and creation process for partnership development requires the use of these different sets of guidelines in order to identify an initial assortment of countries and schools which satisfy a basic level of compliance. Using this initially selected group, the process can begin to identify specific sets of schools which qualify as partners. In this phase it is important to not only employ the above guidelines but to distinguish the schools that, as pairs, will maximize the advantages of diversity while ensuring the highest probability of sustainability. In other words, partner schools should differ in strengths so that they compliment each other but should not be so different that they lack any basis for collaboration.

In order to illustrate this process we can review IHPP's experience with the selection and pairing process. Initially, the IHPP had a total of 26 different schools interested in participating in a partnership or a total of 160 possible combinations of international and US schools. By employing the previously discussed guidelines an initial set of 12 schools were identified as satisfying a basic level of compliance with the guidelines. Table 1 illustrates the criteria used for this exercise, Table 2 describes how the six identified international schools ranked, and Table 3 indicates how the six US schools fared. The analysis was based on information provided by the schools in responding to "request for proposals" disseminated by the IHPP or information obtained from site visits to the international schools.

Table 1

(1) Expressed and demonstrated commitment of faculty
(2) Expressed and demonstrated commitment of the school
(3) History and experience with international programs
(4) Existing programs with community-based services
(5) Exposure to diverse cultures and underserved populations
(6) Transportation and housing
(7) Partnership would have meaningful impact on medical education and health care services
(8) Political stability and safety (international schools).

PAIR A

School J		School C
Strong international experience which included programs in Africa and Asia.	1	Young school which previously had one international program.
Philosophical background in the biopsychosocial approach to medical student training with need to expand opportunity for students.	2	A strong commitment to service to the underserved through community-based programs which required systematization and development of the curriculum.
Worked with Latino migrant-workers.	3	Worked with foreign refugee groups.

In pair A the first characteristic provided complimentary experiences with international health, through which school C would greatly benefit. In terms of experience and philosophy with community-based programs, school J's background could assist school C in developing an appropriate curriculum in community-based training. And the commonality of the target populations provided a basis for collaborative research and service programs.

PAIR B

School K		School A
Individual student international programs but no institutional sponsored program.	1	Ample experience with grants and programs from international foundations and UN agencies.

Table 2

School/Criteria	1	2	3	4	5	6	7	8	Rank
School in Country A	VG	VG	VG	VG	VG	VG	VG	F	1
School in Country B	VG	F	F	G	VG	VG	F	G	5
School in Country C	G	G	P	F	VG	G	VG	G	4
School in Country D	VG	VG	VG	VG	F	VG	F	G	3
School in Country E	G	P	G	P	G	F	VG	P	6
School in Country F	VG	VG	VG	G	VG	VG	VG	G	2

VG-Very Good, G-Good, F-Fair, P-Poor

Proposed community-based program for student training in varied rural areas.	2	Proposed community-based program with urban organizations.
Proposed to work through Family Medicine Department.	3	Proposed to work through Family Medicine Department.

In this second pair the schools' international experience provided a good complement. The difference in the sites for the community-based program offered the participants alternative community experiences, thus enhancing their medical training. The commonality around the Family Medicine approach provided a basis for a collaborative research and service program.

PAIR C

School L		School B
Individual student international programs.	1	Experience with foreign students and programs from international UN agencies programs.
Institutional experience in evaluating students in community programs, but limited community-based sites.	2	Historically strong program in community-based program, with interest in developing student evaluation protocols.
Department of community health had sound research component.	3	Community health department had sound student service programs.

Table 3

School/Criteria	1	2	3	4	5	6	7	Rank
School G	VG	G	F	VG	VG	G	VG	4
School H	G	F	VG	G	G	VG	F	6
School I	VG	F	VG	G	G	VG	F	5
School J	VG	VG	VG	VG	VG	G	VG	1
School K	G	G	G	VG	VG	VG	VG	2
School L	VG	VG	F	G	G	G	G	3

VG-Very Good, G-Good, F-Fair, P-Poor

With this group of twelve the final phase of selecting and pairing was based on pairing schools that complimented one another's programmatic features: location of student training sites, participating academic departments, proposed target populations, and ethnic mix of the schools' population.

In this partnership the schools had similar levels of international experience and had good complimentary approaches to their community-based training programs.

It is important to highlight that the selected schools were not those which ranked highest in terms of compliance with the guidelines. Instead they were those schools with distinctive characteristics that could offer the most novel experience for the participants. The guidelines are only a general set of parameters.

Partnership Implementation and Maintenance

Once the partnerships have been created and are operational, the nature of the relationship between the schools must be modified. Where the initial contact was inquisitive in nature, the new relationship must be based on mutual collaboration and full reciprocity. The partnership defines its own operational structure so as to enhance each partner's educational capacity, making equal use of the resources and expertise of the two linked institutions. Reciprocity, a "balance of benefits," must be ensured; a relationship which is viewed as "parasitic" by one institution will not be sustainable. In order to ensure this reciprocity, special attention is required in delineating expectations, establishing lines of communication, organizing program administration, and designing the monitoring and evaluation plan.

Clear Expectations

After the selection and creation of the partnership, it is essential that the development of the partnership program begin with a clear delineation of expectations. Expectations must be developed on two different levels: goals and objectives must be developed for the partnership as a program; and the expectations of the individual schools, departments, and faculty need to be articulated. While the goals and objectives will be vital in the process of continuous monitoring and evaluation, the individual expectations will ensure agreement on how to accomplish the programmatic objectives. The IHPP provides a good example of how agreement on goals and objectives is not the same as clear expectations.

The IHPP had developed as its goal:

> the development of a critical mass of physicians who value primary health care, community health, and service to the world's medically underserved by means of strengthening the skills of medical educators in and increase the commitment of medical schools to community oriented medical education worldwide.

and as two of its objectives:

> A. To equip students and medical educators with an insight to the specific medical aspects of community health care delivery in Latin America, the Caribbean, and the US; and,
>
> B. To orient students to the common health issues in Latin America, the Caribbean, and the US.

The faculty persons in the IHPP agreed to these goals and objectives, but a problem arose around their implementation because of diverging expectations. US students had been told that they would develop an academic program when they arrived at their partner schools based on a medical topic of personal interest. One of the faculty persons at an international school had been undertaking a COPC research program and thought the visiting students could work on the research project for at least 50% of the time in order to give it a boost. When the students arrived, they were insistent that they wanted a program that was of their personal interest and were not inclined to work on the faculty member's research project. Although the issue was finally resolved by promising the faculty person that subsequent students were to be informed that they would work on the established program, the incident increased the tension between the two schools and created an initial unpleasantness which could have been prevented by clearly delineating expectations.

In this example, the faculty person's COPC research project would have satisfied IHPP objectives, but did not address the expectations of the students. In addition to demonstrating the need for clear expectations, this incident reveals the need for reciprocity. If the partnership, to the best of its ability, accommodates and enhances the professional development of the participating faculty and school it increases the chances of its sustainability. At the same time, it is important that the partnership not become the vision of just one faculty person or school. With the modifications that followed this incident, the IHPP benefitted from allowing the faculty person and school to use the IHPP to advance professional development while not adversely affecting the integrity of the program objectives.

At an institutional level the partnership needs to define clear expectations through a Memorandum of Agreement in which each institution agrees to a set of responsibilities. This is crucial in terms of financial responsibility, legal liability, and ownership of finished projects produced by the partnership. The IHPP included the following language in its Memorandum of Agreement with partner schools.

> A. School X agrees to appoint a qualified faculty member to coordinate the development and implementation of the activities that are undertaken as part of the program. This individual will receive an honorarium in recognition for his or her services throughout the three years.
>
> B. School X shall maintain written records on the activities carried out at the institution, over the course of the period covered by this agreement. School X shall furnish a final partnership program activity report to the IHPP outlining and assessing their experiences during each exchange cycle.
>
> C. The IHPP agrees to fund the institution within the limitations set by the financing agency. The IHPP will fund all air travel arrangements for students and faculty arranged by the IHPP, faculty stipend, student stipend, a payment to School X every year in recognition of School X's participation in the exchange program.
>
> D. The IHPP and School X agree that any results obtained as a result of the performance of this agreement will be jointly published by the IHPP and School X.

Good Communication

As in any relationship a lack of communication will destroy a partnership's chances for sustainability. Communication should be based on equal levels of accountability. The individuals who are ultimately responsible for the partnership in both institutions should be considered equals in a professional sense. If the individuals at each school are at different professional levels, the partnership will often produce one-sided agreements.

When working with international partners, the logistical side of the communication process may be the most trying. Regular mail is slow and unreliable, special delivery couriers are expensive, and telephone lines are unreliable. Even when the lines are reliable, other external factors can inhibit communications.

> In one instance the IHPP worked with an international partner school which was changing deans; it was a contentious process and the outgoing dean had not been a supporter of the incoming dean. Just before leaving, the outgoing dean refused to pay a telephone bill for the previous six months, so the school was left without telephone service during the four-week transition period.

Most international institutions have telex numbers to which telegrams may be sent. An increasing number of institutions also have telefax capabilities, although these suffer the same ills as telephone communication, and finally as international computer networks become more available, e-mail communication becomes increasingly feasible with international partners. Until e-mail becomes commonplace, international couriers, although expensive, are the most reliable.

> Cross-cultural factors will also affect communication. Edward Hall, in his 1959 book *The Silent Language*, summarizes the subtleties surrounding cross-cultural communication: As a country [the US] we are apt to be guilty of great ethnocentrism. Consequently, we manage to convey the impression that we simply regard foreign nationals as 'underdeveloped Americans.' Most of our behavior does not spring from malice but from ignorance, which is a grievous sin in international programs. . . . Most Americans are dimly aware of this silent language. . . .. They are not conscious of the elaborate patterning of behavior which prescribes our handling of time, spatial relationships, our attitude toward work, play and learning.[15]

The IHPP has been forced to resolve many misunderstandings that arose as a result of this "silent language."

> In one case a US faculty person after having visited the international partnership, sent a thank you letter to her host expressing her gratitude. She included with the letter a list of general suggestions which she intended as helpful advice. The international school held numerous faculty meetings to discuss these suggestions and incorporated them as part of their curriculum reform because of fears that they would be dropped from the program if they were not complied with.
>
> In a second incident, during the first week of an exchange, the US participants were given an itinerary of events they were expected to attend. The group was baffled that at the first events to which they arrived late, knowing of the tendency for things to start late, they found everybody waiting and upset. They later realized that the organizers had printed two sets of agendas, one for the visitors and one for the local officials which showed the event beginning 30 to 45 minutes earlier, and so by being sensitive they actually had insulted their host.

The large number of anecdotes concerning cultural miscommunications indicates how difficult they are to prevent. Nevertheless, it is both possible and important to develop the necessary tools and skills to recognize their occurrence and to attempt to resolve them immediately. One can observe one's host's reactions, inquire from the host if there might have been a misunderstanding, and communicate clearly with one's hosts about one's intentions. There are myriad books and courses designed to help the cross-cultural traveler to prepare for his or her need to communicate.

The anecdotal evidence also points to the need to "roll with the punches." One can be certain that no matter how clear one's communication, some event will catch one unprepared. The secret lies in not panicking. One should prepare all those who will be involved with the partnership to expect to have to deal with a crisis at any moment; the better the lines of communication that have been established, the easier it will be to manage.

During the IHPP student exchange component, students receive visa forms, letters of endorsement from the IHPP and their host medical school, and instructions on how to get the needed visas including time requirements. During the first exchange year, two students coming to the United States began their visa request procedure the day before they were supposed to depart. They did not have all the required papers, and they were going to have to wait one or two weeks. When they were given this news they called the IHPP requesting help. It took over twenty telephone calls to different individuals in the US State Department, Immigration Service, and the US Embassy within a period of three hours to get approval for their visas.

Program Administration

Each institution should assign a person to be responsible for all administrative activities. The administrative person should be responsible for all logistical aspects of the program activities, such as lodging, transportation, financing, etc. This individual also needs to be responsible for program communication, including, but not limited to: responding to inquiries, communication with the partner institution, and promoting the program. In the IHPP's seven year experience with international partnerships there has always been a direct correlation between partnerships with assigned administrative personnel and successful program activities.

Consistent Monitoring and Evaluation

Continual monitoring is key to the partnership's ability to respond quickly with necessary program adjustments. It is important to monitor the following components of an international program:

- *The partnership formation process*, which will give insight into the clarity of the process, the communication system between institutions, and the level of ease with which the institutions feel the process was managed. This information will prove useful for future program development activities, partner recruitment, and selection process.
- *The program mechanics*, which include logistical arrangements, participant selection process, technical support offered to institutions and participants, communication efficiency, program costs, and participant's performance.
- *Program quality*, which includes evaluating the value of the information being provided and the experiences of the participants during the orientation, the required program, the elective program and the debriefing seminars.
- *Program impact*, which should be surveyed in the long-term, through a longitudinal tracking system to monitor the impact of the partnership experience on career choices and professional interests of the participants.

The tools for evaluating the components listed above include open-ended interviews administered during site visits to the institutions, faculty reports, student reports, and pre-post self-assessment questionnaires administered to the participating students and faculty. For longitudinal follow-up to assess impact, it is vital that student careers be tracked and periodic surveys be administered.

Partnership Activities and Learning Processes

The goal of academic partnerships is to allow the participating institutions to share educational resources and enable them to offer enhanced educational opportunities to their students. To take maximum advantage of the enormous potential of these partnerships, an appropriately designed structure is required; in the absence of an adequate structure, this potential is lost.

The first structural element is an orientation period in which the goals and objectives of the program should be delineated and discussed. The participants should be broadly informed about what they can expect and how to interpret what they experience. Personal expectations and learning plans should be developed. Participants should be informed of legal liabilities and responsibilities and be provided with information on how to ensure their safety and personal well-being.

The orientation should then provide a general introduction to areas that the participants will encounter. The IHPP has concentrated on the following core topics:

- *Culture and Health*. The orientation has included a discussion on the importance and the skills necessary for cross-cultural communication and delineated some of the local cultural parameters for the health and the health care of the population.
- *Politics, Economics, and Health*. The students are provided with a quick view of the economic base of the country, and how international circumstance affect its economy and hence its health care system. The students are also provided with an overview of the country's political situation and how health is viewed within the political system.
- *Local Community Health Practices*. The students are provided with the technical knowledge of community health care delivery, which include the immunization cold chain, epidemiology and diagnosis of common ailments, technique employed in health education, and the use of lay health care providers.

The next program element consists of required academic assignments and elective activities exposing students to a cross-section of the health care delivery in the region, services to underserved populations, and local medical education. It should also include a longitudinal, in-depth examination of a specific delivery program.

> During one IHPP program, students observed health care delivery in an emergency room, rural hospital, private practice clinic, and community health care center. They attended a weekly seminar on health policy issue and were encouraged to attend Grand Rounds.

They observed weekly medical school classes, and had a five-week rotation in a community health clinic with exposure to primary care clinical services and health education programs.

The elective component should be structured as an independent study program. This component should commence once the participants have familiarized themselves with local opportunities. The participants, working with their assigned tutors, should define their own areas of interest and then design, develop and implement research or intervention programs. These programs should complement the students' academic work so as to expand their understanding of a topic which is consistent with the objectives of the partnership. Students should have an opportunity to publicly present their work to their colleagues.

Debriefing seminars are an essential component that help participants consolidate and internalize their experiences. Much of the learning in a partnership and an international proram is based on spontaneous and unstructured exchanges that fall outside the formal curriculum. During weekly or biweekly debriefing sessions, the participants should be assisted in reflecting on their experiences and comparing notes with other participants. Some examples of insights that emerged from previous debriefing sessions are: the need to conceptualize one's role as "colleague" rather than as "outside benefactor" when working with local counterparts to improve a community's health; the importance of working not only with local health professionals but also with individual lay persons with practical knowledge about the target community; the power of health promotion and disease prevention methodologies in resource deficient conditions; and the relevance of cultural factors in health care.

Program Outcomes

Evaluations from the IHPP indicate that at both the student and institutional levels, the partnerships derive most of their educational potential from the person to person communication. The strength of the partnership at the student level resided in the opportunities students had to exchange perceptions and misconceptions about health care and medical education in other parts of the world. Students were most impacted by reading articles written by their counterparts, discussing their common complaints about medical education, observing first-hand the current trends in career choices, and attending student organization activities. In two cases, participating students from one institution formed new student organizations based on organizations they observed at their partner schools. Again, this was an educational program which required very little planning, needing only the creation of forums to allow individuals to meet and discuss their ideas. As one student participant wrote, "it was not just the technical abilities that were important to learn, it was the frank, non-apologetic discussions I had with student colleagues that were the most memorable and informative."

At the institutional level, medical school administrators were initially unsupportive when they were informed that the IHPP carried very little financial backing and that it concentrated in the area of international community health. It was not until the partnership developed and administrators from partner schools visited one another that they found that their international counterparts shared many of their concerns, and offered solutions from an alternative perspective. In those partnerships that involved administrators in the exchange, the partnerships were expanded into additional areas and the administration began to encourage expanded international contact. The program planners had only to create an appropriate forum for contact between administrators; all the rest emerged spontaneously from there.

The partnerships have benefitted the participating departments enormously, enhancing their funding, staffing, and attention from international sources. In one case the institution created a permanent position for the IHPP coordinator to develop and implement additional student oriented community-service projects. In a second institution the department received additional funding from the university to establish a special research and intervention program, and in the third a special office of academic promotion for other international collaborative programs was established.

For the faculty, the partnerships provide new opportunities for academic and professional development. The partnerships allowed medical faculty the opportunity to observe diverse teaching protocols, curriculums and student evaluation methodologies. In the case of the IHPP, the faculty observed diverse community-based preceptor sites and diverse teaching philosophies with respect to COPC. Comments from the faculty indicated that the partnership increased their understanding of the similarities between medical education issues worldwide, while gaining a greater appreciation of their own academic culture by observing the academic culture in a different country. The exchanges allowed for the individual faculty members to expand their experience of medicine into the international arena, expanding their research and teaching interests. Faculty also benefitted from colleagial assistance in publishing materials in countries other than their own.

Conclusion

It would be a daunting task for a single institution to assemble a conventional curriculum exposing students to the complexities of international health, the intricacies of social and economic relationships between nations, and the challenges and effectiveness of international institutional collaboration. The International Health Partnerships program offers all this without a major investment in facilities or staff. Its essential feature is a structure to promote dialogue and relationship; to stimulate new personal and professional experiences (and reflection on those experiences); and to integrate and distribute the new knowledge and insights that emerge spontaneously from these activities. The IHHP is an extension of the student-student, teacher-student, and teacher-school partnerships that already exist within medical schools on a global scale.

References

1. World Health Organization. *Changing Medical Education: An Agenda for Action*. Geneva, Switzerland, 1991.

2. Kaufman A. International Medical Education. Academic Medicine 1993;68:5:336.

3. Dyal WM. Foreword—Aid for Just Development. Boulder, Colorado: Lynn Rienner Publishers, 1988.

4. Pan American Health Organization. *International Health: A North South Debate*. Washington, DC: Pan American Health Organization, 1992.

5. Hall E. *The Silent Language*. Garden City, NJ: Doubleday & Co., 1959.

XXII Addressing Patient Complaints: A Consultation Service for HMO Physicians

Margaret Blandin-Clark

As a family therapist and faculty member in the Department of Family Medicine at the University of Maryland School of Medicine, I teach behavioral science, communication skills, and family issues to medical students and residents. These skills help physicians-in-training develop partnerships with their patients that promote health and satisfaction with health care delivery. My expertise in this area has led me to work with physicians employed by a local Health Maintenance Organization (HMO).

I was contacted by the Chief of the Family Medicine Department (a former faculty member) at a local HMO where they collect written and verbal patient complaints about physicians on a monthly basis. He was concerned about a physician on his staff who was well respected for his biomedical expertise, but who had recurring complaints from patients about how he related to them. The chief asked if I would help this physician deal with the issues identified in these complaints. He arranged for me to consult with this physician over a period of time (six patient care sessions over a five-month period). The consultation process improved the physician's interactions with patients and reduced patient complaints.

Since 1993, I have worked with 13 board-certified physicians employed by this HMO from the following departments: family medicine, OB-GYN, internal medicine, ENT, neonatology, dermatology, gastroenterology, and pediatrics. The number of sessions can range from one encounter to fifteen encounters, but typically involves five to seven sessions over a four- to five-month period. The HMO continues to contact me episodically as an outside consultant to work with physicians who have repeated complaints from their patients.

Patients' complaints fell into two categories: a) administrative or procedural concerns, and b) physician-patient interactions. The patients' complaints regarding physician interactions included: "I am being treated like a child"; "The doctor was condescending, arrogant, patronizing"; "My thoughts and information are not taken seriously"; "The doctor has a rude, abrupt, authoritative, blaming manner"; "I was seen as a medical condition, not a person with physical symptoms"; "The doctor had an uncaring, cold, distant, business-like manner"; "The doctor was cursory, lacked follow-through, was defensive and not informative"; and "The doctor was

---

I would like to express my deepest appreciation to Richard J. Botelho, MD, for his encouragement and valuable editorial support.

difficult to reach by phone and would not facilitate a referral to specialist." In addition, some physicians had difficulty working well with the health assistants and staff. Health assistants complained that the physician was demanding, spoke abruptly, and had an authoritarian manner; also, they did not feel respected or valued in the provision of patient care. Most physicians were shocked and/or upset when informed about these complaints.

This chapter will describe:

- an educational consultation service for physicians about whom there have been repeated complaints;
- the challenges and issues in providing such a service; and
- the outcome of these consultations.

Overview of the Consultation Process

I apply the concept of parallel process (isomorphism)[1–3] in my work with physicians. I use skills in my educational consultations with physicians so that they can use the same skills with their patients. In other words, I model for them how to establish effective partnerships that they can develop with their patients. In this way, I help them constructively reduce their frustration and annoyance about patients' complaints, which are often preceded by frustrating and annoying encounters, by helping them learn how to prevent these problems from arising in the first place.

Some medical schools and residency training programs provide longitudinal training programs to enhance interpersonal skills with patients.[4–7] Although physicians can participate in experiential training opportunities at professional conferences, practicing physicians rarely have the opportunity to participate in a longitudinal training program.[8,9] Yet it may be just this group of physician who, having grown in their medical expertise, are most open to professional development of this kind.[10]

I will describe the five chronological steps that characterize the facilitation of an effective consultation process. In Step I (Enhancing the Partnership between Administration and Physician), the chief of the department helps to make physicians feel respected and valued by identifying their areas of competence while also addressing the need to resolve issues related to complaints from patients. The chief encourages the physician to consider professional development in the area of patient interactions, and to consider various methods.

In Step II (Initiating a Partnership between Physician and Consultant: Initial Contact), I ask physicians what they understand about the reasons for referral before providing a description of what I may offer and how I can work with them. I ask physicians to think about these options and to suggest any other ideas about how to work together. I also set the tone for collaboration that lets physicians actively decide how to work together, and when and how often to meet.

In Step III (Developing the Consultant-Physician Partnership: Creating a Contract), I assist physicians in clarifying their goals for the consultations and work out a plan to achieve these goals. I try to summarize my understanding of the physician's view of his/her situation and needs. I explore with physicians how we might collaborate toward achieving the goal(s). We also try to anticipate areas of conflict, tension, and differences of opinion in working together, and address whether to set "homework assignments" to prepare for the next session.

In Step IV (Observation and Feedback), I check whether physicians did any "homework" from the prior session. I observe and provide feedback to physicians working with patients and staff. I ask them to identify what worked well to achieve their goals and what did not work so well. We discuss ways of achieving his/her goals. At the end of the half-day work session, we discuss the patient encounters and design "homework assignments" to focus efforts until the next work session.

In Step V (Terminating the Consultation), we review progress made toward the goals, identify future issues or challenges, and discuss ways of maintaining and continuing to enhance their communication skills. I make a telephone follow-up one month after the session to ask for feedback about the consultation process.

Challenges and Issues: A Case Example

A case example will illustrate a typical educational consultation. Dr. A is a composite of the thirteen physicians with whom I have worked. Table 1 summarizes the communication skills that I use during the educational consultation with physicians. Physicians may then use the same skills with their patients.

Several frameworks of communication are available to use in consultation with physicians. Communication skills enhance physicians' ability to develop partnerships with their patients and to empower them to change.[11–16] For physicians who are the focus of repeated complaints, I find it helpful to use an explicit framework to help them improve their communication skills.

The chief spoke with the physician chairing the Patient Relations Committee about patients' complaints about Dr. A. The chief contacted and informed me about referring Dr. A for a consultation. I have instructed the chief to introduce the idea of referring physicians to me in the following way. Dr. A did not contact me by phone nor did he get back to his chief, necessitating a direct inquiry from the chief regarding his decision. Dr. A had expressed reluctance about the referral. I suggested to the chief that he:

- inquire about what Dr. A decided to do, what his plans were to prevent these patients' complaints (going with the resistance), and to acknowledge his affect.
- allow Dr. A to attempt his own plan for making a change for a prescribed amount of time and then together evaluate the progress at the end of this time.
- recommend that if he is unable to reduce the number of patient complaints, he reconsider a referral to a consultant.

Table 1. The Consultation Process

Step 1:	Enhancing the Partnership Between Administration and Physician	Goal:	Physician feels respected and valued, and encouraged to consider Professional Development in the area of physician-patient interactions.	Methods:	- Department chief identifies physician's areas of competence. - Department chief encourages physician to consider Professional Development focused on physician-patient interactions. - Department chief encourages physician's input on where to seek this professional development.
Step 2:	Initiating a Partnership Between Physician and Consultant: Initial Contact	Goal:	Begin to develop rapport (join) and create an atmosphere of collaboration *(inviting the physician's involvement).*	Methods:	- Physician contacts consultant; not the reverse. - Consultant describes how the educational consultation is structured. - With physician having input in time, place, or other details, first appointment is set.
Step 3:	Developing the Consultant-Physician Partnership: Creating a Contract	Goal:	Continue developing rapport and partnership with the physician. Establish goals: specific desired outcomes derived from consultation experience. "Plant seeds" about each contributing in different ways towards reaching the goals.	Methods:	- Brief "social stage." - Open-ended questions designed to enhance consultant's understanding of the physician's view of his situation and needs. - Identification of physician's goal(s). - Speculate what both consultant and physician might contribute in order to successfully reach goal(s). - Draw out for discussion areas of conflict, tension, differences of opinion. - Design "homework" in preparation for next work session.

Table 1. The Consultation Process, cont'd.

Step			Goal		Methods
Step 4:	Observation and Feedback	Goal:	Assist physician in identifying what works well for him, and new techniques for achieving his goals with patients and staff.	Methods:	- 15–20 minutes at beginning of work session reviewing "homework" and prior patient encounters. - Accompanying physician as he goes about his clinical work. - At end of half-day work session, discuss encounters and design "homework" to focus efforts until next work session.
Step 5:	Terminating the Educational Consultation Process	Goal:	Evaluation and support for the maintenance of acquired skills.	Methods:	- Review progress made. - Discuss challenges that might lie ahead and ways of managing them. - Consultant makes available a telephone follow-up in one month.

Table 2. Communication Techniques

Technique	When Use	Examples
Acknowledging affect—Commenting on one's feelings/affect.[27–29]	Throughout the consultant-physician and physician-patient relationship	• "I can imagine that might have been . . . scary, aggravating, embarrassing, deflating, wonderful, a huge relief. • "It sounds like you are very worried about your . . . " • "Who would have guessed that healthcare would change so drastically" (consultant-physician)
Inviting the patient's/physician's involvement[18,22–26,30–31] Open-ended questions designed to increase understanding of the patient's/physician's situation.	Throughout the physician-patient and consultant-physician relationship	• "What do you think your symptoms have to do with?" • "What have you tried thus far?" "What do you think will help you feel better?" "What can we do to help you feel better?" "I think you are suffering from . . . and my recommendation is . . . What do you think?" "What would you like to accomplish? How do you think I can be helpful? What do you think this will require of you?" (Consultant-physician)
Going with the resistance—Reduce the tension by focusing on the person's affect: finding a feeling you can identify with. Avoid a struggle, debate, battle.[12, 32, 33]	When one is hesitant about recommendations and resisting your efforts, leaving you more interested in his situation and working harder at improving it. When one is angry about something and needs to be heard.	• "As important as I think it is for you to quit smoking, I am beginning to get the message that maybe now is not the best time for you to focus on this." (I too have delayed important actions.) • "It sounds like it is OK with you to run the risk of possible future health problems." (I too have made risky decisions before.) • "Given the potential losses you have identified should you start exercising regularly (employing new communication techniques), are you sure you want to proceed?" • "I don't blame you for being upset about waiting so long to be seen (my having to make last minute changes in our scheduled time). I am sorry."

Table 2. Communication Techniques, cont'd.

Technique	When Use	Examples
Taking a pro-active posture[15]—Open and closed questions and statements designed to invite discussions of controversial issues that might interfere with movement toward a goal.	When individuals or circumstances might oppose efforts made by you or your patients. When tension is present and causing a potential impasse.	• "What might your spouse think about your stopping smoking, drinking, etc.? How will it change his/her life?" • "Who might not think it such a good idea if you . . . ?" • "What do you think will be the hardest part?" • "Is it really OK with you that people not agree with your decision to . . . ? Where will you turn for support now?" • "What will you lose if you increase your repertoire of communication techniques?" (Consultant-physician)
Commenting on the process—Physician tells patient or consultant tells provider how he's feeling about their work together.*	When the physician-patient or consultant-physician relationships are working well. When one is concerned/frustrated with the recipient's opposition or lack of concern.	• "It is a pleasure working with you. We do not always agree, and we let each other know." • "I am very worried about you, Mr. Smith. I feel frustrated that I have not found the perfect thing to say to you that will convince you to . . . " • "I have had a few suggestions of techniques you might want to consider using, but none of them have appealed to you. I am beginning to wonder if I will be able to be helpful to you." (Consultant-physician)

*I have not located literature with the focus on the provider (consultant) constructively and strategically sharing their reactions/affect with a patient, as opposed to focusing on the patients reactions.

Table 3. Referring a Physician for Educational Consultations

1. Meet with the physician to discuss his/her strengths, expertise, and issues related to the patient (and sometimes staff) dissatisfaction.
2. Express interest in assisting the physician in how to address these issues.
3. Ask the physician how he/she would like to address these issues.
4. Inform the physician about a consultation service with a behavioral medicine specialist to assist in improving the situation.
5. Explain that the consultant works individually with the physician during patient care sessions.
6. Encourage the physician to take advantage of arranging several sessions over many months.
7. Explain that the HMO will pay for half of the fee, and the other half can come out of the physician's continuing education allowance.

Dr. A attempted to improve his interactions with patients on his own. Several months later, he contacted me at the suggestion of his chief. Dr. A was angry, defensive, and embarrassed. I acknowledged his discomfort before briefly introducing myself. I presented a plan for how we could work together, but explained to him that I was open to work with him in whatever way he would feel comfortable. I suggested the following for our first meeting: 1) talk for an hour about how to work together, 2) work alongside him as he goes about his daily work with patients, and 3) debrief at the end of the patient care session for about 20–30 minutes. (This method of presenting a consultation plan to physicians is similar to what physicians might use when they try to develop a partnership with patients by asking them what they think about the treatment recommendations.)

Dr. A wanted to meet first at my office before deciding whether to go further. I sensed his need to become better acquainted with me before working together. I accepted his preference, rather than insisting that we first meet at his office, to avoid evoking his resistance or enhancing his reluctance to see me. (Similarly, when patients are not ready to follow a treatment recommendation, physicians can initially accommodate patients' preferences rather than imposing their own agenda prematurely, to avoid having patients complain about them, or leave their practices.) Change is a process, not an event; and unless things are imminent, acknowledging this process is helpful, if not essential.

At our first meeting, a brief "social stage" (three minutes or less) is important for putting both of us at ease and for beginning a process of establishing rapport with Dr. A and promoting his active involvement in the consultation process. I introduced myself, smiled, engaged him with direct eye contact, and shook his hand. I followed this by offering him a cup of coffee and asking whether he had difficulty getting to my office.

To develop a rapport with Dr. A, I tried to understand his goals and what concerns and issues he might have in working together. To effectively *join*[17–21] with him and convince Dr. A in a nonjudgmental, genuine, and empathetic way that I was interested in his unique situation, I used a variety of open-ended questions. This fostered an understanding of his concerns and feelings about participating in these consultations. For example, "How did you come to decide to participate in this endeavor?" "What do you hope to gain?" "What are your goals?" "How do you see the situation that prompted us to work together?" "Who have you told?" "What did they think about your participating?" "Who might your colleagues and staff think I am?" These questions gave me opportunities to acknowledge Dr. A's affect and invite his involvement: two effective methods of joining.

As I began to understand his learning goals, concerns, and feelings, I also tried to clarify and negotiate how we might work together, with the purpose of establishing a partnership that would assist him in achieving his goals. (Physicians can use the same process and skills when working with patients. A brief social exchange followed by open-ended questions about patients' symptoms and their feelings, concerns, and expectations about the situation may help physicians establish partnerships with their patients. This patient-centered approach is different from a physician-centered approach in which the physician takes control over the interaction and asks predominantly closed-ended questions to confirm a diagnosis.)[22–27]

Dr. A wanted to reduce the number of patient complaints that were of concern to his chief. He attributed the repeated complaints to the unreasonable mess of patients, office and policy procedures, or other system issues over which he had no influence. I responded by saying: "Maybe we can find some techniques that you can use to help patients be more reasonable and make the system work better for your patients" (going with the resistance). I tried to understand his unique perspective about the situation, without necessarily agreeing with it. I did not push him to verbalize how he may have contributed to these patient complaints. (As with patients, it is often helpful to hear and understand their perspective before trying to present another view or diagnosis.) Dr. A seemed to expect that I, the "expert" consultant, would set the goals and determine the ways to achieve them with little input from him. He seemed taken aback by this approach.

To develop a partnership with Dr. A and clarify how we could best work together, I asked additional open-ended questions to enhance my understanding of his perspective and goals. For example, "What specifically have you tried thus far to improve the situation?" "What do you think or hope I could do for you?" "What do you think will be required of you?" "Do you think the gains will be worth the efforts required?" These questions were designed to elicit affect and information in a way that would reduce defensiveness and enhance Dr. A's involvement. This approach helped to actively involve Dr. A in achieving his goals. (Physicians can ask similar question with their patients: "Do you have ideas about how I could be helpful?" "What have you tried already in an effort to help your symptom go away?" " Is there something you think would be helpful for you to do, but you have not yet been able to do?" "What do you think others, including yourself and myself, think you should do to improve the situation?")

This approach helped me identify early on any potential areas of tension or conflict in working with Dr. A. This proactive approach in identifying and addressing any conflicts and tensions also helped to preempt impasses between Dr. A and myself in later sessions. It was important for both of us to understand how our differences might interfere with his achieving his stated goals.

At the end of our first meeting, I continued to pursue my objective of understanding Dr. A's perspective. I asked him to identify those communication skills that he was already using effectively and to consider doing the following homework assignment for our next session: think about and jot down 1) any additional learning objectives or skills that he would like to learn, 2) his goals for his relationships with patients, staff, and colleagues, and 3) what he will gain and lose if he is successful in reaching his goals. The implied messages underlying these questions and homework assignment were that:

- his current style was "helpful" in some ways;
- successful change in expanding his repertoire of skills was mainly within his control; and
- we would both play active roles in achieving the desired changes.

For example, he may feel more comfortable and competent medically when he is in charge and giving directives, particularly during times of uncertainty; or he may feel less defensive or anxious if the hierarchy is clear and unchallenged. However, some patients may want to deal with uncertainty for themselves, relate to their physician in a nonhierarchical way, and make decisions entirely for themselves. These questions helped Dr. A deal more constructively with his anger about being involved in this consultation process. Otherwise, he could consciously or unconsciously sabotage the work by not truly participating in this process.

In this instance, Dr. A's wife was very angry that his chief thought it necessary for Dr. A to participate in this endeavor. Dr. A appreciated his wife's allegiance and support. As part of the homework assignment, I asked Dr. A to check with his wife how she might respond if he were able to reduce the number of patient complaints and how she might then view the situation. Dr. A subsequently discussed this issue with his wife who thought that it was worthwhile to explore how he could reduce the number of patient complaints. This approach helped Dr. A and his wife align themselves more toward the goals of change, rather than opposing what the Chief did. (Similarly, physicians who do not consider how a spouse disagrees with the patient's treatment plan may experience difficulties in working with the patient to implement the plan over time.)

This one-hour session helped set the stage for me to observe five subsequent half-day sessions with Dr. A interacting with patients and staff at his medical center and returning phone calls to patients. Each session started with 15–20 minutes of conversation that followed up on "homework assignments" and/or observations that Dr. A had made regarding patient interactions since the last session. Dr. A introduced me to patients saying, "Mrs. B, this is Margaret Blandin-Clark. She is working with me today; is that OK with you?" Patients had the right to refuse, but most patients consented to this arrangement without further explanation.

At the end of these sessions, Dr. A and I spent 20–30 minutes discussing the interactions. I began the debriefing session by asking questions. For example: "Were those typical encounters? What did he think about the encounters? Were there aspects of the encounters he wished he had handled differently? Had there been examples of things he wanted to look at in our work together? Was there anything specific he would like me to comment on? Which were the easiest encounters; which were the more difficult medically and/or interpersonally?" Following this inquiry, I positively reinforced interactions that were particularly helpful or skillful. For example, "Your warm smile and handshake at the beginning of each encounter seemed very inviting to your patients." Or "Your ability to sit still with consistent eye contact and listen to your patients' concerns seemed to convey total interest and that they had your undivided attention." Or, "Your style of saving your questions and asking them after Mr. B had finished his description of his problem seemed to convey that you needed his information and were respectful of his style of providing it."

When introducing new techniques, I asked Dr. A what he thought about one or two of his questions that he asked of his patient that focussed a nodal point of the interaction. After listening to his response, I asked for his ideas about how he could handle the situation differently before I introduced a technique that he could have used with the patient. I asked him which of these ideas he would most like to practice in using with patients. I commented on how such questions might help him enhance his partnership with patients. When Dr. A thought that a technique was inappropriate, I asked him for his opinion about what he thought would be more effective. This approach gave us the opportunity to explore our differences of opinion or conflicts. At one session, I made a number of suggestions that Dr. A did not think would work. I took a proactive posture by saying, "It seems that these suggestions are not very helpful to you. Do you think you could use these suggestions in other situations? What do you think is not helpful about these suggestions for this particular patient and situation? Do you have other suggestions for dealing with this situation?" In effect, I tried to understand his resistance to these suggestions rather than convince him of what I thought were the merits of the ideas.

I promoted the idea of Dr. A developing partnerships with patients by asking, "What do you think your patients see as your contribution to their staying in good health? And what about the patient's contribution?" Dr. A responded that he did not think his patients took an active role in their health. I asked him to reflect on the importance of his patients contributing to their own health by asking, "Is it important to you that they contribute to their own health? How would that be helpful or not? And if it were helpful, what might be some ways to convey the importance of this to your patients?"

Dr. A began to use this approach in patient encounters. He asked his patients their views about his treatment plan, and thus involved them in the decision making process. If patients agreed with his advice, Dr. A would say, "Together we'll see how this works for you." If patients did not agree with his recommendation, he would ask them what about this recommendation would not work for them. This approach helped Dr. A understand how to deal with patient resistance to his recommendations. Just as I recruited Dr. A's active involvement to enable us to share power and responsibility in working together towards his goals, I encouraged Dr. A to take a similarly proactive posture in asking his patients what they thought was not appropriate or helpful about his recommendations. Taking a proactive posture enhances understanding of the patients' perspective; going with the resistance by incorporating the patients' ideas furthers the patients' involvement, and builds partnership. When Dr. A was hesitant to abandon his recommendation, he could then express his frustration about not providing the best possible care for them and his inability to convince them about the benefits of the recommendations (commenting on the process). Furthermore, Dr. A could ask his patients if there is anything that might convince them to reconsider their decision (inviting their involvement).

In some encounters, physicians face time pressures and a sense of urgency that limits their time to form partnerships with patients. When faced with such situations, Dr. A could *comment on the process* by saying, "Mrs. B, I think I understand why you don't want to take your insulin, and I know it is a decision only you can make for yourself *(going with resistance)*. I am worried, however, about your decision not to take insulin, and frustrated that I have not found anything that might convince you to take it. I imagine you may sense my frustration and hope that I have successfully conveyed to you my concern about what may happen to you." When Dr. A *commented on the process*, he abandoned talking about medical content and why insulin is needed. Instead he calmly commented on his feelings, commented on the process of their interaction, and talked about the impact on himself rather than focus on Mrs. Jones's noncompliant behavior.

I worked with Dr. A for five such work sessions. During the 20–30 minutes of debriefing session, I pointed out those techniques that worked well for Dr. A, and used nonconfrontational ways of addressing interactions that did not work well for him. I formulated questions in lieu of making statements, went with the resistance, took proactive approaches when tension or conflict seemed evident, and commented on the process when my emotions were stimulated.

At our last session, I asked Dr. A to think about what he had or had not accomplished, what had and had not been helpful (proactive posture), what additional learning goals he may have had, and how he might continue to improve his skills. Dr. A said he was able to: 1) regularly acknowledge patients' affect, which he did not do before, 2) understand patients' roles more clearly and see how his patients were helpful to him, 3) recognize his frustration and disappointment with patients, and 4) feel more comfortable commenting on the process. He was pleased and satisfied with this progress. Dr. A decided to write on an index card how to use three communication techniques; he would later have the card laminated and keep it in his coat pocket. He then agreed to call me in four weeks to let me know his progress in maintaining these skills. When I did not hear from Dr. A, I called him. He was continuing to use these particular techniques and felt more positive about his encounters with patients.

Outcome of Consultations

Several factors shaped physicians' experiences of these consultations: professional and personal developmental stage, ability to identify the gains and losses in learning how to expand their repertoire of communication skills, genuine upset that their patients and /or staff perceive them as "uncaring," and the reactions of their spouses. Of the eleven physicians who have completed participation for at least six months, four had significant decreases in patient complaints, two showed no change in the frequency of patient complaints, one had an increase in the frequency of patient complaints, and four lacked data from which conclusions could be drawn. This information was obtained by comparing the number of patient complaints per six-month intervals prior to the consultation process with the number of patient complaints per six-month intervals following participation in the educational consultation. Depending on when the physician participated, the follow-up data ranges from six months post intervention to four years. For the four physicians whose patient complaints decreased, there were reductions by 50–90% in frequency of complaints.

Three physicians who participated in this learning opportunity were stimulated to think about their personal and professional development, the changing work place, and the balance between their personal and professional lives. These physicians had been practicing for over twenty years and had seen great changes in the provision of healthcare. Paying attention to these issues combined with my acknowledging their feelings about the changes in healthcare improved my ability to provide educational consultations to these physicians.

I encouraged physicians to explore and speculate about the potential gains and losses of expanding their repertoire of communication skills. When physicians actively participated in this exercise, they seemed more likely to achieve their goals. For example, two physicians were never able to identify any gains other than appeasing their superiors, or losses they might experience if they broaden their repertoire of communication skills. Despite my efforts, I never successfully established a partnership with them, and no real change took place. However, three physicians who were able to identify specific gains and losses actively participated in the consultation process; their interactions with patients and staff improved tremendously.

Several physicians were upset to learn that patients and/or staff perceived them as "not caring" as they efficiently went about the business of practicing medicine. They were eager to understand how others got this impression. They were surprised about the power of using new approaches: for example, interrupting patients less and working at a slower pace. Balancing efficiency with new approaches became a new challenge for them. In particular, two young physicians who were upset and embarrassed by the patient complaints were eager to enhance their professional skills, to try new approaches, and to have less tense and more comfortable working relationships with their patients. They were able to identify losses and gains in trying new ways, and became more focused on how they practiced medicine. They commented on never having thought about how subtle interventions can have a powerful impact on others. Initially, their attention to details, words used, timing, and body language increased their anxiety, but it focused their energy on enhancing the art of practicing medicine. As they increased their confidence in their interpersonal skills, they lessened their anxiety about working on their professional goals.

With three physicians, the reactions of their spouses were very influential. Two spouses were very angry about the chief's referral suggestion to an educational consultant. These physicians appreciated the allegiance of their spouse against the chief's suggestion, in effect, unifying against a "common enemy." I was unable to create an effective partnership with one physician, and this experience stimulated my thinking about how to deal with this issue. For example, I could have suggested to him that the loss of his spouse's support might outweigh his gains. This may have brought this issue into better focus in a way that would have assisted him in more clearly understanding losses. Having learned from this experience, I was more effective in assisting the other physician and his spouse to address this issue.

I provided only one consultation to two of these physicians. One physician was not interested in the learning opportunity, yet felt it was important to follow his chief's suggestion. I was unable to increase his interest. The other physician had a totally different experience. He was humbled by his superior's suggestion that he contact me; he had not even told his wife of our meeting. Nevertheless, during our one encounter which included Steps III and IV, he was intrigued by the notion that he might gain and lose something by using certain communication techniques. He tried new techniques during our only work session. He decided to think about the broader concepts on his own and contact me at a later date if he wanted to participate further.

Anecdotally, one patient commented that he felt discounted and patronized when his physician interrupted him and ignored his opinions regarding his symptoms. Later when he needed to return to see his doctor, he contemplated seeing another physician. For practical reasons, he returned to his original physician, who in the meantime had participated in this educational consultation. The patient commented to another physician in the practice that he noticed a change in his doctor—he listened attentively to him and made him feel important during their discussion together.

I contacted two physicians to inquire about their experience retrospectively and the current impact of the consultations on their work. Both physicians expressed an increase in professional satisfaction and skill, and a reduction in patient complaints. They valued the educational consultation and suggested that, in addition to the consultation process, normal professional maturation and personal and spiritual development played a part in the changes they made.

One physician said that he regularly employs acknowledging affect, inviting patients' involvement, going with the resistance, and commenting on the process. In addition, he spoke of originally seeing his relationship with patients and staff as follows: he provided the knowledge that they needed and expected them to use it appropriately to improve the situation. He now sees the relationship as an opportunity to focus on learning what patients or staff members identify as their needs and goals, instead of trying to figure that out for them, and then working with them to achieve their goals. He describes the relationship as more equal and satisfying. The losses for this physician are the additional time required and guilt. "I learned enough to now feel guilty when time pressures persuade me to return to old ways and 'do to' patients."

Another physician was hesitant to work with me, but his spouse's support and encouragement helped him to accept the referral for a consultation. Initially, he felt threatened and uncomfortable, but he ended up enjoying our work together. He described how his prior "autocratic, patronizing, dictating to patients" style was originally easy because he was not comfortable with patients. Now that he listens better to patients, he notices barriers going down, as has his self-imposed pressure to talk, explain, and fix. "Acknowledging patients' ideas and reactions creates better connections, and patients feel that their needs are being met." He also pays similar attention to the staff with whom he works. This physician reports an increase in professional satisfaction, and that the greatest loss was his initial embarrassment and discomfort!

In summary, the physicians' gains have been: developing an appreciation for the complexities in the art of practicing medicine; an increased sense of control over some of the frustrating aspects of practicing medicine; and valuing their patients' positive responses to changes they have made in their interactions with patients, thus increasing patient and physician satisfaction. The biggest challenge for me was helping some physicians speculate and identify the gains and losses that might arise from actively participating in the consultations. However, this process helped me develop more effective partnerships with physicians. I believe that advanced skills are most effectively taught with the experiential, longitudinal component that allows me to apply the concept of parallel process more effectively by modeling skills in the educational consultation that physicians can use in their encounters with patients. I cannot tell which approach works better: the modeling of these preventive and therapeutic communication techniques in my consultations with physicians (unconscious learning), or the teaching of how and when to employ them directly with patients (conscious learning). However, I am convinced that the combination works better than one approach alone, and is applicable not only to HMOs, but also to small medical and faculty practices.

References

1. Lipkin, Jr. M, Williamson PR. Teaching Interviewing Using Direct Observations and Discussion of Actual Interviews. In: Lipkin Jr. M, Putnam SM, Lazare A, ed. *The Medical Interview: Clinical Care, Education, and Research*. New York: Springer-Verlag, 1995:416.

2. Getz HG, Protinsky HO. Training Marriage and Family Counselors: A Family-of-Origin Approach. Counselor Education and Supervision 1994;33(3):183–90.

3. Botelho RJ, Seaburn DB, Harp JJ. Isomorphism in Resident Consultations: Implications for Education and Patient Care. Family Systems Medicine 1991;9(2):137–49.

4. Branch WT, Arky RA, Woo B, Stoeckle JD. Levy DB, Taylor WC. Teaching Medicine

as a Human Experience: A Patient-Doctor Relationship Course for Faculty and First Year Medical Students. Annals of Internal Medicine 1991; 114(6):482–9.

5. Wilkes MS, Slavin SJ, Usatine R. Doctoring: A Longitudinal Generalist Curriculum. Academic Medicine 1994; 69(3):191–3.

6. Bird J, Hall A, Maguire P, Heavy A. Workshops For Consultants on the Teaching of Clinical Communication Skills. Medical Education 1993; 27(2):181–5.

7. Lipkin Jr. M, Kaplan C, Clark W, Novack DH. Teaching Medical Interviewing: The Lipkin Model. In: Lipkin Jr. M, Putnam SM, Lazare A, ed. *The Medical Interview: Clinical Care, Education, and Research*. New York: Springer-Verlag, 1995:432.

8. Levinson W, Roter D. The Effects of Two Continuing Medical Education Programs On Communication Skills of Practicing Primary Care Physicians. Journal of Internal Medicine 1993;8(6):318–24.

9. Gordon GH, Rost K. Evaluating Faculty Development Course on Medical Interviewing. In: Lipkin Jr, M, Putnam SM, Lazare A, ed. *The Medical Interview: Clinical Care, Education, and Research*. New York: Springer-Verlag, 1995:436–47.

10. Carroll JG, Lipkin Jr. M, Nachtigall L, Weston WW. A Developmental Awareness of Teaching Doctor/Patient Communication Skills. In: Lipkin Jr. M, Putnam SM, Lazare A, ed. *The Medical Interview: Clinical Care, Education, and Research*. New York: Springer-Verlag, 1995:396.

11. Quill. Recognizing Barriers to Patient Communication. Annals of Internal Medicine 1989; 111:51.

12. Benarde MA, Mayerson EW. Patient-Physician Negotiation. JAMA 1978; 239(14):1413–5.

13. Carter WR, Inui TS, Kukull WA, Haigh VH. Outcome-based Doctor-Patient Interaction Analysis:II. Identifying the Effective Provider and Patient Behavior. Medical Care 1982;20(6):550–66.

14. Heszen-Klemens I, Lapinska E. Doctor-Patient Interaction, Patients' Health Behavior and Effects of Treatment. Social Science & Medicine 1984;19(1):9–18.

15. Martin AR. Exploring Patients' Beliefs. Steps to Enhancing Physician-Patient Interactions. Archives of Internal Medicine 1983;143(9):1773–5.

16. Bartlett EE, Grayson M, Barker R, Levine DM, Golden A, Libber S. The Effects of Physician Communication Skills on Patient Satisfaction: Recall and Adherence. Journal of Chronic Diseases 1984;37(9–10):755–64.

17. Engel L, The Clinical Application of the Biopsychosocial Model. American Journal of Psychiatry 1980; 137(5):535–44.

18. Suchman AL, Matthews DA. What Makes The Patient-Doctor Relationship Therapeutic? Exploring the Connexional Dimension of Medical Care. Annals of Internal Medicine 1988;108(1):125–30.

19. Lazare A, Putnam SM, Lipkin Jr. M. Three Functions of the Medical Interview. In: Lipkin Jr. M, Putnam SM, Lazare A, ed. *The Medical Interview: Clinical Care, Education, and Research*. New York: Springer-Verlag, 1995:11–14.

20. Lipkin Jr. M, Frankel RM, Beckman HB, Charon R, Fein O. Performing The Interview. In: Lipkin Jr. M, Putnam SM, Lazare A, ed. *The Medical Interview: Clinical Care Education, and Research*. New York: Springer-Verlag, 1995:76.

21. Minuchin S, Fishman HC. *Family Therapy Techniques*. Cambridge, Massachusetts: Harvard University Press, 1981:28–49.

22. Neighbour R. Paternalism or Autonomy? The Practitioner 1992;236:860–4.

23. Sharf BF. Teaching Patients to Speak Up: Post and Future Trends. Patient Education Counseling 1988; 11:95–108.

24. Levenstein JH, McCracken EC, McWhinney IR, Stewart MA, Brown JR. The Patient-Centered Clinical Method. 1. A Model for the Doctor-Patient Interaction in Family Medicine. Family Practice 1986; 3(1):24–30.

25. Smith RC, Hoppe RB. The Patient's Story: Integrating The Patient-and Physician-Centered Approaches to Interviewing. Annals of Internal Medicine 1991;115(6):470–7.

26. Brody DS. The Patients Role in Clinical Decision-Making. Annals of Internal Medicine 1980; 93(5):718–22.

27. Conrad P. The Non-Compliant Patient In Search of Autonomy. Hastings Center Report 1987;17(4):15–7.

28. Lazare A. Shame, Humiliation, and Stigma in the Medical Interview. In: Lipkin Jr. M, Putnam SM, Lazare A, ed. *The Medical Interview: Clinical Care, Education, and Research.* New York: Springer-Verlag, 1995:342.

29. Cohen-Cole SA. Function 2: Building Rapport and Responding to Patient's Emotions (Relationship Skills). In: Cohen-Cole SA. *The Medical Interview: The Three Function Approach.* St. Lewis, Mo: Mosby Year Book, 1991a:21–7.

30. Lazare A, Putnam SM, Lipkin Jr. M. Three Functions of The Medical Interview. In: Lipkin Jr. M, Putnam SM, Lazare A, ed. *The Medical Interview: Clinical Care, Education, and Research.* New York: Springer-Verlag, 1995:13.

31. Delbanco TL. Enriching The Doctor Patient Relationship By Inviting the Patient's Perspective. Annals of Internal Medicine 1992;116:414–8.

32. Lazare A. The Interview As a Clinical Negotiation. In Lipkin Jr. M, Putnam SM, Lazare A, ed. *The Medical Interview: Clinical Care, Education, and Research.* New York: Springer-Verlag, 1995:50–62.

33. Lazare A. Shame, Humiliation, and Stigma In The Medical Interview. In Lipkin Jr. M, Putnam SM, Lazare A, ed. *The Medical Interview: Clinical Care, Education, and Research.* New York: Springer-Verlag, 1995:343.

XXIII Family Medicine/Family Systems Case Consultation: An Educational Partnership

Geoffrey R. Swain and Eric L. Weiner

In this chapter, we will describe the interdisciplinary approach that we use for conducting the Family Medicine/Family Systems Case Consultation method (FFCC) in our Family Practice Residency setting. We will include:

1. The history and rationale for our collaboration;
2. A description of our case consultation method;
3. An abridged transcript of consultation; and
4. An evaluation of this method.

History

We were hired into a family practice residency setting as faculty members, coincidentally about one month apart, in the fall of 1990. ELW was originally drawn into the field of family medicine because of similarities in values between family systems therapists and family physicians. Family Medicine is mandated to incorporate the psychosocial aspects of care into its training curriculum. However, he became disillusioned about the actual realities of this mandate. GRS was likewise drawn into family medicine because of a concordance of personal and professional values. He considered himself to be a "big picture" person, concerned more with "high touch" than "high tech," and more with establishing working connections than with isolating and managing "pathology" out of context.

We were both surprised to note that the realities of the clinic and the staffing room seemed uncomfortably discordant with our values. We were struck by the residents' tendency to focus exclusively on the biomedical aspects of healthcare. They seemed to view themselves as separate from the dynamics of the patient and patient's family. Residents seemed cavalier about prescribing a course of action for

We are grateful to all of our residents for their patience in helping us to meet their needs, and particularly to Dr. Michelle Eich, who allowed us to use a transcript of one of her cases for this chapter. This project was funded at least in part by the Milwaukee Area Health Education Center, Inc., and we are grateful to them as well.

patients to follow, became easily frustrated when patients didn't "comply," and were puzzled about possible barriers to patient compliance.

As frustration levels mounted, residents seemed to respond by becoming more authoritarian with patients, staff, and faculty alike. This was apparent even in the language that the residents used, with an obvious tendency toward dichotomous descriptors as frustration grew. Patients were labeled as either "interesting" or "not very interesting," "good" or "bad," "compliant" or "noncompliant." Further increases in anxiety levels led to calling patients names such as "gomer" or "dirtball," and to telling derogatory stories about specific patients, and sometimes even about faculty and staff.

Resident physicians seemed to be most interested in quick fixes and technological approaches. They had varying levels of interest in the psychosocial aspects of care and generally lacked the patience needed for a long-term, process-oriented approach to patient care.

Both of us initially responded to this situation in a way similar to the way our residents might have responded to "noncompliant" patients; we talk "at" the residents in order to try to change their behavior. GRS would urge and cajole frustrated residents to be patient with their frustrating clients, only to find residents responding with statements like, "Just forget it, I'm just not going to see them any more!" Meanwhile, ELW was trying diligently to get the residents to obtain genogram information from their patients; but the more he pushed for that, the more obstinate residents became, simply ignoring him. Fortunately, with some self-reflection, ELW came to an important conclusion: he was not applying his cherished family systems principle of remaining well-differentiated when dealing with the residents. In his zeal to teach family systems medicine to residents, ELW had begun to push, pursue, and overfunction in ways he never would have with his patients.

Rationale for a New Method

ELW decided to discuss his newly discovered insight with his colleague (GRS). Together we spent many hours sharing frustrations and insights and trying to develop solutions. Eventually, we came to believe that we could not hope to influence residents to value a biopsychosocial approach to patient care until we ourselves understood and respected the residents' issues.

Consequently, we began to try to look at the situation from the residents' perspective. They had been placed in a situation where they would have high levels of responsibility but little control over their lives, schedules, and the demands placed on them. They adapted to a system that reinforced the need for quick assessment and treatment but allowed little time for reflection. They struggled to work in a system that is overwhelming, chaotic, and seemed out of control—an experience that many of their patients also encounter in dealing with health issues.

This insight helped us to take a more systemic view of the situation. However, we struggled to develop an approach that would deal with both the residents and their patients. The work of Wolkenstein's group offered one model they called The Family Medicine Chart Review. We liked their idea of a collaborative approach from both the biomedical and the psychosocial perspectives. We were concerned that a chart review format could potentially keep the residents in a defensive, "one-down" position, when they already seemed to be struggling with issues around autonomy. Similarly, we did not want to continue our early approach of pursuing the residents; instead, we wanted to try to get them to move toward us.

The question was, how could we help the residents change their perspective? We started with some observations and hypotheses. We observed that the most popular learning venue among the residents was the weekly "case conference," a biomedically oriented conference in which a puzzling case was presented and input was obtained from a variety of medical subspecialists as well as generalists. We hypothesized that this format was popular for several reasons. First, it was case-driven. The vast majority of residents' time is spent in patient care, and they have a high level of responsibility for managing their patients. "Cases" are what residents' professional lives revolve around, so the case-conference was a relevant educational form as well as a highly recognized one. Second, the residents received practical suggestions on how to manage the case more effectively, with specific ideas that could be implemented right away. Third, it was integrated into the residents' schedule. It was not considered to be an "extra," something to be squeezed in, in addition to their myriad other responsibilities. Fourth, it provided clear communication and feedback. Residents are indeed very interested in providing high-quality patient care, and they value constructive input from more experienced colleagues, if delivered clearly and collegially.

The Family Medicine/Family Systems Case Consultation (FFCC)

How, logistically, could we organize a program that would be case-driven, practical, integrated, and provide collegial feedback? ELW was in charge of a two-month behavioral medicine rotation for second-year residents. This rotation allowed one second-year resident to learn various aspects of the psychosocial side of medicine during sessions (with ELW) held 2.5 days per week over a two-month period. Since the behavioral medicine rotation was already integrated into the residents' schedule, we decided to use some of that time for our FFCC. In this way, residents might view the FFCC as a legitimate, formal educational offering, something that time had been set aside for, not as something incidental or "extra."

We set aside 1–1.5 hours, twice per month, for the resident to meet with both of us. We elected to keep the formal structure of the FFCC to a minimum in order to give the resident plenty of latitude to let us know what needs to address. All that we required of residents was that they bring us cases that were troubling, frustrating, confusing, or otherwise bothersome to them.

We have found that this approach allows residents to seek us for consultation because they take control of choosing the case and presenting what concerns them.

We stay focused on residents' concerns and give clear, practical feedback to them. At the consultation, we then review the written goals of the FFCC:

- to provide a supportive environment to reflect on a difficult patient situation without the time pressures of the usual clinical encounter;
- to demonstrate how to apply the biopsychosocial model to patient care, and how to use family systems medicine concepts in the physician-patient relationship; and
- to reduce physician frustration and enhance residents' ability to deal with difficult situations.

Case Presentation and Problem Identification

The resident brings the chart of the patient, meets with us in a quiet room away from the exam room area, and presents the case. During the 15–20 minute presentation of the patient's genogram, medical and psychosocial history, we listen and ask questions to clarify the purpose of the consultation: questions such as, "What bothers you the most about this case?" or "How can we help you with this situation?" In most instances, residents are surprisingly clear about what they would like to get out of the consultation. Responses range from the biomedical (e.g. "Is it OK to prescribe more narcotic pain meds for this patient?"), to patient-focused psychosocial (e.g., "How can I get this patient to be less anxious?"), to self-focused questions (e.g., "How can I be less frustrated by this patient?").

We now have clearly in mind a list of between two and five items with which the resident wishes to have help. Generally, we have also identified one or two other issues that we feel merit addressing as well. Sometimes these latter issues are biomedical (e.g., appropriate prescribing of antidepressant medications); but often they are psychosocial, such as parallel process, overfunctioning vs. underfunctioning, or overinvolvement vs. cutoff. We will generally contract with the resident at this point to address his/her concerns and discuss some of ours as well. Residents have been uniformly agreeable to this.

We spend the remainder of the time discussing these identified issues, helping the resident develop new approaches, and sometimes making specific suggestions. GRS usually simultaneously types a summary into a laptop computer; this time-efficient method allows us to provide residents with a written summary at the end of each consultation. Thus they leave with a consultation note to assist them with the ongoing care of the patient.

FFCC Techniques

Throughout each FFCC, we try to model collaborative process with each other. Sometimes we switch roles. ELW may ask biomedical questions, for example, or

GRS psychosocial ones. We may have a brief discussion between the two of us to try to clarify a point, and we are comfortable demonstrating to the resident that we learn something from each other and from the resident in every FFCC.

We also try to model the value of a systems perspective. When appropriate, we share our own genograms or ask the resident to share his/hers. Occasionally, the connection between the physician-patient relationship and the physician's own family-of-origin is striking. In one such case, the resident was frustrated by a drug-abusing patient, and in fact had a drug-abusing sibling of her own. Another consultation, in particular, stands out in regard to our own genograms; the resident was a younger sister of a brother, and we are both older brothers of sisters. When we found ourselves tending towards rescuing and overfunctioning, a quick glance at our genograms made a reason for this tendency clear. We all had a good laugh about it, but it was also powerful modeling to the resident that we take systems concepts very seriously.

Because the residents are involved in very serious, even life-and-death decision-making, humor in the FFCC can facilitate an atmosphere more conducive to learning by relieving tension. We frequently tease each other in front of the resident. Occasionally we offer paradoxical feedback to the resident, suggesting for example that residents should sacrifice anything and everything for the good of their patients. The message to the residents behind the humor is that we don't take *ourselves* too seriously; this allows the residents to be more relaxed, and perhaps to be more open in discussing some issues about which they may feel anxious or vulnerable.

Common Themes

Retrospective review of our written summaries shows that some themes predictably recur. A typical scenario is that residents get into a trap of overfunctioning for their underfunctioning patients, and then become so frustrated by the patients' behaviors that the only solution they can see is to cut off from these patients by transferring their care to another clinic physician or simply dismissing them from the clinic system. We try to guide the resident to a more balanced position that is neither over- nor under-functioning, and not cut off either.

For example, an "under-functioning" diabetic who is noncompliant with the prescribed insulin regimen might engender high levels of frustration in an overfunctioning resident, who believes that it's the physician's responsibility to make the patient better. A more balanced position, with more appropriate boundaries delineating the physician's and patient's responsibilities, not only allows the physician to stay connected to this patient with less frustration, but may even allow the patient to step out of the under-functioning role.

Residents identify several common characteristics of patients which are difficult for them to manage. The over-arching characteristic seems to be that of chronicity. Specific characteristics include noncompliance, anger, passivity, and psychological disturbances such as anxiety, depression, and poor self-esteem. Residents also tend to identify several common areas with which they would like help. These include assessment and management of depression, management of chronic pain, stress management techniques for patients, and how best to deal with demanding or noncompliant patients.

As consultants, we have noted several other themes common to most FFCCs. We generally try to help the resident with his/her own tendencies related to over- or under-functioning (almost always over-functioning), limit-setting and appropriate availability, seeing self as part of a system, pressure to fix and cure others, and the need to treat patients in context. We strive to remain aware of our own tendencies in these areas as well. Residents predictably feel a sense of relief when they examine these themes as they relate to being a "good" doctor.

Abridged Example of an Actual Case Consultation

> This patient is a 49-year-old black female with asthma, obesity, osteoarthritis, hypertension, tic douloureux, and fatigue. Our resident had seen her for the past year and a half on about five occasions; she'd been a patient at our clinic for about ten years. She's weighed approximately 270 lbs ± 10 lbs. for the past twelve years, and has seen a dietician on numerous occasions. She will hardly ever leave her house because she's so embarrassed by her weight.
>
> She has frequent asthma exacerbations. She takes Azmacort and Ventolin inhalers, and continues to smoke one-half pack per day. Her hypertension is well controlled, and her osteoarthritis (DJD) doesn't seem to be a big factor. Her tic douloureux is well controlled on baclofen and tegretol. She complains of fatigue and decreased energy recently, but hasn't been able to make it to the lab to get a TSH level checked. She occasionally uses oxygen at night because a sleep study showed some obesity hypoventilation or sleep apnea. Often she will stay awake late into the night watching TV.
>
> Her social history is significant for smoking a half pack per day and occasional alcohol use. She has used marijuana in the past, about one joint per week. She is not married nor currently involved in a similar relationship. She lives with her mother and her daughter. She had had a 3-month trial of Prozac several months ago for some of the various symptoms of fatigue and some crying spells, but it didn't help. She had been referred to a counselor, but didn't show for her appointment.

Weiner: What questions do you have about working with this patient?

Resident: I feel stuck in several respects. Her obesity seems to be the major problem. She's isolating herself from society because of her perception of her obesity. In ten years, she hasn't been able to change her weight. I think that contributes to many of her problems: fatigue, sleeping disorder, DJD. Her smoking doesn't help her asthma. I don't feel like I'm making progress in any of these areas with her.

Weiner: You said that in ten years no one's really made much progress? Or has there been some progress?

Resident: Not with her weight. There's been progress with her other problems. At least she hasn't been hospitalized this past year.

Swain: So you feel kind of stuck, like you're not making the improvement that you'd like to see. What are the things that we can maybe help you with?

Here we are directly addressing one of our explicit goals: to reduce physician frustration relative to a difficult patient situation. We are also trying to follow principles of adult learning by elucidating and focusing on the resident's own concerns.

Resident: I think trying to figure out more about her fatigue, or is this depression? Should I try another course of antidepressants?

Weiner: She doesn't admit to being depressed?

Resident: Often it seems that people who have crying spells may not say "I'm really depressed," but present with a lot of other symptoms.

Weiner: What's it like, being stuck with her?

Resident: I don't know—I just feel frustrated.

Weiner: And who would you say is more frustrated, you or the patient?

Resident: The patient.

Swain: What is she most frustrated about?

Resident: That she doesn't get out of the house, that she sits around and watches TV, and that she's tired all the time.

Swain: I'd like to get a little more of a feel for what it's like for you when the patient's in that room. Does she say, "You know doctor, it's been ten years since I've been coming to this clinic, and nobody's been able to make me any better; and, you're not any better than anybody else?"

Resident: She actually isn't putting blame back on any of the providers. She recognizes that a lot of it is herself, but she continues to come with the same complaints.

Swain: And how about you? Do you think that there should be something else that we could do for her, or is this it?

Resident: I don't know. Sometimes I feel like maybe I'm missing something with her fatigue or something, and that's why I can't figure out why this is going on.

At this point, we have several issues to address. The resident reports feeling stuck, and a bit frustrated. She also admits to fear that she's missing something important, probably something biomedical. However, from the history, we are reasonably certain that the patient's weight, sleep, fatigue, and isolation all stem primarily from psychosocial issues. Since the medical diagnoses of hypertension, asthma, and tic douloureux do not seem to be standing out for the resident or for us, we focus the discussion section of the FFCC on the psychosocial aspects of weight and social isolation, and try to tie it in with the resident's sense of frustration and stuckness.

Weiner: When you see her with her mother, does her mother give you any kind of insight? Does she talk about it; is she frustrated? Does her mother get much involved in what's going on here with the patient?

Resident: No.

Swain: I was wondering if maybe that was a nice thing for the mother, that she's always got someone at home with her.

Resident: That's true. Probably very reassuring for her that her daughter's there.

Weiner: And it might just be that giving up weight would also be a way of giving up one aspect of that relationship. Either one may not be ready for that. To the extent that this might be the case, then it might be a matter of showing more respect for the difficulty in making this change, that it's more than just losing weight. It may have something to do with defining their relationship with one another. If being overweight gives her an "excuse" to stay home to take care of her mother, it maintains their closeness to one another. Just one way to think about it.

At this point, we are using reframing techniques with at least three different goals in mind. First, we hypothesize that the resident is being a bit over-responsible for the patient's situation, and hope our reframe helps the resident to be able to see things from a more balanced perspective. Second, we are inviting the resident to consider the patient's illness perspective. Third, we are making a hypothesis about the adaptive nature of a symptom within a system. This could have been done with other reframes, such as hypothesizing about the patient's possible agoraphobia or other phobias or anxieties. The specific hypothesis is not as important as the point that there may be any number of reasons (besides the physician's inadequacy) why the patient's symptoms persist.

Swain: There are all kinds of possibilities. Your patient may think that to be "good" she must stay home and take care of her mother, so the part of her that wants to get out of the house and do stuff might be in conflict with the part of her that thinks being good means staying home with mother. It might be very, very difficult for her. These are just some hypotheses, but I think that the theory underneath is that there are probably some good reasons why she's not losing weight, good reasons why she's staying home. We don't necessarily know what they are yet. After all, you and other people have offered her other options to help her with these problems, setting up counseling for example, which she didn't show up for. It may be very scary for her to make a change, and that may be why she didn't go to counseling. Part of her wants to change, and part of her is stuck, really stuck. I don't know the reasons exactly why, but we can speculate.

Weiner: Is this making some sense? Does it fit with the picture you've been seeing now for a year and a half? This has been going on for ten years, and no change? I think it speaks to what your predecessors have been up against as well, how difficult it's been to try to impact this particular system.

Swain: When we first asked you what issues you'd like to talk about, you said that you felt stuck. It's striking to me because I don't see you as being the one who is stuck; I see the patient as being the one who is stuck. I appreciate that it's frustrating as a physician to see all these other problems: hypertension, degenerative joint disease, asthma, fatigue, all related to her weight. It's as if you could just get this patient to lose some weight, she'd feel a lot better. But it's the patient who's stuck, because she has a number of options that she could do to feel better: go to counseling, wear her oxygen every night, you name it. You tried to get her in for a TSH because she might be hypothyroid, but she won't go to the lab. So, it's a common thing, even for practicing physicians and certainly for residents, for the doctor to feel like they're the stuck one. "She won't go to the lab, I'm stuck." But really *she's* stuck then, because she might be hypothyroid and you could help her with that fairly easily, but she won't go to the lab. It might make it feel a little different for you to have that view. That doesn't mean you don't care about what's happening.

At this point, we are beginning to move from redirecting responsibility and helping the resident to process for greater understanding, to giving her some specific practical suggestions for management.

Weiner: So it kind of has the effect of turning around, putting it back onto the patient. For example, "If this continues for another ten years, what will your life be like?" Or, you could take more of the following path: "Well, I'm not quite sure what to tell you. We've been trying to help with this for ten years now, and really haven't made much progress. I'm not quite sure what else to tell you at this point." Another comment that could be made would be, "Tell me what you think of this: maybe it's important for you to have your weight? Maybe it's more important for you to be overweight than it is for you to lose weight?"

Resident: I could say that, but I just can't imagine that someone could actually feel that way. Why would somebody want to be overweight? I know what you're saying is that this may be a way of life for her. I know some overweight people, but I just can't imagine it.

Swain: Assuming that she doesn't have a biomedical problem that makes her overweight, the fact that she *is* overweight virtually proves that it's more important for her to be overweight than not. It doesn't tell you why, but what it does tell you is that there's some reason. It's too difficult for her to lose weight. Right, or not?

Resident: Well, yeah, it makes sense.

The resident is now, slowly, able to express some understanding of the potential for the patient's symptoms to serve some sort of adaptive function.

Swain: There are some other options about how to approach her that might just jar the relationship a little bit, and get you both in a little different space. What will it be like for her after her mom dies? You could ask, "What's scary about it," because there must be something there. What would it mean to her? Or to lose weight at all? What would her mom think? She might come up with something. Basically, what you're doing here is some counseling, some psychotherapy. Since she's not willing to come to a counselor *per se*, but she does come to you, you have your foot in the door to deal with and explore some of these things.

Weiner: My guess would be that her behavior with being overweight serves some sort of adaptive function within that family system. Otherwise, I think we would have seen something different by now, certainly after all this time.

For a number of reasons, we have focused this FFCC on the psychosocial aspects of the situation. However, there are biological explanations for obesity as well, and there is no guarantee that psychosocial explanations predominate in this case. In addition, since the resident has expressed a fear of missing something important medically, a few nonpsychosocial comments are in order.

Swain: Regarding the biomedical aspects, I do think that getting the TSH would be a very reasonable thing to do. If she's got normal thyroid function, normal electrolytes, nothing abnormal on her chemistry screening panel, and she's not anemic, you've covered 98–99% of all metabolic things that could cause her obesity and fatigue. I think then you don't have to worry that you're missing something serious medically, particularly since she's been stable for years. If you want to try another antidepressant, my recommendation would be a trial of something like nortriptyline at bedtime, because its slight drowsy effect might help her sleep better, and then she might be more awake during the day. I'd probably try to work with her from a psychosocial level first, but it's an option. Talking to her about sleep hygiene might help too.

The FFCC is wrapping up, and we want to be sure that we have addressed all the resident's important issues and that we get some feedback as to the effectiveness of the consultation.

Weiner: Do you have any other questions at this point?
Resident: No.
Weiner: Had I asked you earlier when you said you feel stuck how stuck are you on a scale of 1–10, with 10 being as stuck as a person could be, where would you have put your stuckness?
Resident: Six.
Weiner: And now, on a scale of 1–10, are you still at a six?
Resident: Four.
Weiner: Anything else?
Resident: It was good to take an hour to review her chart. I didn't realize that I didn't even really know her whole social situation; she actually has three children. I've only heard her talk about one. She doesn't really talk about her kids, how old they are, or anything.
Weiner: And that might be another thing to do next time. Get a little more information so you could put that together. It's interesting that after ten years of visits, you had to work real hard just to dig that up in the progress notes. Maybe this should have been done a long time ago.
Swain: Let us know if we can help you further with this.

Evaluation

At the conclusion of each consultation, we ask the resident if there are any other issues that we need to address; we try to be sure that we have satisfied the resident's needs. As mentioned earlier, we give the residents each a written summary that they can keep and refer back to in the future when providing further care to the patient. Finally, we solicit verbal and written evaluative feedback from the resident.

The immediate feedback that we receive is overwhelmingly positive. Residents consistently indicate that they either agree or strongly agree that we met each of our stated goals. The single exception to this was one occasion when we videotaped the consultation; that particular resident felt that we did not meet the "provide a supportive environment" goal. Other residents have not found videotaping to detract from the sessions.

We have continued to videotape our sessions from time to time, but only with resident's consent. The main purpose for videotaping was to have a dynamic illustration of our model for various conference presentations, such as the University of Rochester's 1993 "Partnerships in Health Care" symposium. We have also found that videotaping provides us with a convenient way to reflect on our performance during the FFCC, and to discuss means of future improvement.

Representative comments made by residents on their evaluation forms are reflected in the table below.

Representative Comments by Residents Immediately After FFCC

- "This gave me a whole new perspective and understanding of why this patient has been difficult for me."
- "The format is excellent."
- "This experience [was] very helpful, especially regarding objective #4 [optimize functioning/reduce frustration], which I need lots of help with."
- "This was good, as often what we need is reassurance/feedback and don't have time for it in the staffing room."
- "Get one or two other residents involved to share other points of view."
- "The two of you provide a well-balanced view."
- "We developed several strategies to help me manage this patient; I have developed a more systems approach and hope to use this in my management."

Results of a Retrospective Survey

In addition to our request for immediate verbal and written feedback from each resident, we recently undertook a retrospective survey of residents who participated in the FFCC over the past three years. The survey consisted of eight Likert-scale questions (Strongly Agree, Agree, Disagree, or Strongly Disagree) and two fill-in-the-blank spaces with stem-question prompts. We received responses from 15 of 18 residents, for a response rate of 83%. The responses are summarized in the following two tables.

Representative Responses to Comment Section of Survey

- "It [the FFCC] helped me to step back and evaluate my behavior and options."
- "Helped give me a new outlook and re-energize myself."
- "Allowed an intensive review of a problem patient and one-on-one staffing with MD and therapist."
- "A plan was generated which I could implement."
- "I still use techniques learned at the FFCC process."

Question "The FFCC . . ."	Strongly Agree	Agree	Disagree	Strongly Disagree	N.A. or Can't Remember
helped me give biopsychosocially integrated care	20%	67%	0%	0%	13%
increased my confidence about the case	33%	40%	7%	0%	13%
helped me identify appropriate level of functioning	27%	53%	0%	0%	20%
helped me to value collaborative consultation	20%	60%	7%	0%	13%
helped me to manage relevant personal issues	20%	33%	27%	7%	13%
had a helpful written case report	13%	40%	7%	0%	40%
was conducted in a supportive environment	47%	47%	0%	0%	7%
would be recommended by me to my peers	33%	40%	13%	0%	13%

When asked to comment on how the FFCC might be more helpful, the responses were unanimous that the process should be done more frequently, and be more consistently spread throughout the three-year residency program.

Evaluation by Faculty

We have given careful thought to this form of consultation and to what benefits have accrued to us, the residents, and the residency program as a result of this program. Several benefits are apparent. The program benefits by having a family physician and family therapist collaborating in the training of residents. Such collaboration is powerful role-modeling, which the residents can observe and incorporate into their practices. In fact, one of our FFCC graduates is now on faculty at this residency program and participates in the FFCC process with our current group of residents.

The residents' comments (above) summarize their perception of how FFCC benefits them. One additional point is worth making. The FFCC creates a supportive, positive space for residents where they can be reflective, because they know that they won't be grilled, "pimped," or even have to return immediately to the exam room to face the patient and his/her demands. In the chaotic world of the residents' lives, the FFCC provides a safe place where support leads to significant growth on the part of the physician-in-training.

As faculty, we have ourselves benefitted extensively from the FFCC process. We have learned much from each other, and from our residents. We enter each FFCC not knowing exactly where it will lead, and leave each FFCC having learned something new. The mutual support that we provide to each other is also very positive, as is the pleasure of participating in an effective educational intervention. When we can watch as residents become less frustrated and more comfortable, when we know that our collaborative effort is more effective than if we would provide similar feedback independently, then we, too, feel less frustrated and more fulfilled as educators.

Summary

Despite disparate educational backgrounds, we were both struck early in our faculty careers by residents' tendencies toward quick technological fixes, frustration and feeling cut off from patients, and dichotomous thinking about biomedical vs. psychosocial issues.

In order to devise a successful collaborative intervention, we first needed to examine the residents' roles in a larger system, and our own roles as well, including parallel process issues. We then tried to understand what interventional format would be viewed from the residents' perspective as helpful.

We went on to develop our case-consultation model, keeping in mind the residents' needs for practicality, collegiality, and "schedule neutrality." We knew that it would be important for our intervention to be integrated into the residents' schedules without appearing to be extra work, and recognized the residents' desire for a case-driven format.

By collaborating with each other, we have developed a successful educational intervention. We model collaboration to the residents and collaborate directly with them, as well, by frequently and genuinely soliciting feedback from them about how we can best meet their needs, and by contracting with them to allow us to address other issues that we've identified as important.

We feel that residents gain insight into the importance of using biopsychosocial approach. They leave our case consultations feeling recharged and much less frustrated. They develop an appreciation for the benefits of a balanced approach to patients that is not under-functioning, over-functioning, or cut off. They also seem to retain the idea that a review of a challenging case with a colleague of the same or another discipline can be a very helpful experience.

We are pleased to know that graduated residents still remember and appreciate our program, and still use techniques learned from the FFCC in their medical practices.

References

1. Botelho RJ, Seaburn DB, Harp JJ. Isomorphism in resident consultations: Implications for education and patient care. Family Systems Medicine 1991;9:137–49.
2. Bowen, M. *Family Therapy in Clinical Practice*. New York: Jason Aronson, 1978. Crouch M, Roberts L (eds). *The Family in Medical Practice*. New York: Springer-Verlag, 1986.
3. Doherty WJ, Baird MA. Developmental levels in family centered medical care. Family Medicine 1986;18:153–6
4. McDaniel SH, Campbell TL, Seaburn DB. *Family-Oriented Primary Care: A Manual for Medical Providers*. New York: Springer-Verlag, 1990.
5. McDaniel SH, Campbell TL. Physicians and family therapists: The risk of collaboration. Family Systems Medicine 1986:4–8
6. McDaniel SH, Hepworth J, Doherty WJ. *Medical Family Therapy*. New York: Harper Collins, 1992.
7. Mengel M. Physician ineffectiveness due to family-of-origin issues. Family Systems Medicine. 1987;5(2):176–90
8. Searight HR. Psychosocial Knowledge and Allopathic Medicine: Points of Convergence and Departure. The Journal of Medical Humanities 1994;15(4):221–232
9. Weiner EL, Lorenz A. Initiation of a Behavioral Scientist: Lessons Learned From a Boy Who Wouldn't Poop. Family Systems Medicine 1994;12(1):73–9
10. Wolkenstein AS, Lawrence SL, Butler DJ. Teaching "Family": The family medicine chart review. Family Systems Medicine 1985;3(2):171–8

Conclusions

As we look back upon the places we've visited in this intellectual excursion—from community-based healthcare in Africa to an HMO in the United States concerned about patient complaints; from the feelings of fear and trust deep within the hearts of individuals to the linkages between health policy and community economic development in rural Alabama; from improving the health and future social prospects of children to easing the fear of death of elders and their loved ones—we can recognize that partnership is "self similar across scale." This term, as used in fractal geometry, describes the remarkable consistency of angles, shapes and proportions as one proceeds up or down the levels of complex systems (e.g., the trunks, limbs, branches, and twigs of a tree, or the ever finer ramifications of blood vessels from the aorta down to capillaries). Whether we are looking at the level of the patient and clinician working one-on-one; at the level of families or of colleagues on a healthcare team working in small groups; at hospitals working in collaboration with churches or schools, institution to institution; or at counties working together, community to community, the principles of partnership are the same.

Partnership begins with the willingness and the ability to listen. From this foundation, relationship, dialogue, and understanding emerge in a self-reinforcing spiral. We try to communicate respect and invite collaboration through our personal behavior and organizational process, although we do not always succeed; after all, we are not perfectly behaved nor are we possessed of unfailing intuition. Fortunately, perfection is not necessary to make relationships work. In fact, as we saw in the chapter on international partnerships for education, recognizing and repairing mistakes probably enhances trust and builds more robust relationships than does seamless, unerring process (which, by the way, is impossible). Within the context of relationship, we explore goals, strategies, and scenarios. New ideas and possibilities emerge from the interaction—from the "group mind" of the participants—that we would not have developed by ourselves. Ultimately we arrive at a mutually agreeable plan of action: shared decision-making in a supportive climate, the conditions that maximize motivation and commitment.

Relationship, understanding, shared decision making—what could be simpler? And what could be more difficult? We must surrender control to experience the synergistic properties of partnership, to appreciate that the whole is indeed greater than the sum of the parts, but we have been taught in any number of ways to prize control. The thought of entrusting vital decisions to an ineffable process that no one can direct is frightening. Here we arrive at the fuzzy metaphysical edge of partnership. "If *I'm* not driving, and *you're* not driving, then *who is*!!?" How can one trust a process that eludes explanation? The only answer is empirical: it works better.

Learning to trust the process involves a personal transformation, a change that is categorical, not incremental. The strong and growing research literature that supports the efficacy of relationship-centered process is not enough to bring this change about. Cognitive knowledge is no match for the existential fear that accompanies uncertainty; trust can be gained only through direct personal experience. Only by facing the fear, taking a risk, letting go and then experiencing the creativity, joy and communion of partnership will we be convinced that larger wholes really exist and are wiser than individual parts. One success makes the next round easier. One failure . . . well, that happens, too. There are no guarantees.

The theme of letting go to become open to new and better possibilities is played out in many of the partnership stories in this book. Margaret Blandin-Clark allowed her reluctant complaint-prone physician-student to determine the setting and goals for their initial meetings. She could not have known ahead of time what learning agenda and experiences would be needed to engage this individual in improving his communication skills; rather, she structured a process that allowed them to emerge. The staff members of the Alliance of Black Churches Health Project approached their community groups with a public health goal in mind, but they let go of all preconceptions about how to achieve it. Different plans were invented within each community; each one succeeded beyond all expectations. In response to their initial inability to interest family medicine residents in exploring the psychosocial dimensions of care, Geoff Swain and Eric Weiner did not push harder using the same methods. They relinquished their traditional teaching approaches, stepped back to take a fresh look at the situation, and developed a successful new teaching model. In these stories and others, abandoning efforts to control the outcome—letting the process of partnership find its own path—enhanced the enthusiasm of the participants and mobilized creative energy.

And so we close by returning to the original purpose of this book: to improve the quality of partnership process in healthcare. We can now express this goal in more concrete terms. We hope we have encouraged you, our readers, to take the risk of letting go of specific desired outcomes and instead to trust the process—to listen deeply to the patients, families, colleagues, and organizations with whom you seek to partner; to build relationships that encourage them to be active, empowered partners; to share decision making with them (within the limits of their ability and your responsibility)—and then see what happens. This volume is filled with the experiences of people who have tried this approach in many settings and succeeded. We hope that their experiences have made this path more accessible to you.

Index